Environmental Discourses in S<

T0253146

Volume 7

This series wrestles with the tensions situated between environmental and science education and addresses the scholarly efforts to bring confluence to these two projects with the help of ecojustice philosophy. As ecojustice is one of the fastest emerging trends for evaluating science education policy, the topics addressed in this series can help guide pedagogical trends such as critical media literacy, citizen science, and activism. The series emphasizes ideological analysis, curriculum studies and research in science educational policy, where there is a need for recognizing the tensions between cultural and natural systems, the way language is endorsed within communities and associated influence, and morals and ethics embedded in school science. Conversations and new perspectives on residual issues within science education are likely to be addressed in nuanced ways when considering the significance of ecojustice, defensible environmentalism, free-choice. Book proposals for this series may be submitted to the Publishing Editor: Claudia Acuna E-mail: Claudia.Acuna@springer.com

María S. Rivera Maulucci • Stephanie Pfirman
Hilary S. Callahan
Editors

Transforming Education for Sustainability

Discourses on Justice, Inclusion, and Authenticity

Editors
María S. Rivera Maulucci
Education Program
Barnard College
New York, NY, USA

Stephanie Pfirman
College of Global Futures
Arizona State University
Tempe, AZ, USA

Hilary S. Callahan
Department of Biology
Barnard College
New York, NY, USA

ISSN 2352-7307   ISSN 2352-7315   (electronic)
Environmental Discourses in Science Education
ISBN 978-3-031-13538-5   ISBN 978-3-031-13536-1   (eBook)
https://doi.org/10.1007/978-3-031-13536-1

This work was supported by BARNARD COLLEGE, PROVOST OFFICE

This Springer imprint is published by the registered company Springer Nature Switzerland AG
The registered company address is: Gewerbestrasse 11, 6330 Cham, Switzerland

Preface

This volume was conceptualized in January 2017. At that time, we were all at Barnard College and María reached out to Hilary and Stephanie about advancing the teaching of environmental justice across the curriculum. We sought to run a Willen Seminar at Barnard—a convening of faculty around a common interest—and sent out a call for chapters. In the Willen Seminar, we would explore how engaging in environmental research also means grappling with questions of justice and sustainability. We also hoped to explore the limitations of human inquiry, human impacts upon the environment and how to mitigate them, implications for human and ecosystem health, and, ultimately, the rights of future generations. At a time when science was increasingly being called into contention (the first March for Science was held April 22, 2017), especially with respect to climate science, we believed there was a need for scientists and those whose research connected to environmental issues to explore common ground and consider the implications for teaching the next generation.

The Willen Seminar was held during the 2017–2018 academic year with four meetings per semester. The seminar gathered faculty in the sciences, social sciences, and education from Barnard and Columbia who were working with issues of the environment and/or environmental justice to explore the literature on ecojustice and share research and teaching strategies. Seminar sessions investigated the connections between research and teaching in the classroom, the research laboratory, and in the field and how the faculty members were navigating the complexities of particular environmental science issues, such as climate change, water quality, soil health, urban environments, and biodiversity, in ways that conveyed their complexity, the real sociopolitical context, and their hope for the future. We focused on topics such as what is ecojustice, connecting ecojustice research and teaching, and had presentations from several potential contributors to the volume.

At around the same time, a campus-wide effort from 2016 to 2018 worked on defining the college's climate leadership, and later sustainability leadership, through a 360-degree approach to academics, finance and governance, and campus culture and operations with attention to the role of women, people of color, and low-income communities. A significant commitment was made in the spring of 2017, when

Barnard announced an innovative approach to divestment. "In a first for any college in the nation, Barnard will divest from fossil fuel companies that deny climate science or otherwise seek to thwart efforts to mitigate the impact of climate change."[1] This approach was developed by a Presidential Task Force to Examine Divestment, which was established in response to a student-led campaign by the group Divest Barnard. Grounded in the diverse perspectives of the faculty, students, administrators, board members, and facilities representatives on the Task Force, their recommendations approached the issue from a new direction, one that was authentic to Barnard College: our mission and our context.

Following on the divestment discussions, during the 2017–2018 academic year, the Sustainable Practices Committee at Barnard held a workshop series to develop a shared climate action vision statement. These discussions culminated in Barnard's Climate Action Vision 2019 which recommended that Barnard provide pedagogical support for faculty to create new courses or adapt existing ones with the goal that 100% of students would participate in a sustainability class or event each year.[2]

Coming out of these separate but complementary faculty dialogues across the year, it became apparent that the overarching concept that connected all our work was sustainability, and that justice, inclusion, and authenticity were central to ways in which we approached the work as scientists, social scientists, and educators. Thus, we decided to focus our attention in this volume on teaching sustainability in authentic, inclusive, and just ways—themes that are threaded through the chapters.

As we were gathering manuscripts, life happened. First, one of us needed successive medical treatments beginning in 2016, each of which took a further toll, and one of us took a new position in 2018. Then, in 2020, COVID-19 hit New York City—and one of us very hard—losing a partner to this disease further compelled a focus on protecting life.

2020 was also transformative in raising attention around the country to systemic issues of racism and justice. The Black Lives Matter movement and the disparate effects of the pandemic further reinforced issues of justice as life and death.

2020 also showed unmistakable signs of climate change with intense wildfires in Australia and the USA, catastrophic flooding in China and the USA, the most active hurricane season on record for the Atlantic Ocean with 30 named storms, and record-breaking heat and storms in Europe and the Arctic, with a temperature over 100 degrees in Siberia.

The following year, Barnard convened Sustaining Curricula 2021, a series of four interdisciplinary faculty panels at Barnard College that took place virtually during the Spring semester. The panels encouraged faculty integration of climate action, sustainability, and the environment into their courses, with collaboration and pedagogical exchange across the disciplines.

---

[1] Barnard magazine, Spring 2017 https://barnard.edu/magazine/spring-2017/divesting-deniers; also see Union of Concerned Scientists, 2017 https://blog.ucsusa.org/kathy-mulvey/barnard-college-to-divest-from-fossil-fuel-companies-that-deny-climate-science/

[2] https://barnard.edu/sites/default/files/inline-files/ClimateActionVision-2019-final-5.pdf

These events and experiences strengthened our focus in new ways—not only did we want this Environmental Discourses volume to address teaching the subject matter of sustainability in authentic, inclusive, and just ways, but we also wanted to consider the process and institutional aspects of teaching, and the human side of the people involved—the faculty, students, and staff. To this end, we reached out to additional BIPOC colleagues to gain more diverse perspectives, and included the Barnard Sustaining Curricula panel discussions in this volume.

As we draw this project to completion, we hope that the voices and experiences we bring together in this volume point to the challenges as well as the possibilities of doing authentic and inclusive work for sustainability and justice in the context of our curricula, our students, our institutions, our communities, and our world.

New York, NY, USA   María S. Rivera Maulucci

Tempe, AZ, USA   Stephanie Pfirman

New York, NY, USA   Hilary S. Callahan

Contents

Editors and Contributors

About the Editors

María S. Rivera Maulucci joined the faculty at Barnard College in 2004 and is the Ann Whitney Olin Professor of Education. Prior to Barnard, she served as Director of the Science & Technology Professional Development Center for Region One in the Bronx and as the Director of Urban Forestry for the Environmental Action Coalition. Her expertise in STEM pedagogy and teacher education draws on 16 years of teaching mathematics, science, and technology at the elementary, secondary, and postsecondary levels. She has worked with both school and community-based education programs. Her interdisciplinary scholarship focuses on how inservice and preservice teachers learn to teach for social justice, particularly in STEM fields, and the role of language, identity, and emotions in teacher development. Professor Rivera was the principal investigator for the *Barnard Noyce Teacher Education Scholars Program* (BNTSP) and Co-PI for the *Summer STEM Teaching Experiences for Undergraduates Program*, both funded by the National Science Foundation. She is currently the Co-PI for a Carnegie Foundation grant, *Joint Barnard-American Museum of Natural History Summer STEM Teaching Experiences for Undergraduates*, a project that provides STEM pedagogical training for undergraduates who teach math and science summer enrichment courses for high school youth. She has served on the board of the *National Association for Research in Science Teaching* (NARST) and on the editorial boards of the *Journal of Research in Science Teaching* and *Cultural Studies of Science Education.* She has been a strong supporter of diversity, inclusion, and sustainability initiatives at the college and in professional organizations, including NARST and the *Association for Science Teacher Education.*

Stephanie Pfirman is a Foundation professor in the College of Global Futures and a Senior Sustainability Scientist at the Julie Ann Wrigley Global Institute of Sustainability and Innovation at Arizona State University. Before joining ASU in 2018, Pfirman was Hirschorn Professor of Environmental and Applied Sciences and co-chair of the Department of Environmental Science at Barnard College. She held a joint appointment with Columbia University's Earth Institute and the Department of Earth and Environmental Sciences as an adjunct research scientist at the Lamont-Doherty Earth Observatory. Professor Pfirman focuses on understanding and responding to the changing Arctic, developing innovative approaches to formal and informal education, and exploring the intersection between diversity and interdisciplinarity. Pfirman's Arctic research addresses implications of changes in sea ice origin, drift, and melt patterns, including defining the Last Ice Area, recently established as a protected area by Canada. She is an elected fellow of the American Association for the Advancement of Science. As a former co-PI of a National Science Foundation Advancing Women in the Sciences (ADVANCE) grant, past president of the Council of Environmental Deans and Directors, and chair of the Columbia Earth Institute's Faculty Development Committee, Pfirman has helped to understand and foster the career trajectories of women and interdisciplinary scholars. Pfirman co-designed EcoChains: Arctic Life, a card-game that earned a Parent's Choice award. A longtime advocate for action on climate change, as a senior scientist at the Environmental Defense Fund, Pfirman co-developed one of the first climate change exhibitions, "Global Warming: Understanding the Forecast," produced jointly with the American Museum of Natural History.

Hilary S. Callahan, an Ann Whitney Olin Professor of Biology, joined the faculty at Barnard College in 1999 and today serves as department chair and directs the Arthur Ross Greenhouse; she is also an affiliate of Columbia University's Department of Ecology, Evolution, and Environmental Biology. She learned her early love for nature from knowledgeable relatives, and gained her identity as a botanist and activist during her undergraduate, graduate, and postdoctoral years in New Haven, Connecticut; Madison, Wisconsin; and Knoxville, Tennessee, where she fostered friendships with scholars across disciplines and with labor activists, grassroots conservationists, community journalists, and progressive politicians. Today, she resides in The Bronx. She has received multiple National Science Foundation funding awards for her research projects and mentoring of experiences in STEM research for undergraduates, including UNPAK: Undergraduates Phenotyping *Arabidopsis* Knockouts, described in this volume. Her writings in plant ecological genetics focus on plant responses to environmental change. As part of the Barnard Teaches digital initiative, she created an interdisciplinary course in botany and its history in collaboration with the New York Botanical Garden and the Columbia Center for Science and Society. Over the years, she has contributed to the Barnard Noyce Teacher Scholar Program (BNTSP), founded Barnard's Beckman Scholars Program, and served on many sustainability committees and working groups. Since 2018, she has served as President of the Board of Directors of Black Rock Forest in Cornwall, New York.

Contributors

Laura Arbelaez Barnard College, New York, NY, USA

Shirley-Ann Augustin-Behravesh Thunderbird School of Global Management, Arizona State University, Phoenix, AZ, USA

Rachel Narehood Austin Department of Chemistry, Barnard College, New York, NY, USA

Kadambari Baxi Department of Architecture, Barnard College, New York, NY, USA

Orlando Betancor Department of Spanish and Latin American Cultures, Barnard College, New York, NY, USA

April Bisner Department of Biology, College of Charleston, Charleston, SC, USA

Jamie Bousleiman Jacobs School of Medicine University at Buffalo, Buffalo, NY, USA

Peter Bower Department of Environmental Science, Barnard College, New York, NY, USA

Lisa A. Brundage Macaulay Honors College, City University of New York, New York, NY, USA

Angelo Caglioti Department of History, Barnard College, New York, NY, USA

Hilary S. Callahan Barnard College, New York, NY, USA

Ana Cardenas Barnard College, New York, NY, USA

Linda Chen Barnard College, New York, NY, USA

Elizabeth M. Cook Environmental Science, Barnard College, New York, NY, USA

Andrew Crowther Department of Chemistry, Barnard College, New York, NY, USA

Tal Danino Department of Biomedical Engineering, Columbia University, New York, NY, USA

Yuval Dinoor Barnard College, New York, NY, USA

Arminda Downey-Mavromatis Chemical & Engineering News, Washington, DC, USA

Rachel Elkis Barnard College, New York, NY, USA

Rahma Elsiesy California Institute of Technology, Pasadena, CA, USA

Kassandra Fuiten Barnard College, New York, NY, USA

Nicholas Gershberg Arthur Ross Greenhouse, Barnard College, New York, NY, USA

Ralph Ghoche Department of Architecture, Barnard College, New York, NY, USA

Sandra Goldmark Department of Theatre, Barnard College, New York, NY, USA

Tejashree S. Gopal Barnard College, New York, NY, USA

Mary A. Heskel Department of Biology, Macalester College, Saint Paul, MN, USA

Aastha Jain Barnard College, New York, NY, USA

Joscelyn Jurich Center for Engaged Pedagogy, Barnard College, New York, NY, USA

Manu Karuka Department of American Studies, Barnard College, New York, NY, USA

Laura Kay Physics & Astronomy, Barnard College, New York, NY, USA

Sohee Ki Keck School of Medicine, University of Southern California, Los Angeles, CA, USA

Katrina Korfmacher Environmental Medicine, University of Rochester, Rochester, NY, USA

Terryanne Maenza-Gmelch Department of Environmental Science, Barnard College, New York, NY, USA

Ann McDermott Chemistry, Biological Sciences, and Chemical Engineering, Columbia University, New York, NY, USA

Krista L. McGuire Biology and Institute of Ecology and Evolution, University of Oregon, Eugene, OR, USA

Jennings G. A. Mergenthal Department of Biology, Macalester College, Saint Paul, MN, USA

Courtney J. Murren Department of Biology, College of Charleston, Charleston, SC, USA

Kelly L. O'Donnell Macaulay Honors College, City University of New York, New York, NY, USA

Angelica E. Patterson Black Rock Forest and Miller Worley Center, Mt. Holyoke College, South Hadley, MA, USA

Stephanie Pfirman College of Global Futures, Arizona State University, Tempe, AZ, USA

Suzanne Pierre Critical Ecology Lab, California Academy of Sciences, San Francisco, CA, USA

Jorge Ramos Jasper Ridge Biological Preserve, Stanford University, Stanford, CA, USA

Meena Rao Department of Chemistry, Barnard College, New York, NY, USA

Leslie Raucher Sustainability, Barnard College, New York, NY, USA

Matthew E. Rhodes Department of Biology, College of Charleston, Charleston, SC, USA

María S. Rivera Maulucci Education Program, Barnard College, New York, NY, USA

Sedelia Rodriguez Department of Environmental Science, Barnard College, New York, NY, USA

Jennifer Rosales Center for Engaged Pedagogy, Barnard College, New York, NY, USA

Emma Ruskin David Geffen School of Medicine, University of California Los Angeles, Los Angeles, USA

Matthew T. Rutter Department of Biology, College of Charleston, Charleston, SC, USA

Katherine L. Shek Department of Biology, Barnard College and University of Oregon, Eugene, OR, USA

Angela M. Simms Departments of Sociology and Urban Studies, Barnard College, New York, NY, USA

Jonathan Snow Department of Biology, Barnard College, New York, NY, USA

Martin Stute Department of Environmental Science, Barnard College, New York, NY, USA

Jennifer Jo Thompson Crop and Soil Sciences, University of Georgia, Athens, GA, USA

Unyimeabasi Udoh School of the Art Institute of Chicago, Chicago, IL, USA

Olivia Wang Barnard College, New York, NY, USA

Carl Wennerlind Department of History, Barnard College, New York, NY, USA

Shoshana Williams Stanford University, Stanford, CA, USA

Tanisha M. Williams Biology Department, Bucknell University, Lewisburg, PA, USA

Gisela Winckler Lamont-Doherty Earth Observatory, and Department of Earth and Environmental Sciences, Columbia University, Palisades, NY, USA

Michael Wolyniak Department of Biology, Hampden-Sydney College, Hampden Sydney, VA, USA

Chapter 1
Education for Sustainability: Connecting with Signs of Hope

María S. Rivera Maulucci, Stephanie Pfirman, and Hilary S. Callahan

1.1 Setting the Scene

In 2017 when this project began, hardly a week went by without some headline or another testifying to environmental justice issues. For example, "Storms hit poorer people harder from superstorm Sandy to Hurricane Maria" (Sellers, 2017) and "Urban noise pollution is worst in poor and minority neighborhoods and segregated cities" (Casey et al., 2017). We saw regular headlines about climate change, such as, "Federal scientists call the warming of the Arctic 'unprecedented over the last 1500 years'" (Mooney, 2017) and "2017: The year climate change hit" (Banos Ruiz, 2017). We also saw headlines reporting attacks on science and scientists. "Climate scientists face harassment, threats and fears of 'McCarthyist attacks'" (Milman, 2017). These attacks led to "Tens of thousands march[ing] for science and against threats to climate research" (Cushman, 2017).

Fast-forward 5 years to the completion of this project and the headlines for 2021 show problems still abound, but some signs of hope are recognized. For example, "How much air pollution do you live with? It may depend on your skin color." (McCormick, 2021) contrasts with the "EPA gets serious about environmental justice" (Macfarlan et al., 2021). "Climate change [is] widespread, rapid, and intensifying – IPCC" (IPCC, 2021) contrasts with, "COP26 news: Coal phase-out boosts

M. S. Rivera Maulucci (✉)
Education Program, Barnard College, New York, NY, USA
e-mail: mriveram@barnard.edu

S. Pfirman
College of Global Futures, Arizona State University,
Tempe, AZ, USA

H. S. Callahan
Department of Biology, Barnard College, New York, NY, USA

© The Author(s) 2023   1
M. S. Rivera Maulucci et al. (eds.), *Transforming Education for Sustainability*,
Environmental Discourses in Science Education 7,
https://doi.org/10.1007/978-3-031-13536-1_1

hope for limiting warming to 1.5 °C" (Marshall et al., 2021). "'We're heartbroken. We're overwhelmed' – U.S. hospitals grapple with delta outbreak as omicron takes root" (Kimball et al., 2021) appears alongside, "How Puerto Rico became the most vaccinated place in America" (Narea, 2021).

In the context of a rapidly-warming world, a global pandemic, systemic racism and environmental racism, as science researchers and educators, we recognize an urgent need for comprehensive sustainability and environmental justice education. We also recognize the special role that higher education can and should play in the creation of a sustainable future. In particular, higher education must challenge several prevailing assumptions (Cortese, 2003, p. 17):

- Humans are the dominant species and separate from the rest of nature.
- Resources are free and inexhaustible.
- Earth's ecosystems can assimilate all human impacts.
- Technology will solve most of society's problems.
- All human needs and wants can be met through material means.
- Individual success is independent of the health and well-being of communities, cultures, and the life support system.

To challenge these assumptions and achieve a holistic vision of education for sustainability that attends to viability, feasibility, durability, wholeness, and scale-sensitivity, we must thoughtfully and carefully address issues of justice, inclusion, and authenticity. Justice includes commitments to integrity, civil and human rights, fairness, and equity. Inclusion unpacks how we involve, incorporate, and weave together diverse stakeholders and perspectives and foster a sense of belonging. Finally, authenticity deepens our understanding of the genuineness, originality, credibility, and trustworthiness of our different approaches to sustainability.

In this volume, we gather voices from natural sciences, social sciences and the humanities, educators, administrators, staff, and students. Contributions range from life in the universe, climate change, volcanoes, birds, honey bees, microbiology, and the chemistry of lead, to green roofs, Bio Blitzes, science teacher education, and higher education. The work takes place in the community, at institutions, across institutions, and in the field, the classroom, the lab, and even within personal lives. Time scales range from billions to hundreds of thousands of years, to the historical, decadal, present, and projected and predicted future. Formats include reviews, personal essays, project narratives, curriculum reflections, interviews, panel discussions, and group conversations. The chapters explore sustainability and environmental justice through a focus on the subject matter (what to teach or research), the process (how to teach or motivate students and stakeholders to care about sustainability), and the people (who teaches, who learns, and who matters in education for sustainability). That said, this volume has an admittedly STEM foundation – in part due to the background of the editors – but also because many faculty come to sustainability from the natural sciences. All tell the story of their learning as they grapple with teaching for sustainability with authenticity, inclusivity and justice.

Part I of the book focuses on how we frame and reframe sustainability and environmental justice through a historical lens (María Rivera Maulucci, Chap. 2), the lens of diversity in academia (Shirley-Ann Behravesh, Chap. 3), an intersectional lens (Angelica E. Patterson, Tanisha M. Williams, Jorge Ramos, Suzanne Pierre, Chap. 4), a critique of ecology's white nationalism problem (Ralph Ghoche and Unyimeabasi Udoh, Chap. 5), and the lens of science teacher education (María S. Rivera Maulucci, Chap. 6). Together, these chapters explore the fraught history (Ghoche and Udoh; Rivera Maulucci) and present-day experiences of academics grappling with sustainability in authentic, inclusive, and just ways (Behravesh; Patterson et al.; Rivera Maulucci). Concerning the student experience, "Education for Sustainability can help us answer the timeless question, 'Why are we learning this.' … Students also see that sustainability pathways can expand their ideas about the meaning and relevance…" of the subject (Rivera Maulucci, Chap. 6). On the other hand, a feeling of responsibility for relevance and inclusivity can wear on Black, Indigenous, and People of Color (BIPOC) faculty. Through contributing to numerous department, university, and community committees, along with stepping up to informally mentor students, etc., as the only Black faculty member in her unit, Behravesh found that she lost traction towards her goal of academic advancement. "[B]y appearing to take everything in stride – I also made my efforts, and the toll it was taking on me, invisible." In her discussion with colleagues of color, Williams explains that, "… being the color we are, from the cultures we are from, and having the experiences we have, [means] we are not just scientists. We are actually doing justice work just by showing up" (Patterson et al.).

Part II of the book explores sustainability and ecological perspectives on biodiversity. The chapters delve into sustainability issues surrounding the purpose of ex-situ conservation, specifically in greenhouse collections (Nick Gershberg, Chap. 7), cell biology and saving pollinators (Jonathan W. Snow, Chap. 8) finding the most important places for birds (Terryanne Maenza-Gmelch, Chap. 9), the local ecology of New York City green roofs (Matthew Rhodes, Krista McGuire, Katherine L. Shek, and Tejashree S. Gopal), conducting urban BioBlitzes (Kelly O'Donnell and Lisa Brundage, Chap. 11), and using game-based learning to engage students in learning about the human microbiome (Emma Ruskin and Tal Danino, Chap. 12). These chapters provide classroom and field-based examples of using experiences to engage students authentically.

Rhodes et al. note "… we intended for our students to gain a deeper understanding of these ecological challenges in the city in which they lived and achieve a more personal connection to their laboratory research projects." Both Maenza-Gmelch and Snow discuss teaching reciprocity – that managing the environment for non-humans, also protects it for ourselves. Snow explains "… an environment that is healthy for pollinators is also healthy for people. So you automatically get into aspects of climate justice, and where people live, and how the environment they are in is going to impact their health."

O'Donnell and Brundage and Ruskin and Danino also consider justice, but from a knowledge access perspective, "…we consider providing all students, whether they will become scientists or not, with foundational knowledge to navigate the scientific information they will encounter in day-to-day life to be a pressing social justice matter." (O'Donnell and Brundage).

Part III of the book investigates sustainability and environmental justice perspectives in undergraduate science education. The chapters explore the chemistry and pedagogy of environmental lead exposure (Rachel Narehood Austin, Ann McDermott, Katrina Korfmacher, Laura Arbelaez, Jamie Bousleiman, Arminda Downey-Mavromatis, Rahma Elsiesy, Sohee Ki, Meena Rao, and Shoshana Williams, Chap. 13), Brownfield Action, a web-based active learning simulation (Peter Bower and Sedelia Rodriguez, Chap. 14), integrating ecojustice principles into ecology and environmental science courses (Mary A. Heskel and Jennings G. A. Mergenthal, Chap. 15), and using plant science in the UNPAK project (Undergraduates Phenotyping Arabidopsis Knockouts) to expand critical thinking skills and meaningful personal relationships, both so necessary for journeying from classes and on-campus research toward careers and life's other challenges. (Hilary S. Callahan, Michael Wolyniak, Matthew Rutter, Courtney Murren, April Bisner, and Jennifer Jo Thompson, Chap. 16).

Each chapter addresses experiences with changing perspectives on teaching. Wolyniak of the UNPAK project emphasizes integrating classroom and independent experiences, crucial for allowing college students "to participate in the research regardless of their previous background and experience. This is a powerful tool … [to] inspire many more students than I could have before, and to attract people who may have slipped through the cracks." Heskel and Merganthal, also working on a college campus, discuss a fundamental shift in teacher orientation. While our teaching "… often recognize[s] the importance of human systems and their effects … It is less common for professors to expand into the structural inequalities of human systems and how they shape the science we practice and the content we deliver."

Austin et al. caution,

> …talking about issues of social justice in the classroom requires a skilled facilitator. College science teachers may not be trained or experienced in moderating such difficult conversations. It can be helpful to lay out some guidelines for talking about racism, equity, and justice at the beginning of the class, particularly that their colleagues and community members may have very different life experiences and perspectives… It is important to remind students to remain humble and to acknowledge how much they do not know about diverse communities' experiences and multidisciplinary perspectives. It can be challenging to balance this message while also encouraging students to become as confident of their basic scientific skills as quickly as they can.

Part IV of the book zooms in on climate change, engagement, politics, and action from the perspective of teaching about astronomy (Laura Kay and Kassandra Fuiten, Chap. 17), volcanoes (Sedelia Rodriguez and Kassandra Fuiten, Chap. 18), two decades of evolving approaches to teaching about climate change in environmental science (Stephanie Pfirman and Gisela Winckler, Chap. 19), and the process of building a circular campus at Barnard College (Sandra Goldmark, Leslie Raucher, and Ana Cardenas, Chap. 20). Kay and Fuiten and Rodriguez and Fuiton integrate teaching about climate change into their teaching of astronomy and volcanoes. Kay notes, "It is certainly an environmental justice issue to keep Earth habitable for humans! There is no way we can relocate everyone to another planet in the foreseeable future." Rodriguez talks about intentionality in connecting students with their environments.

> I think that teaching students about the natural world is especially important because hopefully as they look more at their environment, they will appreciate it more and want to preserve it. Knowing what happens naturally helps plan for a sustainable community. We can't control everything – what can we not change? What do we need to work around? When you don't understand the natural world, it's harder to fight for it. Knowledge about the environment becomes a vehicle for activism.

Pfirman and Winckler tackle the political polarization of climate change "…explor[ing] how faculty who taught [a course in] climate over the years responded to [a] shifting landscape of both external context and internal responses." They explain that

> While scientists teaching about evolution have long had to face these issues, the rest of the scientific community was largely distanced from being confronted with questions about their opinions or positions, versus facts or a scientific consensus. For those who were teaching about climate, this was an unusual, and sometimes uncomfortable, situation. … [We] needed to learn how to address these differences with authenticity – congruent with who we are, and our own values – both in the classroom and beyond.

Goldmark et al. bring the need for action outside of the classroom home, stating "building a circular campus becomes about more than recycling, waste, or even greenhouse gas emissions; it is about creating a more equitable and just community."

Finally, Part V brings us up to Spring 2021 with an introduction and series of edited transcripts from sustaining curricula conversations held at Barnard over Zoom due to the pandemic. These conversations brought faculty and students together to explore how sustainability was being integrated into the undergraduate curriculum. The introduction to the panel discussions (Jennifer Rosales, Chap. 20) provides an overview of the campus initiatives leading up to the interdisciplinary faculty panels and an explanation of themes for each panel. The first panel (Orlando Betancor, Carl Wennerlind, Hilary Callahan, Yuval Dinoor, Rachel Elkis, Chap. 22) addresses the theme of theorizing the environment from the perspective of History, Spanish and Latin American Cultures, and Ecology. The second panel (Angelo Caglioti, Andrew Crowther, Jonathan Snow, María S. Rivera Maulucci, Yuval Dinoor, Aastha Jain, Chap. 23) takes up the theme of climate change and the Anthropocene from the perspective of Environmental History, Chemistry, Biology, and Science Education. The third panel (Elizabeth M. Cook, Manu Karuka, Angela M. Simms, Joscelyn Jurich (Ed.) Yuval Dinoor, Rachel Elkis, Chap. 24) focuses on environmental justice. The fourth panel discussion (Kadambari Baxi, Sandra Goldmark, Martin Stute, Joscelyn Jurich (Ed.), Linda Chen, Olivia Wang, Chap. 25) takes up the challenge of decarbonization and design.

Wennerlind (Panel 1) reminds us of the foundation for our historical trajectory. It is

> …essential to deconstruct the notion of scarcity and point out that it is based on assumptions made in the 1870s, grounded in very limited psychological or anthropological research. It was just analytically convenient to assume that human desires are insatiable and

that people are mostly focused on consumption. While this way of thinking fostered behavior and policies conducive to economic growth, our present challenge is to build a social science that is conducive to more sustainable practices.

Karuka (Panel 3) connects a history of consumption to a need for a new emphasis on justice. He notes, "[in] addition to these broader overarching themes of imperialism, racism, and colonialism, I think environmental justice is also a way to think about place and our relationships and responsibilities to place."

Cagliotti (Panel 2) moves from responsibility to empowerment and personal action, explaining,

> When you try to examine such a global issue, an issue that has so many facets and so many angles, it is easy to feel a little bit powerless in a sense, and to say, "Okay, so what can I do?" The goal of teaching a subject like this is to empower students and make them feel that precisely because this is such a global issue with so many facets, that regardless of the discipline that you are studying, regardless of your major, or regardless of what you will be doing after college, for example, or regardless of your locality…or where you are situated geographically, where is your community? You can always make an impact, from any angle if you can think globally and act locally to use a very common expression.

Stute (Panel 4), in reference to tackling climate change, says "… we have to move people's minds, we have to generate some groundswell of support for these massive changes we have to see in the future of our lives."

This volume will point to some of the ways we see hope grounded in the difficult, meaningful work of each of the contributors as they strive for a more sustainable future by teaching sustainability with authenticity, inclusion, and justice.

References

Banos Ruiz, I. (2017, December 28). *2017: The year climate change hit*. DW.Com. https://www.dw.com/en/2017-the-year-climate-change-hit/a-41944142

Casey, J. A., James, P., & Morello-Frosch, R. (2017, October 5). Urban noise pollution is worst in poor and minority neighborhoods and segregated cities. *The Conversation*. http://theconversation.com/urban-noise-pollution-is-worst-in-poor-and-minority-neighborhoods-and-segregated-cities-81888

Cortese, A. D. (2003). The critical role of higher education in creating a sustainable future. *Planning for Higher Education, 31*(3), 15–22.

Cushman, J. H. (2017, April 22). Tens of thousands march for science and against threats to climate research. *Inside Climate News*. https://insideclimatenews.org/news/22042017/march-for-science-scientists-climate-change-donald-trump-climate-denial/

Intergovernmental Panel on Climate Change. (2021, August 9). *Climate change widespread, rapid, and intensifying – IPCC*. IPCC. https://www.ipcc.ch/2021/08/09/ar6-wg1-20210809-pr/

Kimball, S., Rattner, N., & Constantino. (2021, December 15). *"We're heartbroken. We're overwhelmed" – U.S. hospitals grapple with delta outbreak as omicron takes root*. CNBC. https://www.cnbc.com/2021/12/15/covid-delta-drives-surge-of-us-infections-hospitalizations-amid-omicron-fears.html

Macfarlan, T. J., Boden, S. R., & Rothenstein, C. L. (2021, November 15). EPA gets serious about environmental justice. *The National Law Review*. https://www.natlawreview.com/article/epa-gets-serious-about-environmental-justice

Marshall, M., Lawton, G., & Vaughan, A. (2021, November 4). COP26 news: Coal phase-out boosts hope for limiting warming to 1.5°C. *New Scientist.* https://www.newscientist.com/article/2296411-cop26-news-coal-phase-out-boosts-hope-for-limiting-warming-to-1-5c/

McCormick, E. (2021, December 15). How much air pollution do you live with? It may depend on your skin color. *The Guardian.* https://www.theguardian.com/us-news/2021/dec/15/air-pollution-people-of-color-breathe-more-us-study

Milman, O. (2017, February 22). Climate scientists face harassment, threats and fears of "McCarthyist attacks." *The Guardian.* https://www.theguardian.com/environment/2017/feb/22/climate-change-science-attacks-threats-trump

Mooney, C. (2017, December 12). Federal scientists call the warming of the Arctic 'unprecedented over the last 1,500 years'. *Washington Post.* https://link.gale.com/apps/doc/A518483949/AONE?u=columbiau&sid=bookmark-AONE&xid=9ccfcc39

Narea, N. (2021, November 26). *How Puerto Rico became the most vaccinated place in America.* Vox. https://www.vox.com/22761242/puerto-rico-vaccine-covid-hurricane-maria

Sellers, C. (2017, November 19). Storms hit poorer people harder, from Superstorm Sandy to Hurricane Maria. *The Conversation.* http://theconversation.com/storms-hit-poorer-people-harder-from-superstorm-sandy-to-hurricane-maria-87658

María S. Rivera Maulucci is the Ann Whitney Olin Professor of Education at Barnard College. Her research focuses on the role of language, identity, and emotions in learning to teach science for social justice and sustainability. She teaches courses in science and mathematics pedagogy and teacher education for social justice.

Stephanie Pfirman is Foundation Professor, College of Global Futures, and Senior Sustainability Scientist in the Global Institute of Sustainability and Innovation at Arizona State University. Until 2018 she co-chaired Barnard College's Department of Environmental Science, with a joint appointment at Columbia University on the faculties of the Earth Institute and Department of Earth and Environmental Sciences.

Hilary S. Callahan is the Ann Whitney Olin Professor of Biology. Her integrative research examines many different features of plants – from roots to flowers to seeds. She seeks to understand how plants and their traits function in nature, and how external factors affect trait expression and how rapidly traits evolve.

Part I
Framing and Reframing Sustainability and Environmental Justice

Chapter 2
A History of Ecojustice and Sustainability: The Place Where Two Rivers Meet

María S. Rivera Maulucci

2.1 Mamaroneck Land Acknowledgement

I am writing to you from the Village of Mamaroneck, New York, a place that gets its name from the Munsee Lenape language of the people who first lived in the region. Mamaroneck is a coastal village and town on the Long Island Sound in Westchester County, New York. According to local lore, the area that is now the town of Mamaroneck was "purchased" from the Native American Chief, *Wappaquewam*, and his brother, *Manhatahan*, of the Siwanoy tribe, by an Englishman, John Richbell, in 1661. As with many other places, there is a memorial plaque and artwork commemorating this event. There is some disagreement about the historical accuracy of the tribe name, Siwanoy, which may be a corruption of *Siwanak*, or "salt people." Due to removal, intermarriage, assimilation, and erasure of the Native American people from the area, we cannot know for sure if the name is one that was given to people living in this area by settlers rather than one that the people used for themselves at the time.

The name, Mamaroneck, is most likely derived from the Munsee, *maamaalahneek* which means "striped stream," which comes from *maamaaleew*, "to be striped." I wondered if the river may have been called striped because, towards the mouth of the river, where it flows into the Mamaroneck Harbor and the Long Island Sound, both freshwaters from the land and saltwater from the sea meet and mix. Then, I came across a reference that stated that Mamaroneck means, "the place where the sweet waters fall into the sea" (Scarsdale Inquirer, 1936).

I begin with this land acknowledgment because I am about to share a chapter that explores the history of the ecojustice and sustainability movements. I cannot write about this history and these issues without acknowledging the ways in which they

M. S. Rivera Maulucci (✉)
Education Program, Barnard College, New York, NY, USA
e-mail: mriveram@barnard.edu

© The Author(s) 2023

M. S. Rivera Maulucci et al. (eds.), *Transforming Education for Sustainability*,
Environmental Discourses in Science Education 7,
https://doi.org/10.1007/978-3-031-13536-1_2

11

intersect with the place I am writing from. Mamaroneck exists as a result of settler colonialism. Today, parts of the Village that have historically had issues with flooding are under a deeper threat due to climate change. The Village sits on both sides of the Mamaroneck River, which is joined by the smaller Sheldrake River before emptying into the Long Island Sound. In September 2021, the Village suffered catastrophic flooding during the remnants of Hurricane Ida. One Mamaroneck resident died and there were more than 150 water rescues, 535 flooded homes, 1000 people displaced, 310 abandoned cars, and $93 million in Village, residential, and commercial damage (Edwards, 2021). Mayor Tom Murphy said, "As we know, global warming is getting worse. The storms are coming with greater rapidity and greater intensity. And here in the Village of Mamaroneck, especially in some of the neighborhoods that abut the rivers, [we] are at the epicenter, the bullseye, the cutting edge of climate change" (News 12 Bronx, 2021).

Ecojustice is salient to this story because Mamaroneck has about 19,000 residents, 65% of whom identify as White only, 5% Black, 4% Asian, and 25% Hispanic or Latino of whom 15% identify as White (US Census, 2019a). The median household income is $102,000, and the median home value is $623,000. Twenty-five days after the flood, New York State Senator Chuck Schumer announced that they had secured federal approval for the Mamaroneck and Sheldrake River Flood Risk Management Project (News 12 Bronx, 2021). This $100 million project will deepen and widen the rivers to slow flooding and remove bridges that currently constrict river flow.

Climate resiliency is a central ecojustice and sustainable development issue, and we have to ensure that the poor who are most vulnerable have adequate support for mitigation and adaptation efforts. When Hurricane Maria hit the island of Puerto Rico in 2017 with devastating force, a report from researchers at George Washington University's Milken Institute found that there were more than 2975 excess deaths in the 5 months following the hurricane (Santos-Burgoa et al., 2018). Importantly, they compared the number of deaths in each month to the comparable number in prior years and they took into account the number of people who left the island after the disaster. Dean, Lynn R. Goldman (2018), explained:

> We do not know the exact circumstances around each of the 2,975 excess deaths that occurred. Many factors – disruption in transportation, access to food, water, medications, power, and other essentials – may have contributed. In interviews, we heard many heartbreaking stories of families struggling to obtain emergency health care, power for medical devices, prescription drugs, or even food and drinking water. This is why we were not surprised to find that the highest rates of excess deaths occurred among those living in the poorest municipalities, as well as those over the age of 65, especially men.

In Loiza, Puerto Rico, a town of comparable size to Mamaroneck, with a population of 25,400 and a median income of $17,900 (US Census Bureau, 2019b), Amnesty International (2018) reported that a year later, 350 houses still had blue tarps covering the roofs. Thus a year after the disaster, American citizens living in this town still needed roofs! It is not hard to find headlines that show how the poor and marginalized are grievously and disproportionately impacted by climate change, the pandemic, or other environmental issues.

In the rest of this chapter, I provide a history of the sustainability and ecojustice movements, primarily in the United States. This history serves as a basis for understanding the discourses that are taken up in this volume. I begin with a sketch of the roots of the concept of sustainability and the early years of the environmental justice movement from 1962 to the early 1970s when the term ecojustice came into use. The third section picks up in 1979 with the case of *Bean v. Southern Waste Management* and moves through the early 1990s. The fourth part brings us up to the present day. In these sections, I seek to demonstrate how two streams of thought and action: ecojustice and sustainability have come together and inform each other in contemporary times.

2.2 Sustainability: A Concept Rooted in Forest Management

Jeremy L. Caradonna (2014) provides a history of sustainability that begins in Seventeenth-Century Europe when forests were on the decline. Timber and other forest products formed the basis of a pre-industrial economy that needed wood for hearth and home, for cooking and heating, for fruits, nuts, berries, fodder, glue, tanning products, glass-making, metal-work, charcoal, hunting, and importantly, ship-building. Populations were growing, placing further demands on a dwindling resource that was urgently needed for naval and state power. In 1664, John Evelyn, a founding member of the Royal Society in England wrote *Sylva, or a Discourse of Forest Trees, and the Propagation of Timber in His Majesty's Dominions*,[1] in which he decried the loss of woodlands and offered recommendations to replant and reforest the landscape. In 1669, Jean-Baptiste Colbert, the minister of Finances under Louis XIV in France issued *Ordonnance sur le fait des Eaux et Forêts*,[2] "a forest code that became the model of 'rationalized' state forestry for governments all over Europe and beyond" (Caradonna, 2014, p. 34).

Inspired by Colbert's treatise, in 1713 Hans Carl von Carlowitz, a royal mining administrator of Saxony, Germany, wrote a forestry manual, *Sylvicultura oeconomica, oder haußwirthliche Nachricht und Naturmäßige Anweisung zur wilden Baum-Zucht*.[3] This text explored the intensive cultivation of wild trees and forests, including seed collection, planting, and harvesting, and thus, for the first time, proposed the concept of sustained-yield forestry. For example, Carlowitz wrote, "We must aim for a continuous, resilient, and sustainable use, because [forests] are an indispensable thing, without which the country and its forges could not exist" (Carlowitz, 1713, p. 106, in Caradonna, 2014).

---

[1] Evelyn, J. (1664). *Sylva, or a Discourse of Forest Trees, and the Propagation of Timber in His Majesty's Dominions*. London: Allestry and Martyn.

[2] Colbert, J. (1669). *Ordonnance sur le fait des Eaux et Forêts*. Paris: Chez P. Le Petit, 1669.

[3] Carlowitz, H.C. (1713). *Sylvicultura oeconomica, oder haußwirthliche Nachricht und Naturmäßige Anweisung zur wilden Baum-Zucht*. Leipzig: Braun.

It is important to note that during this time, many Indigenous societies were already using forests resiliently and sustainably and had done so for thousands of years. For example, in *Changes in the Land,* William Cronon (1983) explains that Native Americans in New England lived sustainably in part because the people took advantage of seasonal abundances by moving from inland areas where they hunted game and harvested nuts, berries, and many other plants, to the coast where they could take advantage of fish spawns and shellfish beds. Their hunter-gatherer lifestyle reduced human demands on the ecosystem and made sure no species of plants or animals were overused. Taking advantage of the ecological diversity available as the seasons unfolded, including a wide array of medicinal plants, allowed Native Americans to thrive.

In southern New England, where agricultural practices were adopted, grain could be stored through winter making starvation less of a problem and these Native American groups had larger population densities. They worked the ground with clamshell hoes, raising heaps in which they sowed the seeds stored from the previous harvest. Their low-impact methods left the soil more intact and less prone to erosion. In addition, they interplanted corn, squash, and beans, allowing each crop to contribute to the success of the others. The squash kept weeds to a minimum and preserved soil moisture, the corn served as a pole for the beans, and the beans as we now know, fixed nitrogen, enriching the soil. When fields became less productive, in about 8–10 years, they moved to another location. Summer camps were located near the fields and winter camps were located in areas where game animals would be more plentiful. By moving camps, they could take advantage of new fuel sources which were needed for cooking and heating.

Another forest management practice involved burning extensive areas once or twice a year in spring and fall to keep the forests open and park-like. According to Cronon (1983), this practice resulted in "a forest of large, widely spaced trees, few shrubs, and much grass and herbage." With reduced fuel on the ground, the low-temperature fires burned quickly and extinguished themselves, rarely burning out of control. The practice of selective burning made hunting and travel through the forests easier, allowed plants to grow more lushly in the nutrient-rich soils, created conditions favorable for berries and other foods, produced edge and meadow habitat with lush growth to support a variety of wildlife, such as elk, deer, hare, turkey, and quail, and reduced plant diseases, pests, and fleas. Thus, burning purposefully created abundant game and foods for hunting and gathering.

In a similar way, Jared Diamond (2011) describes how people have lived sustainably in the highlands of New Guinea for about 46,000 years, raising pigs and chickens and growing taro, bananas, yams, sugarcane, and sweet potatoes, with terraced gardens, vertical drainage ditches, and other agricultural practices uniquely suited to a region prone to earthquakes, landslides, and as much as 400 inches of rain per year. Central to the New Guineans' success is their practice of silviculture, or the intentional planting and careful nurturing of the casuarina tree on a large scale, from which they obtain the bulk of their wood products for homes, fences, utensils, weapons, and fuel. The tree is a fast-growing hardwood with root nodules that fix nitrogen and a heavy leaf fall that together enrich the soil. The trees reduce erosion, provide shade, and are enjoyed for the sound they make when the wind blows through their leaves. Also during this time, the Tokugawa Shogunate in Japan addressed deforestation by urging people to plant seedlings, conducting forest inventories, regulating the use of wooded areas, developing forest plantations, and engaging in forest practices to address the imbalance between cutting and producing trees (Diamond, 2011). Finally, places like India, China, and Taiwan also developed state-sponsored forest conservation efforts similar to those in Europe (cf. Grove, 1997).

Richard H. Grove (1997) documents further lessons that were learned during the colonial period as rapidly deforested tropical islands such as Jamaica, St. Helena, St. Vincent, Barbados, and Mauritius showed the adverse impacts of deforestation on the local climate, flora, fauna, and hydrology in a relatively short period of time. The denuded landscapes were hotter and more prone to drought and erosion. Although island conservation efforts were driven primarily by economic concerns, such as impacts on sugar production, island bureaucrats were among the first people to realize that human activity can have an impact on the local climate (Grove, 1997). Thus, Caradonna (2014, p. 45) notes that "sustainability [at this time] traces its roots primarily to imperialists…who cared *very little* about nature or social justice and *very much* about state power, industrialization, and profit."

The industrial revolution and colonialism were marked by a rapacious appetite for natural resources and rapid depletion of forests in colonial territories and the United States. Gifford Pinchot, the first chief of the US Forest Service, pushed for sustainable forestry, or the management of forests for natural resource consumption (Caradonna, 2014). Pinchot had studied at the forestry school in Nancy, Germany, and his leadership set the precedent for managing federal lands from a utilitarian and economic perspective which valued forests for their timber and natural resources rather than their beauty and other ecosystem services. By the 1900s, critics such as George Perkins Marsh, who condemned the reckless destruction of the natural world, and John Muir, who advocated for the preservation of wild places, widely denounced the focus on utilitarian and economic perspectives (Caradonna, 2014). As we will see in the next section, these utilitarian and economic perspectives also came into conflict with poor, Latino, Black, and Indigenous communities who were bearing the brunt of the impacts of industrialization on the environment.

2.3   Civil Rights, Ecojustice, and Environmental Protection

In many ways, the ecojustice movement in the US was born out of an ongoing struggle for civil and human rights that coalesced around issues of environmental racism (cf. Garcia, 2016; Bullard & Johnson, 2009). Environmental justice seeks equal treatment with respect to environmental laws and regulations, whereas ecojustice, or ecological justice, recognizes and tries to address the deeper social issues and ecological concerns embedded in environmental problems within communities (Maranda & Bermel, 2020).

In 1962, two different labor organizations formed in rural California to secure better wages, health care, and living conditions for farmworkers, the Agricultural Workers Organizing Committee (AWOC), led by Larry Itliong, a Filipino labor organizer, and the National Farm Workers Association (NFWA), led by César Estrada Chávez and Dolores Huerta, Mexican labor organizers (Garcia, 2016). The passing of the Civil Rights Act in 1964 galvanized Latinx and Filipino communities that had been organizing around themes of racial and social justice. In September 1965, AWOC initiated a series of strikes against grape growers in Delano, CA. A week later, the NFWA began to strike in solidarity with the AWOC. Then, in 1966, the two organizations merged to create the United Farm Workers Organizing Committee (UFWOC). The organization embraced nonviolent means of protest. For example, in the Spring of 1966, Chávez organized a pilgrimage from Delano to Sacramento, CA. While his supporters sang Mexican protest songs, Chávez walked barefoot along the highways to draw converts and media attention to the cause. Dr. Martin Luther King sent Chávez a telegram saying, "Our separate struggles are really one – a struggle for freedom, for dignity and for humanity" (Ott, 2019).

Chávez and his supporters began organizing boycotts of grape products, especially wine and table grapes sold in stores, an approach that was widely regarded as a turning point in the movement (Garcia, 2016). The boycotts spread across the country. For example, in 1968, Dolores Huerta appealed to unions at the Port of New York and New Jersey and they were able to block the delivery of grapes to New York City. They also initiated secondary boycotts of large chain stores selling table grapes, such as the A & P, and were successful in getting the chain to stop selling grapes in all 430 of its stores. Similar boycotts in other cities were also successful. The Toronto mayor declared November 23, 1968, "Grape Day," and announced that the government would no longer purchase any grapes in solidarity with the farmworkers. In Cleveland, OH, Mayor Carl Burton Stokes, the first African American to be elected mayor of a major U.S. city, also ordered all government facilities to stop serving grapes.

As the boycott dragged on and striking farm workers were subject to egregious, racist attacks, some supporters became impatient and began advocating violence (Kim, 2017). Chávez began a 25-day hunger strike following the example of Mahatma Gandhi. His fast stopped talk of violence and attracted further support from Dr. Martin Luther King, who wrote another telegram to him, saying, "I am deeply moved by your courage in fasting as your personal sacrifice for justice through nonviolence" (Ott, 2019).

In 1969, Chávez gave testimony before the Subcommittee on Labor of the Senate Committee On Labor And Public Welfare. He said,

> An especially serious problem in agricultural employment is the concerted refusal of growers even to discuss their use of economic poisons or pesticides. There are signs that several members of Congress are becoming increasingly aware of the dangers posed by economic poisons to human life and to wildlife, to the air we breathe and the water we drink (Chávez, 1969).

The grape boycott stopped the sale of grapes in New York City, Boston, Detroit, Chicago, Philadelphia, Montreal, and Toronto and in 1970 led to the first contract between growers and farmworkers in California. Importantly, the contract included protection from pesticides. Thus, the movement brought national and international attention to the issue of pesticide use and its effects, not only on farmworkers, but on their families and communities, and on the consumers who would eat the tainted grapes and grape products.

In 1968, Black sanitation workers in Memphis began protesting economic and job-safety issues, bringing together labor and civil rights issues (US EPA, 2015a). Focused on the dignity of the Black person, the ongoing strike was not resolved until after Martin Luther King was assassinated and the mayor agreed to engage in collective bargaining. In 1969, the Young Lords, a group of mostly Puerto Rican activists in New York City, organized around the condition of the streets in "El Barrio," a predominantly Puerto Rican neighborhood in East Harlem (Gandy, 2002). Matthew Gandy (2002) explained, "Piles of garbage were…routinely ignored by the city's sanitation department, in stark contrast to the pristine sidewalks of affluent districts in downtown Manhattan." In July 1969, The Lords launched the "garbage offensive." Equipped with brooms and garbage bags, they began cleaning the streets of their community. By August, the Lords began setting the trash on fire and barricading the streets to protest the piles of garbage that remained uncollected. The activists would go on to address other environmental health issues, such as lead paint and tuberculosis rates in the community.

In 1970, the National Environmental Protection Act was passed. This groundbreaking United States legislation is considered the "Magna Carta" of environmental legislation (cf. Mandelker, 2010). The stated goals of the legislation were, "…to create and maintain conditions under which [humans] and nature can exist in productive harmony, and fulfill the social, economic, and other requirements of present and future generations…" The legislation established that an environmental review process would be needed for all federal actions, such as the construction of airports, buildings, military complexes, or highways. Despite its goals, the legislation relied heavily on impact statements and underestimated the ways in which the complexity of ecological systems might make it difficult to predict the impacts of projects (Mandelker, 2010). For example, Lapping (1975) noted

> Ultimately, full compliance with NEPA's Section 102 may be impossible. The ability to measure impact implies the ability to describe an environmental system. It is inherent in the understanding of ecosystem dynamics that change is the most basic and significant quality of real-world ecosystem behavior.

In addition, early impact statements focused mostly on environmental impacts and not as much on impacts on people. For example, Bhatia and Wernham (2008) called for integrated health impact statements developed by public health institutions since aside from toxic exposures environmental impact statements rarely addressed human health in a comprehensive way.

Meanwhile, at a 1970 conference on "War Crimes and the American Conscience," Arthur Galston, an American plant biologist, and bioethicist proposed the term, ecocide, to mean "the willful and permanent destruction of [an] environment in which a people can live in a manner of their own choosing" (Zierler, 2011). Galston and other scientists had been lobbying at home and abroad against the use of Agent Orange as a defoliant during the Vietnam War and considered its use on par with other crimes against humanity. Also in 1970, the first Earth Day was celebrated, and by the end of that year, the Environmental Protection Agency (EPA) would be established and other new environmental laws, including the National Environmental Education Act, the Occupational Safety and Health Act, and the Clean Air Act. The Clean Water Act would follow in 1972. Importantly, these laws were not passed solely for the sake of the environment, rather they were instigated by concerns for the impact of environmental degradation on human well-being (Salkin et al., 2012). The Marine Mammals Protection Act in passed in 1972 and the Endangered Species Act in 1973 began to focus on habitat areas and the function of habitats as ecosystems.

While the environmental movement was expanding, the ecojustice movement also began to grow due to community and ecumenical interest in environmental and social justice issues. Rather than choosing between issues of conservation or poverty, churches and campus ministry programs began to pursue both ecological and economic justice goals (Gibson, 2004). As William Gibson (2004, p. 3) eloquently maintained, "Concern for the earth and its myriad creatures and systems should not, must not, be a turning away from the cause of oppressed and suffering people." Ecojustice has been defined as, "the condition or principle of being just or equitable with respect to ecological sustainability and protection of the environment, as well as social and economic issues" (Oxford English Dictionary). Although not explicitly mentioned in this definition, at its outset, ecojustice was very much an ethic that viewed environmental health and wholeness as "inseparable from human well-being" (Hessel, 2004, p. xii.).

For example, in 1971, Norman J. Faramelli, an Episcopal minister, wrote:

> Environmental quality can be achieved by either expansive applications of pollution control technology, or by a long-range reduction in the production of material goods. In either case, there will be severe repercussions on the poor. The neglect of the poor, and the impact of specific ecology solutions on them, are among the weakest links in the ecology movement. Thus, the relationship of ecological responsibility to economic justice needs to be explored (1971, p. 218).

Faramelli (1971) made it clear that there would be costs to achieving environmental quality and that these costs would not be borne equally by the affluent and the poor. Technological solutions would increase the costs of production, which could be passed on to the consumer through increased prices (Option 1), or costs could be borne by the taxpayer if solutions were funded through government subsidies (Option 2), since reducing corporate profits (Option 3) was not likely. Faramelli explained:

> The American dream is rooted in a three-pronged syndrome - an active process of acquisition, consumption, and disposal. The net result of this process is dissatisfaction which we try to solve by repeating the cycle. Hence, an insatiable material appetite makes the wheels of American progress go around (Faramelli, 1971, pp. 225–226).

Environmental activists increasingly understood that surging rates of consumption, in the nation, and indeed the world, were unsustainable. At the same time, the environmental movement was increasingly criticized for being a predominantly White, affluent movement (Faramelli, 1971; See also Ghoche & Udoh, this volume; Betancor et al., this volume). A growing distinction was also being made between environmentalism, which focused on reducing human impacts to the environment, and environmental justice, which focused on remediating the impacts of pollution and environmental degradation on communities (both human and nonhuman) suffering adverse impacts from human activity (Carder, 2017).

In 1972, The Club of Rome founded by Aurelio Peccei in 1968, published *The Limits to Growth* (Meadows et al., 1972). This text among others at the time, such as Ernst F. Schumacher's (1973) *Small is Beautiful: Economics as if people mattered,* and Ezra J. Mishan's *The Cost of Economic Growth,* highlighted the costs of growth (questioning the idea that all economic growth and development are inherently good), critiqued economic models ignoring the environment (so-called externalities), derided useless metrics and measurements (such as GDP, a crude measure of economic activity rather than the actual health or resilience of the economy), and denounced technology worship (the idea that all technology was essentially benign without determining whether it would allow humans to live within biophysical limits) (Caradonna, 2014).

In 1977, the National Academy of Sciences (NRC, 1977, p. viii) released a report, *Energy and Climate,* in which they stated, "The principal conclusion of this study is that the primary limiting factor on energy production from fossil fuels over the next few centuries may turn out to be the climatic effects of the release of carbon dioxide." The report acknowledged that heat, particulates, and especially carbon dioxide (CO_2) from energy production and use had the potential to modify the global climate. In addition the report called for further research on climate, carbon dioxide, the atmosphere-ocean-biosphere system, world population, energy consumption, potential renewable resources, potential impacts of climate change on food and water resources, and the development of better climate models.

2.4 Ecojustice, the Fight for Healthy Communities and Sustainable Development

In 1979, a group of African American homeowners in suburban Houston protested the building of Whispering Pines Sanitary Landfill in their community (Bullard, 1990). The landfill site would be 1500 feet from a local public school and within two miles of six schools. Although the lawsuit, *Bean v. Southern Waste Management,* led by Linda McKeever Bullard, ultimately failed, it was the first case to charge environmental discrimination. In 1982, a case in Warren County, North Carolina drew national coverage. More than 500 arrests of environmentalists and civil rights activists were made during protests of a polychlorinated biphenyl (PCB) landfill planned for construction in a rural African American community. The protests did not succeed in halting construction.

In 1980, the Comprehensive Environmental Response, Compensation, and Liability Act (CERCLA) was passed (US EPA, 2013a). Superfund sites consist of "uncontrolled or abandoned hazardous-waste sites as well as accidents, spills, and other emergency releases of pollutants and contaminants into the environment" (US EPA, 2013a; See also Bower & Rodriguez, this volume). CERCLA gave the EPA the authority to identify responsible parties and ensure their participation in the cleanup. Meanwhile in 1981, NASA (National Aeronautics and Space Administration) scientists, James Hansen et al. (1981) published their findings regarding global climate trends. They noted that globally, the "temperature rose by 0.2 °C between the middle 1960's and 1980, yielding a warming of 0.4 °C in the past century. This temperature increase is consistent with the calculated greenhouse effect due to measured increases of atmospheric carbon dioxide." Earlier studies had already concluded that the increased CO_2 was primarily from burning fossil fuels (NRC, 1977). The study also established the ways in which climate models and confidence in the findings were improving.

Two studies at this time drew a clear link between race and the siting of hazardous waste landfills. The first, *Solid Waste Sites and the Black Houston Community* (Bullard, 1983), conducted by Dr. Robert Bullard (husband to Linda McKeever Bullard) investigated the location of municipal waste disposal facilities in Houston. Although African Americans only made up 25% of the population of Houston, all five city-owned dumps, 80% of city-owned incinerators, and 75% of privately-owned landfills were located in Black neighborhoods. The second, a 1983 U.S. General Accounting Office study, *Siting of Hazardous Waste Landfills and Their Correlation with Racial and Economic Status of Surrounding Communities*, confirmed that three out of four of the commercial hazardous waste landfills in Region 4, which includes eight Southern states, were located in majority African American communities. In addition, the percent of people with income below the poverty level in the four communities ranged from 26% to 42%, and of those, 90–100% were Black.

Meanwhile, in 1986, César E. Chávez started a new boycott with a famous *Wrath of Grapes* speech. In the speech, Chavez said, "The worth of humans is involved here" (Chávez, 1986). He also noted:

> We farm workers are closest to food production. We were the first to recognize the serious health hazards of agriculture pesticides to both consumers and ourselves…Our first contracts banned the use of DDT, DDE, Dieldrin on crops, years before the federal government acted. Twenty years later, our contracts still seek to limit the spread of poison in our food and fields (Chávez, 1986).

Chávez (1986) called for an immediate ban on the use of Parathion, Phosdrin, Dinoseb, Methyl Bromide, and Captan in grape production; a testing program for pesticide residues on grapes; free and fair elections for farmworkers to decide whether to organize and negotiate contracts; and good faith bargaining.

A third study of toxic wastes, completed in 1987, *Toxic Wastes and Race in the United States,* by the United Church of Christ (UCC) Commission for Racial Justice, expanded the documentation of unequal and discriminatory siting of toxic waste facilities across the United States. This study concluded that "Race proved to be the most significant among variables tested in association with the location of commercial hazardous waste facilities. This represented a consistent national pattern (UCC, p. xii)." Paul Mohai et al. (2009, p. 406) explain, "…hundreds of studies conclude that, in general, ethnic minorities, [I]ndigenous persons, people of color, and low-income communities confront a higher burden of environmental exposure from air, water, and soil pollution from industrialization, militarization, and consumer practices."

In 1987, the report of the Brundtland Commission, Our Common Future (WCED, 1987), was released by the United Nations. This report famously defines sustainable development as "development that meets the needs of the present without compromising the ability of future generations to meet their own needs." It is important to note that up to this point, development focused on three pillars, peace and national security, economic prosperity, and human rights; however, with the addition of the modifier, sustainable, the idea took hold that development must also protect and restore the environment (Salkin et al., 2012). At the same time, the term, sustainable development, was seen as problematic because development generally implied economic growth; however, such growth often happened at the expense of ecological sustainability goals. (See Robinson, 2004).

In 1988, West Harlem Environmental Action (WE-ACT) was founded to address local environmental justice issues, including the construction of the North River Sewage Treatment Plant and a bus depot across from an intermediate school and a large housing development. Planners originally sited the sewage plant near 72nd street, a predominantly white, upper-middle-class neighborhood, but due to technical problems and community resistance, they relocated the site to West Harlem. The plant spanned eight blocks along the Hudson River, from 137th street to 145th street. From the beginning, West Harlem residents dealt with foul odors that were particularly potent during the summer months, itchy eyes, shortness of breath, asthma, and other respiratory conditions. WE-ACT sued the NYC Department of Environmental Protection for operating the plant and in 1993, they won a $1.1 million settlement. To this day, driving by if the plant is belching remains unpleasant. Bullard (1990) made it clear that NIMBY, "not in my backyard," often resulted in PIBBY, "put in Black's backyards."

Also in 1988, The United Nations Environment Programme (UNEP) and the World Meteorological Organization (WMO) established the Intergovernmental Panel on Climate Change (IPCC), which published its first assessment report in 1990. This global group joined prominent US scientists, such as NASA scientist James Hansen, to form a working group of several hundred scientists from 25 countries. They concluded that "emissions resulting from human activities are substantially increasing the atmospheric concentrations of the greenhouse" and that "Business-as-Usual (Scenario A) emissions… [could result in] a rate of increase of global mean temperature during the next century of about 0.3 °C per decade" (IPCC,

1990, p. xii). Although uncertainties were recognized regarding the specific timing, magnitude, and regional patterns of climate change, nevertheless, warming, sea-level rise, and climatic changes were expected. At the same time, Pearce et al. (1989, p. 10) laid out *A Blueprint for a Green Economy* for the United Kingdom in which they noted:

> When our demand for resources and environmental services starts to outstrip the planet's capacity to provide them, then the problems we are storing up for ourselves become exceptionally serious…The underlying cause is a way of life which is out of step with the long-term health of the planet. The solution requires us to dig deep into our reserves of human ingenuity: to challenge our own cultural beliefs, economic assumptions, and policy frameworks.

According to the authors, a green economy requires a system of environmental accounting that respects: "the four interdependent 'securities' of nature – energy security, water security, food security, and climate security" (p. 22). They explained,

> All overlap in complex ways. For example, if we put huge areas of fertile land over for production of biofuels to gain energy security or increase climate security, what will be the effects on food and water security? Failing to understand how these things mesh together ultimately damages us all (p. 22).

Pearce et al. (1989) recognized the inherent tensions regarding what to prioritize among the four interdependent securities–energy, water, food, and climate–and the concomitant desire to foster economic and social development.

2.5 EcoJustice and Indigenous Communities: Shared Concerns

In 1990, the Indigenous Environmental Network (IEN) was formed with the goal of "building the capacity of Indigenous communities and tribal governments to develop mechanisms to protect [their] sacred sites, land, water, air, natural resources, the health of both [their] people and all living things, and to build economically sustainable communities (IEN, 2012)." During this time, as with the above cases, many tribal communities were being targeted for large, toxic, municipal, and hazardous waste dumps or nuclear storage facilities. At the same time, industrial and mineral development on tribal lands was contaminating the soil and water. Indigenous communities began organizing to resist these incursions on tribal lands. Also in 1990, Dr. Robert Bullard published *Dumping in Dixie*, a book that documented the efforts of five grassroots efforts of Black communities in the South to address environmental justice issues including, the siting of municipal landfills and incinerators, a lead smelter, and hazardous waste facilities (See also Cook et al., this volume).

That same year, Richard Moore, director of the SouthWest Organizing Project, a grassroots advocacy group in Albuquerque, N.M., sent a letter to the Executive Directors of the top ten environmental conservation groups, including the Sierra Club and the National Resources Defense Council (Durlin, 2010). "Signed by 100 cultural, arts, community and religious leaders – all people of color – …the letter charged the organizations with a history of 'racist and exclusionary practices,' a lack of in-house diversity, and an all-around failure to support environmental justice efforts" (Durlin, 2010). The letter caught many of the directors by surprise but was pivotal in helping some groups begin incorporating environmental justice advocacy, such as the Sierra Club.

A year later, delegates from Puerto Rico, Canada, Central, and South America, and the Marshall Islands drafted and adopted 17 Principles of Environmental Justice at the First National People of Color Environmental Leadership Summit in Washington, D.C. (Alston, 2010). Reporting on the summit, Alston (2010, p. 15) noted, "Delegates detailed numerous examples where the unilateral policies, activities, and decision-making practices of environmental organizations have had a negative impact on the social, economic, and cultural survival of communities of color in the United States and around the world." The principles affirmed the sacredness of Mother Earth, demanded that public policy be based on mutual respect and justice for all peoples, called for ethical and responsible use of land and resources and protection from nuclear testing, toxic/hazardous wastes, and poisons, as well as the right to political, economic, cultural, and environmental self-determination and participation in decision-making. Importantly, the gathering and discussion of ongoing struggles "dispel[led] the myth that people of color are not interested in or active on issues of the environment" (Alston, 2010, p. 15; also see Patterson et al., this volume).

In 1992, a report from the Environmental Equity Group of the EPA, *Environmental Equity: Reducing Risk for All Communities*, concluded, "The evidence indicates that racial minority and low-income populations are disproportionately exposed to lead, selected air pollutants, hazardous waste facilities, contaminated fish tissue, and agricultural pesticides, in the workplace" (US EPA, 1992). The report defined environmental equity as an examination of the differential risk burden borne by low-income and racial minority communities and how governmental agencies respond. As a result of the recommendations in this report, the office of Environmental Equity was established and later renamed the Office of Environmental Justice in 1994.

Also in 1992, Bunyan Bryant and Paul Mohair, who had established an Environmental Justice program at the University of Michigan, published *Race and the Incidence of Environmental Hazards*, a collection of 16 articles that exposed examples of environmental inequity from consumption of toxic fish from the Detroit River, the fallout from hazardous waste incineration in Louisiana, pesticide exposure among farmworkers, and the effect of uranium production in Navajo communities. The authors pushed back against the idea that laws and environmental regulations equally protected all communities from harmful pollutants.

That same year, the United Nations Conference on Environment and Development, also known as the Earth Summit, was held in Rio de Janeiro in June 1992. As part of this summit, nations endorsed *Agenda 21*. The report of the summit clearly noted:

> Poverty and environmental degradation are closely interrelated. While poverty results in certain kinds of environmental stress, the major cause of the continued deterioration of the global environment is the unsustainable pattern of consumption and production, particularly in industrialized countries, which is a matter of grave concern, aggravating poverty and imbalances (UNCED, 1992).

The recommendations of Agenda 21 were non-binding but they called for nations to act locally and in partnership with other nations to foster more sustainable development (UNCED, 1992). Also in 1992, William E. Rees and Mathis Wackernagel (Rees, 1992; Wackernagel & Rees, 1996) developed the concept of an "ecological footprint," or how much land is required to support the standard of living of a city's population. Rees (1992) made it clear that the ecological footprint of a city is several orders of magnitude larger than the geographical footprint.

In 1993, the National Environmental Justice Advisory Council (NEJAC) was established as an advisory committee to the EPA, with representatives from the community, academic institutions, industry, and environmental, Indigenous, and government groups. A central objective of NEJAC was "to improve the environment or public health in communities disproportionately burdened by environmental harms and risks" (US EPA, 2015b). In 1994, President William Clinton issued Executive Order 12898, which required federal agencies to incorporate environmental justice as part of their missions (US EPA, 2013b), and in 1995 the first Interagency Public Meeting on Environmental Justice was held.

Globally, the first United Nations Framework Convention on Climate Change (UNFCCC) Conference of the Parties (COP) took place in April 1995 in Berlin, Germany (United Nations Climate Change, n.d.-a). These meetings would be held annually to review implementation and emissions inventories submitted by the Parties (nations). Later that year, the IPCC Second Assessment Report was issued in December 1995 which affirmed the science of climate change (IPCC, 1995). The report noted that "carbon dioxide remains the most important contributor to anthropogenic forcing of climate change..." (p. xi). In addition, "projections of future global mean temperature change and sea level rise confirm the potential for human activities to alter the Earth's climate to an extent unprecedented in human history" (p. xi). In 1996, The Kyoto Protocol, adopted in 1997, set emissions targets for six greenhouse, carbon dioxide (CO_2); methane (CH_4); nitrous oxide (N_2O); hydrofluorocarbons (HFCs); perfluorocarbons (PFCs); and sulfur hexafluoride (SF_6); however, the United States did not ratify the Kyoto protocol (UNCC, n.d.-b).

While COP6 was being held at the Hague in 2000, an alternative meeting of community activists from around the world gathered for the first Climate Justice Summit (Karliner, 2000). Karliner reported, "Speaker after speaker described the human rights violations and environmental devastation wrought by the fossil fuel industry as well as the industry's responsibility for the global dynamic of climate change." Meanwhile, in 2001, Warren County, North Carolina finally secured state and federal funding to remediate the PCB landfill they had protested unsuccessfully back in 1982 (see above), and in 2002, the Second National People of Color Environmental Leadership Summit was held in Washington, D. C. A report following the summit explained, "Summit II should have been a proud display of what had been accomplished by the original networks, organizations, and leaders of color that had come together for the First Summit;" however, the overall feeling was that as environmental justice became more bureaucratized, and part of the EPA, grassroots voices and concerns were sidelined (Martinez, 2003). Tools and resources for environmental justice, established by President Clinton's Executive Order 12898, were used to grow the EPA, and fund research and environmental justice programs at universities rather than address local issues. The Indigenous Caucus at the summit highlighted the ways in which the United States continued to violate legal and moral obligations to Indigenous peoples and called for an end to unsustainable energy policies, mining practices, and oil development, as well as greater protection from pollution and destruction of waters, forests, and sacred sites in Indigenous territories (SRIC, 2003).

In 2005, the World Summit on Social Development affirmed three goals of sustainable development: economic development, social development, and environmental protection. The first two goals, economic and social development were not supposed to come at the expense of the third, environmental protection. A report of the 59th session of the General Assembly of the United Nations (UN, 2005, p. 51) noted that both halves of the term, sustainable development, should be given "their due weight." A year later, former Vice President, Al Gore's documentary, *An Inconvenient Truth* (Guggenheim, 2006) aired. The documentary explained the science behind global warming, the relentless rise of CO_2 in the atmosphere, the ongoing loss of ice caps on mountains worldwide, such as Kilimanjaro in Africa, the Himalayas in Nepal, and the Alps in Italy, and the retreat of glaciers in Glacier National Park and the Columbia Glacier in Alaska. Gore noted, "There are good people who are in politics, in both parties, who hold [global warming] at arm's length because if they acknowledge it and recognize it, then the moral imperative to make big changes is inescapable." The film is credited with waking many people up to the issue, helping young people grow up knowing the facts about global warming and taking action, spurring individuals and businesses to change and become more sustainable, and engaging more people in speaking up about climate change and teaching others (The Climate Reality Project, 2016; see also Pfirman & Winckler, this volume).

2.6 Coming Up to the Present: Climate and Indigenous Environmental Justice

In 2007, the report, *Toxic Race and Wastes at Twenty: 1987–2007* (Bullard et al., 2007, p. viii) concluded that "people of color are found to be more concentrated around hazardous waste facilities than previously shown." Key findings from the report were:

- Racial disparities for people of color as a whole exist in nine out of 10 U.S. EPA regions (all except Region 3, the Mid-Atlantic Region).

- Forty of the 44 states (90%) with hazardous waste facilities have disproportionately high percentages of people of color in circular host neighborhoods within three kilometers of the facilities.
- In metropolitan areas, where four of every five hazardous waste facilities are located, people of color percentages in hazardous waste host neighborhoods are significantly greater than those in nonhost areas (57% vs. 33%) (pp. x–xi).

The authors concluded, "[t]he current environmental protection apparatus is 'broken' and needs to be 'fixed.' The current environmental protection system fails to provide equal protection to people of color and low-income communities" (p. xii).

One of the difficulties with securing legal remedies and environmental justice is that it can be difficult to prove racist intent. For example, Section 2 of the 1965 Voting Rights Act focused on discrimination that "results in the denial or abridgment of the right of any citizen to vote" on the basis of race and ethnicity, and a 1972 amendment included membership in a language minority group (US DOJ, 2021). While voter registration rates surged, problems remained due to gerrymandering, annexation, and other practices that limited Black people from voting (US DOJ, 2017). Also, "the Supreme Court required that any constitutional claim of minority vote dilution must include proof of a racially discriminatory purpose, a requirement that was widely seen as making such claims far more difficult to prove" (US DOJ, 2017). Thus in 1982, "Congress amended Section 2 to provide that a plaintiff could establish a violation of the Section without having to prove discriminatory purpose" (US DOJ, 2017). Nevertheless, since the same story (discrimination), different context (environmental justice issues) was repeated over and over, there was a dire need for a concerted effort to better enforce existing laws, to remediate toxic sites, and address the disparate impacts that continued to affect Black, Latinx, and Indigenous communities.

Also in 2007, the United Nations Declaration on the Rights of Indigenous Peoples (UNDRIP) called for "Free, Prior and Informed Consent (FPIC)" (UN, 2008). Article 19 declares:

> States shall consult and cooperate in good faith with the [I]ndigenous peoples concerned through their own representative institutions in order to obtain their free, prior and informed consent before adopting and implementing legislative or administrative measures that may affect them. (p. 8)

According to the UN, FPIC allows Indigenous people "to give or withhold consent to a project that may affect them or their territories." The US was one of four nations, including Canada, Australia, and New Zealand to vote against the declaration and the last of the four nations to change their position (UN, n.d.). President Barack Obama announced support for the declaration in 2010 (US DOS, n.d.). While the announcement affirmed the moral and political force of the UNDRIP, it also stated that the document was not "legally binding," that it was "aspirational" and would be "achieve[d] within the structure of the U.S. Constitution, laws, and international obligations, while also seeking, where appropriate, to improve our laws and policies" (p. 1).

Progress with sustainability was made in 2009 when the Federal Partnership for Sustainable Communities was formed. The six "livability principles" of this partnership between the US EPA, the Department of Housing and Urban Development, and the Department of Transportation were to: (1) provide more transportation choices; (2) promote equitable affordable housing; (3) enhance economic competitiveness; (4) support existing communities; (5) coordinate and leverage federal policies and investment, and (6) value communities and neighborhoods (US EPA, 2013c). The partnership mobilized funding for sustainability projects across the country, streamlined federal regulations and policies to support sustainable development, and aligned and coordinated agency priorities. On the other hand, the 2009 UN Climate Change Conference was considered disastrous, with no agreements or meaningful plans to reduce greenhouse gases adopted by the attendees (Caradonna, 2014).

In 2010, a *Symposium on the Science of Disproportionate Environmental Health Impacts* explored the factors that contribute to disparate and "adverse health outcomes [such] as elevated blood lead, asthma, preterm births, and morbidity and mortality from cardiovascular diseases" (Nweke et al., 2011, p. S19) for racial and ethnic minority, Indigenous, and low-income populations in the United States. The symposium explored the state of scientific knowledge for seven factors that contribute to disparate health outcomes, including unique exposure pathways, multiple and cumulative environmental burdens, physical infrastructure, diminished capacity to participate in decision-making, vulnerability and susceptibility, proximity to sources of environmental hazards, and chronic psychosocial stress. Methods to assess the factors and incorporate the data into community-based decision-making processes were presented. Importantly, they recognized that collaborative and holistic approaches were needed to address the root causes of environmental justice issues as well as their disparate health impacts. Also in 2010, the Global Alliance for the Rights of Nature (GARN) was formed. The alliance states, "Rather than treating nature as property under the law, the time has come to recognize that natural communities have the right to exist, maintain and regenerate their vital cycles" (GARN, 2021).

The Rio +20 Conference on Sustainable Development in 2012 and its report, *The Future We Want* (UNCSD, 2012), reaffirmed the Rio Principles of 1992, committed to poverty eradication, and urged all Parties to fulfill their pledges to sustainable development goals. The report acknowledged "that climate change is a cross-cutting and persistent crisis," and emphasized that "combating climate change requires urgent and ambitious action, in accordance with the principles and provisions of the United Nations Framework Convention on Climate Change" (p. 8). The report also committed to a green economy in the context of sustainable development goals. Although the term, green economy, was not defined in the report, the UN has defined it elsewhere as low carbon, resource-efficient, and socially inclusive (Pearce et al., 1989; UNEP, 2011).

In 2014, the EPA launched a mandatory *Introduction to Environmental Justice Online* training course for all agency staff and managers. The course provided an overview of environmental justice and the EPA's role in fostering healthy communities. Over 17,000 EPA employees completed the training in the first year (US EPA, 2015a). Also in 2014, the EPA Office of Environmental Justice collaborated with the American Indian Environmental Office to create and issue a document, *EPA Policy on Environmental Justice for Working with Federally Recognized Tribes and Indigenous Peoples* (US EPA, 2014). The report served to "clarify and integrate environmental justice principles in a consistent manner in the Agency's work with federally recognized tribes and [I]ndigenous peoples" (US EPA, 2014, p. 1). In that same year, *Indian Country Today* reported that an incredibly disproportionate number, 532 out of 1322, of Superfund sites in the US were located on Indian lands (Hansen, 2014).

In 2015, Pope Francis released an encyclical, *Laudato Si, On Care for Our Common Home,* in which he stated,

> The continued acceleration of changes affecting humanity and the planet is coupled today with a more intensified pace of life and work which might be called "rapidification". Although change is part of the working of complex systems, the speed with which human activity has developed contrasts with the naturally slow pace of biological evolution. Moreover, the goals of this rapid and constant change are not necessarily geared to the common good or to integral and sustainable human development (p. 15).

Pope Francis cited an urgent need to address pollution, waste, a "throwaway culture," lack of access to clean drinking water, loss of biodiversity, a decline in the quality of human life, the breakdown of society, and global inequality while recognizing the climate as a "common good." Importantly, common goods are those that cannot be had without cooperation, which in the case of local and global climate must happen across social, cultural, and ethnic differences, borders, nations, and regions.

Also in 2015, The American Heart Association issued a report prepared by Echo Hawk Consulting, *Feeding Ourselves: Food Access, Health Disparities, and the Pathways to Healthy Native American Communities.* The report enumerated the challenges Native communities face with health disparities and food insecurity resulting from their forced separation from traditional lands and U.S. Indian policies that have led to inordinately high rates of obesity and diabetes among other health concerns. For example, childhood obesity exceeded 50% in many communities, and 80% of adults aged 20–74 were obese. As a comparison, recent studies in New York City place childhood obesity rates at 22–27% (Columbia University Mailman School of Public Health, n.d.).

The Dakota Access Pipeline (DAPL) project also made the headlines in 2015 (See McKibben, 2016). Originally, the pipeline was routed to cross the Missouri River near Bismarck, North Dakota. Concerns that an oil spill could damage the water supply for the State's capital prompted a move of the pipeline south of the capital and just north of the Standing the Rock Sioux Reservation. The Standing Rock Sioux Tribe passed a resolution regarding DAPL, which was to be constructed on historically unceded land and under the Lake Oahe reservoir that provides their drinking water (Camp of the Sacred Stones, 2015). The tribe invoked the 1868 Fort Laramie Treaty (US National Archives, n.d.) and cited the harms, including threats to public health and welfare, destruction of valuable cultural resources, and abrogation of their right to "undisturbed use and occupation" of their lands. The Standing Rock Sioux Tribe cited multiple federal violations in the permitting process (Camp of the Sacred Stones, n.d.). The disproportionate impacts of rerouting the pipeline went against Executive Order 12898. Not designating the Missouri aquifer as a "high-consequence area" went against the Pipeline Safety and Clean Water Acts. Finally, a lesser environmental assessment was conducted instead of an in-depth environmental impact statement, which went against NEPA, and the fact that historical ceremony and burial grounds would be impacted went against Executive Order 13007 on Protection of Sacred Sites. The resolution called on the Army Corps of Engineers (ACE) to reject the permit for the pipeline.

Despite the Standing Rock Sioux Tribe's resolution, plans moved forward for the pipeline's construction, and a wide-scale, grassroots movement emerged in response with a camp erected at Standing Rock, youth rallies, petitions, relays, and protests in Washington D.C. (Hersher, 2017). Locally, these protests were met with an increasingly militarized response. In 2016, the EPA issued another report, *Promising Practices for Environmental Justice Methodologies in NEPA Reviews* (US EPA, 2016a), which ironically cited the following:

> Meaningful engagement efforts with potentially affected minority populations, low-income populations, and other interested individuals, communities, and organizations are generally most effective and beneficial for agencies and communities when initiated early and conducted (as appropriate) throughout each step of the NEPA process (p. 8).

The Standing Rock Sioux tribe argued they had not had opportunities for free, prior, informed consent to the pipeline project (UN, 2008). Thus, clear contradictions between stated policy intentions and actual implementation of environmental justice guidelines continued to exacerbate instances of environmental injustice.

For example, the EPA publication of an *Environmental Justice 2020 Action Agenda* (US EPA, 2016b) in August 2016 laid out three main goals:

- Deepen environmental justice practice within EPA programs to improve the health and environment of overburdened communities;
- Work with partners to expand positive impact within overburdened communities; and
- Demonstrate progress on significant national environmental justice challenges.

Despite the promise of this document, the incoming Trump administration did not support environmental justice, ecological justice, or sustainability goals. The EPA Environmental Justice timeline (US EPA, 2015a), which had seen a flurry of activity in the years prior, came to a sudden halt in 2016. In January 2017, in one of his first actions as President, Donald J. Trump issued a memorandum for the Army Corps of Engineers to expedite the environmental review of the DAPL project, and 2 weeks later, the easement was granted and construction resumed (Naylor, 2017; Hersher, 2017).

The DAPL case highlights several fundamental differences between mainstream environmental justice or ecojustice and Indigenous environmental justice. One of the main differences hinges on Indigenous beliefs about land and water as sacred, not to be despoiled, but held in trust and care for present and future generations (Gilio-Whitaker, 2019). Unlike Western conceptualizations that require sacred things to be set apart, as with the National Park System (the majority of which came at the expense of Indian removal) Indigenous perspectives advocate for respect and responsibility toward the natural world and relationships that foster reciprocity between humans and nature, positioning humans as part and parcel of nature, not in dominion over it (Gilio-Whitaker, 2019). A rallying cry for the DAPL struggle was, "Water is life. *Mní Wičóni*" (Weston, 2017), which was literally understood as, water is alive. Indigenous environmental justice also encompasses distributive justice (equal protection), corrective justice (addressing damages), social justice (fairness in meeting basic needs), and procedural justice (meaningful participation in decision-making) (Jarratt-Snider & Nielsen, 2020). Karen Jarratt-Snider and Marianne O. Nielsen (2020) provide three additional aspects that frame Indigenous environmental justice, (1) Native American tribes are governments, not ethnic minorities, with a unique political and legal status; (2) Native peoples have deep connections to traditional homelands that are central to their identities and cultural survival; and (3) Native peoples suffer from the ongoing effects of colonization, including land dispossession, destruction of sacred sites, loss of subsistence rights, and the detrimental effects of environmental contamination including the land and water.

Sadly, DAPL was only the first of many actions during this period that placed people and the environment in harm's way. The Trump administration also oversaw more than 100 rollbacks of environmental protection rules, including withdrawing from the Paris Climate Agreement in 2017, weakening limits on carbon dioxide emissions from power plants in 2019 and cars and trucks in 2020, as well as removing protections from more than half the nation's wetlands and removing restrictions on mercury emissions from power plants in 2020 (Popovich et al., 2020). In 2019, the Department of the Interior opened federal land for oil and gas leasing by weakening the Endangered Species Act and the Department of Energy loosened efficiency standards for products (Popovich et al., 2020). In September 2021, Trump officials relocated the Environmental Justice Office from the Office of Enforcement and Compliance Assurance to the Office of Policy, purportedly to improve efficiency; however, critics were concerned that the move would politicize the office (Perls, 2020). The EPA also changed the name of the Office of Sustainable Communities to the Office of Community Revitalization (Perls, 2020).

In January 2021, in one of his first acts as President, Joseph R. Biden wrote a letter to rejoin the Paris Climate Agreement (Biden, 2021). President Biden followed this action with Executive Order 14008 (2021), which lays out a bold vision for addressing both climate change and environmental justice issues:

> It is the policy of my Administration to organize and deploy the full capacity of its agencies to combat the climate crisis to implement a Government-wide approach that reduces climate pollution in every sector of the economy; increases resilience to the impacts of climate change; protects public health; conserves our lands, waters, and biodiversity; delivers environmental justice; and spurs well-paying union jobs and economic growth, especially

through innovation, commercialization, and deployment of clean energy technologies and infrastructure.

Importantly, Biden notes, "Successfully meeting these challenges will require the Federal Government to pursue such a coordinated approach from planning to implementation, coupled with substantive engagement by stakeholders, including State, local, and Tribal governments." The timing of these actions was crucial since the sixth IPCC report was issued in August 2021 which noted, "It is unequivocal that human influence has warmed the atmosphere, ocean, and land. Widespread and rapid changes in the atmosphere, ocean, cryosphere, and biosphere have occurred (IPCC, 2021, p. 5)." In November 2021, COP 26 was held in Glasgow which affirmed four pillars of climate action: adaptation, finance, and mitigation, and collaboration and kept the goal of limiting climate change to 1.5 °C within reach (UNFCCC, 2021). The Glasgow Climate Pact was agreed to by 200 countries in attendance and the rulebook for the Paris Climate Agreement was finalized, which includes the norms for establishing carbon markets (UNFCCC, 2021).

While a full history of ecojustice, environmental justice, sustainability, and their intersections with climate justice would take a volume of its own, this select history of key events and ideas in these movements makes clear the need to develop more comprehensive and just approaches to economic, ecological, and social development. Ecojustice emphasizes the role of just participation in decision-making and free, prior, informed consent, as well as repairing the ecological harms and disparate health effects of economic development and pollution on the communities that have been impacted. History shows that these communities are predominantly Black, Latinx, Indigenous, and low-income. Environmental justice focuses on rules, regulations, and laws, but history also shows that litigation can be a slow, incremental approach to ensuring just outcomes and mitigating environmental justice issues. Sustainability brings forward the ideas of ecological conservation, preservation, and restoration as necessary aspects of economic development. Furthermore, sustainability comprises both a process and a goal that communities and nations have to engage in and work towards (NRC, 2011). Increasingly, sustainability acknowledges the need to balance economic growth with planetary limits in mind with respect to energy, sustainable design and green building, urbanization, transportation, higher education and research, business and finance, as well as equality, democracy, social justice, and well-being (Caradonna, 2014). Indigenous framing and values make it clear that we must consider the seventh generation as we plan for the present (Gilio-Whitaker, 2019).

The COP 26 agreements provide signs of hope that, together, we will become more climate-resilient and build towards a more sustainable and just future. Furthermore, they position ecojustice and Indigenous environmental justice as frameworks that have the potential to bring together concerns for social and environmental justice, including ecological sustainability; social and economic justice; inclusive, healthy, and safe communities; enough food; and potable water in ways that affirm the interdependence and relatedness of all people and all kinds of biodiversity.

References

Alston, D. (2010). The Summit: Transforming a Movement. *Race Poverty & the Environment, 17*(1), 14-17.

Amnesty International. (2018). *Puerto Rico a year after Hurricane Maria.* https://www.amnesty.org/en/latest/research/2018/09/puerto-rico-a-year-after-hurricane-maria/#death

Betancor, O., Wunnerlind, C., Callahan, H. S., Dinoor, Y., & Elkis, R. (this volume). Panel One: Theorizing the environment. In M. S. Rivera Maulucci, S. Pfirman, & H. S. Callahan (Eds.), *Education for sustainability: Discourses on authenticity, inclusion, and justice.* Springer.

Bhatia, R., & Wernham, A. (2008). Integrating human health into environmental impact assessment: An unrealized opportunity for environmental health and justice. *Environmental Health Perspectives, 116*(8), 991–1000. https://doi.org/10.1289/ehp.11132

Biden, J. R. (2021, January 20). *Paris Climate Agreement.* Retrieved December 28, 2021, from The White House website https://www.whitehouse.gov/briefing-room/statements-releases/2021/01/20/paris-climate-agreement/

Bower, P. & Rodriguez, S. (this volume). In M. S. Rivera Maulucci, S. Pfirman, & H. S. Callahan (Eds.), *Education for sustainability: Discourses on authenticity, inclusion, and justice.* Springer.

Bullard, R. D. (1983). *Dumping in Dixie: Race, class, and environmental quality.* Westview Press.

Bullard, R. D. (1990). Solid waste sites and the black Houston community. *Sociological Inquiry, 53*(2–3), 273–288. https://doi.org/10.1111/j.1475-682X.1983.tb00037.x

Bullard, R. D., & Johnson, G. S. (2009). Environmental justice grassroots activism and its impact. In L. King & D. McCarthy (Eds.), *Environmental sociology: From analysis to activism.* Rowman & Littlefield Publishers.

Bullard, R. D., Mohai, P., Saha, R., & Wright, B. (2007). *Toxic race and wastes at twenty: 1987–2007.* Justice & Witness Ministries, The United Church of Christ. Retrieved December 21, 2021, from NRDC website https://www.nrdc.org/sites/default/files/toxic-wastes-and-race-at-twenty-1987-2007.pdf

Camp of the Sacred Stones. (2015). *Excerpts from the Standing Rock Sioux tribe [DAPL] Resolution No. 406-15 September 2, 2015.* Retrieved December 27, 2021, from Sacred Stone Camp – Iŋyaŋ Wakháŋagapi Othí website http://sacredstonecamp.org/resolution/

Camp of the Sacred Stones. (n.d.). *Dakota access is violating Federal laws.* Retrieved December 28, 2021, from Sacred Stone Camp – Iŋyaŋ Wakháŋagapi Othí website http://sacredstonecamp.org/federal-violations/

Caradonna, J. L. (2014). *Sustainability: A history.* Oxford University Press.

Carder, E. F. (2017). The American environmental justice movement. *Internet encyclopedia of philosophy.* Accessed November 22, 2017, from https://iep.utm.edu/enviro-j/

Chávez, C. E. (1969). *Statement of Cesar E. Chávez before the subcommittee on labor of the Senate Committee on Labor and Public Welfare.* Retrieved December 28, 2021, from https://libraries.ucsd.edu/farmworkermovement/essays/essays/MillerArchive/031%20Statement%20Of%20Cesar%20E.%20Chavez.pdf

Chávez, C. E. (1986). *Wrath of Grapes Boycott - Cesar Chavez 1986.* http://www.emersonkent.com/speeches/wrath_of_grapes_boycott.htm

Climate Reality Project. (2016, May 17). *Four ways an inconvenient truth changed peoples' lives.* The Climate Reality Project Blog. Retrieved December 23, 2021, from https://www.climaterealityproject.org/blog/four-ways-inconvenient-truth-changed-lives

Columbia University Mailman School of Public Health. (n.d.). *Obesity | Columbia Public Health.* Retrieved December 29, 2021, from Columbia Center for Children's Environmental Health website https://www.publichealth.columbia.edu/research/columbia-center-childrens-environmental-health/obesity

Cook, E. M., Karuka, M., Simms, A. M., Jurich, J., Dinoor, Y., & Elkins, R. (this volume). Panel Three: Environmental Justice. In M. S. Rivera Maulucci, S. Pfirman, & H. S. Callahan (Eds.), *Education for sustainability: Discourses on authenticity, inclusion, and justice.* Springer.

Cronon, W. (1983). *Changes in the land: Indians, colonists, and the ecology of New England*. Hill and Wang.

Diamond, J. M. (2011). *Collapse: How societies choose to fail or succeed*. Penguin Books.

Durlin, M. (2010, February 1). *The Shot Heard Round the West*. Retrieved December 23, 2021, from https://www.hcn.org/issues/42.2/the-shot-heard-round-the-west

Echo Hawk Consulting. (2015). *Feeding ourselves: Food access, health disparities, and the pathways to healthy Native American communities*. Echo Hawk Consulting.

Eco-justice English Definition and Meaning | Lexico.com. (n.d.). Lexico Dictionaries | English. Retrieved December 16, 2021, from https://www.lexico.com/en/definition/eco-justice

Edwards, J. (2021). *Mamaroneck flood mitigation project to move forward: Schumer*. Patch Media. Retrieved November 3, 2021, from https://patch.com/new-york/larchmont/mamaroneck-flood-mitigation-project-move-forward-schumer

Executive Order 12898. (1994). Federal actions to address environmental justice in minority populations and low-income populations. *Federal Register, 59*(32). Retrieved December 27, 2021, from https://www.archives.gov/files/federal-register/executive-orders/pdf/12898.pdf

Executive Order 14008. (2021). Tackling the climate crisis at home and abroad. *Federal Register, 86*(19), 7619–7633. Presidential Documents. Retrieved December 27, 2021, from https://www.govinfo.gov/content/pkg/FR-2021-02-01/pdf/2021-02177.pdf

Faramelli, N. J. (1971). *Perilous links between economic growth, justice and ecology: A challenge for economic planners, 1 B.C. Envtl. Aff. L. Rev. 218*. Accessed 28 November, 2017, from http://lawdigitalcommons.bc.edu/ealr/vol1/iss1/11

Gandy, M. (2002). Between Borinquen and the barrio: Environmental justice and New York City's Puerto Rican community, 1969–1972. *Antipode, 34*, 730–761.

Garcia, M. (2016, May 9). Cesar Chavez and the United Farm Workers Movement. *Oxford Research Encyclopedias*. Retrieved December 28, 2021 from https://doi.org/10.1093/acrefore/9780199329175.013.217

Ghoche, R., & Udoh, U. (this volume). Ecology's white nationalism problem. In M. S. Rivera Maulucci, S. Pfirman, & H. S. Callahan (Eds.), *Education for sustainability: Discourses on authenticity, inclusion, and justice*. Springer.

Gibson, W. E. (Ed.). (2004). *Eco-justice – the unfinished journey*. State University of New York Press.

Gilio-Whitaker, D. (2019). *As long as grass grows: The indigenous fight for environmental justice, from colonization to Standing Rock*. Beacon Press.

Global Alliance for the Rights of Nature. (2021, August 6). *About GARN – Global Alliance for the Rights of Nature (GARN)*. Accessed June 23, 2022, from https://www.garn.org/about-garn/

Goldman. L. (2018). *Opinion: We calculated the deaths from Hurricane Maria. Politics played no role*. Washington Post. Accessed November 4, 2021, from https://www.washington-post.com/opinions/we-calculated-the-deaths-from-hurricane-maria-politics-played-no-role/2018/09/15/2b765b26-b849-11e8-94eb-3bd52dfe917b_story.html

Guggenheim, D. (2006). *An inconvenient truth*. Paramount Vantage.

Grove, R. (1997). *Green imperialism: Colonial expansion, tropical island Edens and the origins of environmentalism, 1600–1860*. Cambridge University Press.

Hansen, T. (2014, June 17). *Kill the land, kill the people: There are 532 Superfund sites in Indian Country!* Retrieved December 29, 2021, from Indian Country Today website https://indiancountrytoday.com/archive/kill-the-land-kill-the-people-there-are-532-superfund-sites-in-indian-country

Hansen, J., Johnson, D., Lacis, A., Lebedeff, S., Lee, P., Rind, D., & Russell, G. (1981). Climate impact of increasing atmospheric carbon dioxide. *Science, 213*(4511), 957–966. https://doi.org/10.1126/science.213.4511.957

Hersher, R. (2017, February 22). *Key moments in the Dakota Access Pipeline fight*. NPR. Retrieved from https://www.npr.org/sections/thetwo-way/2017/02/22/514988040/key-moments-in-the-dakota-access-pipeline-fight

Hessel, D. T. (2004). Foreword. In W. E. Gibson (Ed.), *Eco-justice – the unfinished journey*. State University of New York Press.

Indigenous Environmental Network (IEN). (2012). *About | indigenous environmental network.* https://www.ienearth.org/about/

Intergovernmental Panel on Climate Change (IPCC). (1990). In J. T. Houghton, G. J. Jenkins, & J. J. Ephraums (Eds.), *Climate change: The IPCC scientific assessment.* Cambridge University Press.

Intergovernmental Panel on Climate Change (IPCC). (1995). *IPCC second assessment climate change.* https://www.ipcc.ch/site/assets/uploads/2018/05/2nd-assessment-en-1.pdf

Jarratt-Snider, K., & Nielsen, M. O. (Eds.). (2020). *Indigenous environmental justice.* The University of Arizona Press.

Karliner, J. (2000, November 21). *Climate justice summit provides alternative vision | corpwatch.* https://www.corpwatch.org/article/climate-justice-summit-provides-alternative-vision

Kim, I. (2017, March 7). *The 1965-1970 Delano Grape Strike and Boycott.* UFW. https://ufw.org/1965-1970-delano-grape-strike-boycott/

Lapping, M. B. (1975). Environmental impact assessment methodologies: a critique. *Boston College Environmental Affairs Law Review, 4*(1), 123–134. Retrieved December 23, 2021, from https://lawdigitalcommons.bc.edu/ealr/vol4/iss1/8

Mandelker, D. (2010). The National Environmental Policy Act: A review of its experience and problems. *Washington University Journal of Law & Policy, 32*(1), 293–312. https://openscholarship.wustl.edu/law_journal_law_policy/vol32/iss1/9

Maranda, J., & Bermel, L. (2020, September 24). *What is ecological justice?* Retrieved December 23, 2021, from Future Proof website https://www.futureproofreviews.com/articles/what-is-ecological-justice/

Martinez, S. (2003). The second national people of color environmental leadership summit: WHAT NEXT? *Voices from the Earth, 4*(1). Southwest Research and Information Center. Retrieved October 21, 2021, from http://www.sric.org/voices/2003/v4n1/summit.php

McKibben, B. (2016, September 6). *A pipeline fight and america's dark past.* The New Yorker. https://www.newyorker.com/news/daily-comment/apipeline-fight-and-americas-dark-past

Meadows, D. H., Meadows, D. L., Randers, J., Behrens, W., & Club of Rome. (1972). *The limits to growth: A report for the Club of Rome's project on the predicament of mankind.* Universe Books.

Mohai, P., Pellow, D., & Roberts, J. T. (2009). Environmental justice. *Annual Review of Environment and Resources, 34*(1), 405–430. https://doi.org/10.1146/annurev-environ-082508-094348

Naylor, B. (2017, January 24). *Trump gives green light to Keystone, Dakota Access Pipelines.* NPR. Retrieved from https://www.npr.org/2017/01/24/511402501/trump-to-give-green-light-to-keystone-dakota-access-pipelines

NEPA. The National Environmental Policy Act of 1969, as amended. (1969). Retrieved December 23, 2021, from https://www.energy.gov/sites/prod/files/nepapub/nepa_documents/RedDont/Req-NEPA.pdf

News 12 Bronx. (2021). *Project aims to prevent future flooding in Mamaroneck following Ida damage.* Accessed March 11, 2021, from https://bronx.news12.com/project-aims-to-prevent-future-flooding-in-mamaroneck-following-ida-damage

NRC (National Research Council). (1977). *Energy and climate: Studies in geophysics.* The National Academies Press. https://doi.org/10.17226/12024

NRC (National Research Council). (2011). *Sustainability and the U.S. EPA.* The National Academies Press. https://doi.org/10.17226/13152

Nweke, O. C., Payne-Sturges, D., Garcia, L., Lee, C., Zenick, H., Grevatt, P., Sanders, W. H., Case, H., & Dankwa-Mullan, I. (2011). Symposium on integrating the science of environmental justice into decision-making at the environmental protection agency: An overview. *American Journal of Public Health, 101*(Suppl 1), S19–S26. https://doi.org/10.2105/AJPH.2011.300368

Ott, T. (2019, October 3). *Martin Luther King Jr. praised Cesar Chavez for his "indefatigable work."* Retrieved December 30, 2021, from Biography website https://www.biography.com/news/cesar-chavez-martin-luther-king-jr-telegram

Patterson, A. E., Williams, M. T., Ramos, J., & Pierre, S. (this volume). Urban-rural sustainability: Intersectional perspectives. In M. S. Rivera Maulucci, S. Pfirman, & H. S. Callahan (Eds.), *Education for sustainability: Discourses on authenticity, inclusion, and justice*. Springer.

Pearce, D. W., Markandya, A., & Barbier, E. B. (1989). *Blueprint for a green economy* (pp. 1–47). Earthscan. Accessed October 13, 2018, from https://www.researchgate.net/publication/39015804_Blueprint_for_a_Green_Economy/link/53e2274a0cf2235f352c1324/download

Perls, H. (2020, November 12). *EPA Undermines its Own Environmental Justice Programs - Environmental & Energy Law Program*. Harvard Law School. https://eelp.law.harvard.edu/2020/11/epa-undermines-its-own-environmental-justice-programs/

Pfirman, S., & Winckler, G. (this volume). Perspectives on teaching climate change: Two decades of evolving approaches. In M. S. Rivera Maulucci, S. Pfirman, & H. S. Callahan (Eds.), *Education for sustainability: Discourses on authenticity, inclusion, and justice*. Springer.

Pope Francis. (2015, May 24). *"Laudato Si:"* On care for our common home. Retrieved December 27, 2021, from https://www.vatican.va/content/francesco/en/encyclicals/documents/papa-francesco_20150524_enciclica-laudato-si.html

Popovich, N., Albeck-Ripka, L., & Pierre-Louis, K. (2020, October 16). The Trump administration rolled back more than 100 environmental rules. Here's the full list. *The New York Times*. Retrieved from https://www.nytimes.com/interactive/2020/climate/trump-environment-rollbacks-list.html

Rees, W. E. (1992). Ecological footprints and appropriated carrying capacity: What urban economics leaves out. *Environment and Urbanization, 4*(2), 121–130. https://doi.org/10.1177/2455747117699722

Robinson, J. (2004). Squaring the circle? Some thoughts on the idea of sustainable development. *Ecological Economics, 48*(4), 369–384. https://doi.org/10.1016/j.ecolecon.2003.10.017

Salkin, P. E., Dernbach, J. C., & Brown, D. A. (2012). Sustainability as a means of improving environmental justice. *Journal of Environmental & Sustainability Law, 19*, 1–34. https://ssrn.com/abstract=2183260

Santos-Burgoa, C., Goldman, A., Andrade, E., Barrett, N., Colon-Ramos, U., Edberg, M., & Zeger, S. (2018). *Ascertainment of the estimated excess mortality from Hurricane Maria in Puerto Rico*. Project Report; Milken Institute School of Public Health, George Washington University.

Scarsdale Inquirer. (1936). How Scarsdale was named and its early history are described in work on Westchester. *Scarsdale Inquirer, 18*(6). Accessed February 11, 2021, from https://news.hrvh.org/veridian/?a=d&d=scarsdaleinquire19360313.2.59&e=%2D%2D%2D%2D%2D%2D-en-20%2D%2D1%2D%2Dtxt-txIN%2D%2D%2D%2D%2D%2D-

Schumacher, E. F. (1973). *Small is beautiful: Economics as if people mattered*. Harper & Row.

SRIC (Southwest Research and Information Center). (2003). Excerpts from the Indigenous Peoples Caucus statement. *Voices from the Earth, 4*(1). Retrieved from http://www.sric.org/voices/2003/v4n1/caucus.php

UN (United Nations). (2008). *United Nations Declaration on the Rights of Indigenous Peoples*. Retrieved December 30, 2021, from https://www.un.org/esa/socdev/unpfii/documents/DRIPS_en.pdf

UN (United Nations). (n.d.). *United Nations declaration on the rights of indigenous peoples | United Nations For Indigenous Peoples*. Retrieved January 8, 2022, from https://www.un.org/development/desa/indigenouspeoples/declaration-on-the-rights-of-indigenous-peoples.html

UNCC (United Nations Climate Change). (n.d.-a). *Conference of the parties (COP)*. Retrieved December 23, 2021, from https://unfccc.int/process/bodies/supreme-bodies/conference-of-the-parties-cop

UNCC (United Nations Climate Change). (n.d.-b). *Kyoto Protocol – Targets for the first commitment period*. Retrieved December 23, 2021, from https://unfccc.int/process-and-meetings/the-kyoto-protocol/what-is-the-kyoto-protocol/kyoto-protocol-targets-for-the-first-commitment-period

UNCED (United Nations Conference on Environment and Development). (1992). *Agenda 21, Rio Declaration, Forest Principles*. United Nations.

UNCSD (United Nations Conference on Sustainable Development). (2012). *The future we want – Outcome document*. Retrieved December 31, 2021, from https://sustainabledevelopment. un.org/futurewewant.html

UNEP (United Nations Environment Program). (2011). *Working towards a balanced and inclusive green economy: A united nations system-wide perspective*. Retrieved from https://wedocs. unep.org/xmlui/handle/20.500.11822/8065

UNFCCC (United Nations Framework Convention on Climate Change). (2021, November 13). *COP26 reaches consensus on key actions to address climate change*. Retrieved December 28, 2021, from https://unfccc.int/news/cop26-reaches-consensus-on-key-actions-to-address-climate-change

US Census Bureau. (2019a). *Quick Facts Mamaroneck Village, New York*. Retrieved November 3, 2021, from https://www.census.gov/quickfacts/mamaroneckvillagenewyork

US Census Bureau. (2019b). *Quick facts Loiza, Puerto Rico, New York*. Retrieved November 3, 2021, from https://www.census.gov/quickfacts/loizamunicipiopuertorico

US DOJ (Department of Justice). (2017, July 28). *History of federal voting rights laws*. Retrieved December 30, 2021, from https://www.justice.gov/crt/history-federal-voting-rights-laws

US DOJ (Department of Justice). (2021, November 8). *Section 2 of the Voting Rights Act*. Retrieved December 30, 2021, from https://www.justice.gov/crt/section-2-voting-rights-act

US DOS (United States Department of State. (n.d.). *Announcement of U.S. support for the United Nations declaration on the rights of indigenous peoples*. Retrieved January 6, 2022, from U.S. Department of State website https://2009-2017.state.gov/s/srgia/154553.htm

US EPA (Environmental Protection Agency). (1992). *Environmental Justice: Reducing Risk For All Communities Volume 1*, Washington D.C.

US EPA (Environmental Protection Agency). (2013a, February 22). *Summary of the Comprehensive Environmental Response, Compensation, and Liability Act (Superfund) [Overviews and Factsheets]*. Retrieved December 30, 2021, from https://www.epa.gov/laws-regulations/ summary-comprehensive-environmental-response-compensation-and-liability-act

US EPA (Environmental Protection Agency). (2013b, February 22). *Summary of Executive Order 12898 – Federal actions to address environmental justice in minority populations and low-income populations [Overviews and Factsheets]*. Retrieved December 23, 2021, from https://www.epa.gov/laws-regulations/ summary-executive-order-12898-federal-actions-address-environmental-justice

US EPA (Environmental Protection Agency). (2013c, February 22). *Creating equitable, healthy, and sustainable communities [Reports and Assessments]*. Retrieved December 26, 2021, from https://www.epa.gov/smartgrowth/creating-equitable-healthy-and-sustainable-communities

US EPA (Environmental Protection Agency). (2014). *EPA Policy on Environmental Justice for Working with Federally-Recognized Tribes and Indigenous Peoples*. https://www.epa.gov/sites/ default/files/2017-10/documents/ej-indigenous-policy.pdf

US EPA (Environmental Protection Agency). (2015a, April 15). *Environmental justice timeline [Collections and Lists]*. Retrieved December 27, 2021, from https://www.epa.gov/ environmentaljustice/environmental-justice-timeline

US EPA (Environmental Protection Agency). (2015b, February 17). *National Environmental Justice Advisory Council [Overviews and Factsheets]*. Retrieved December 23, 2021, from https://www.epa.gov/environmentaljustice/national-environmental-justice-advisory-council

US EPA (Environmental Protection Agency). (2016a, March 31). *EJ IWG promising practices for EJ Methodologies in NEPA Reviews [Reports and Assessments]*. Retrieved December 27, 2021, from https://www.epa.gov/environmentaljustice/ ej-iwg-promising-practices-ej-methodologies-nepa-reviews

US EPA (Environmental Protection Agency). (2016b, May 20). *Environmental justice 2020 action agenda [Reports and Assessments]*. Retrieved December 28, 2021, from https://www.epa.gov/ environmentaljustice/environmental-justice-2020-action-agenda

US National Archives & Records Administration. (n.d.). *Our documents – Transcript of treaty of Fort Laramie (1868)*. Retrieved December 29, 2021, from https://www.ourdocuments.gov/doc.php?flash=false&doc=42&page=transcript

Wackernagel, M., & Rees, W. (1996). *Our ecological footprint: Reducing human impact on the Earth*. New Society Publishers.

Weston, J. (2017). *Water is life: The rise of the Mní Wičóni movement*. http://www.culturalsurvival.org/publications/cultural-survival-quarterly/water-life-rise-mni-wiconi-movement

World Commission on Environment and Development. (1987). *Our common future*. Oxford University Press.

Zierler, D. (2011). *The invention of ecocide: Agent orange, Vietnam, and the scientists who changed the way we think about the environment*. University of Georgia Press.

María S. Rivera Maulucci is the Ann Whitney Olin Professor of Education at Barnard College. Her research focuses on the role of language, identity, and emotions in learning to teach science for social justice and sustainability. She teaches courses in science and mathematics pedagogy and teacher education for social justice.

Chapter 3
Diversity in Academia and Sustainability Science: The STEM Blindspot

Shirley-Ann Augustin-Behravesh

3.1 Change, Without the Change

Universities across the US are currently pushing to increase faculty and student diversity because of recent, pressing events that once again, unearthed the racial inequalities and injustices that exist within our institutions. Yet, the majority of interventions adopted are short-sighted and do not consider the fundamental reasons why these problems persist in our society. These issues are critical to address in order to transform education for sustainability in just, inclusive and authentic ways.

One strategy used by universities is to hire Black, Indigenous, People of Color (BIPOC) faculty who work in fields related to justice, equity, diversity, and inclusion (JEDI), thus working on the assumption that historically excluded persons *should be* working to eradicate the injustices that they had no part in initiating or perpetuating. This misconception that people of color are all activists or vocal on the issues affecting their identified groups, is unfair and furthers the current stereotyping behaviors that exist today. Many BIPOC and women faculty, though aware and unhappy with the current system, cope in ways that do not demonstrate their frustrations and merely want to be recognized for their contributions to their field of study, as their colleagues would. The expectation that they should be more interested or active on issues of JEDI, places an unfair and extra burden – including stress and invisible labor – that others do not have to contend with (Jimenez et al., 2019). This practice also continues the cycle of isolating BIPOC and women in fields where they are already present; a form of epistemic exclusion (Settles et al., 2019).

The STEM fields have been traditionally higher paid and highly regarded when compared to other fields such as the Humanities. Yet, these are the very fields with

S.-A. Augustin-Behravesh (✉)
Thunderbird School of Global Management, Arizona State University, Phoenix, AZ, USA
e-mail: satab2@asu.edu

© The Author(s) 2023
M. S. Rivera Maulucci et al. (eds.), *Transforming Education for Sustainability*,
Environmental Discourses in Science Education 7,
https://doi.org/10.1007/978-3-031-13536-1_3

the lowest recruitment and highest attrition for BIPOC and women (Li & Koedel, 2017). Women and BIPOC faculty are the lowest paid in academia and the barriers to entry in the STEM fields perpetuate this. This disparity further enables the cycle of keeping students who are BIPOC and female, out of STEM fields, as one major determinant of one's field of study, is the ability to find a role model or mentor who shares their characteristics (Arnim, 2019; Beech et al., 2013). It also means that the perspectives and experiences of a large and important group have been mostly absent from the STEM fields, resulting in unrealized solutions and innovations.

3.2 Systemic Stereotype Threats

Many of the interventions used by universities ignore the underlying reasons why historically excluded persons of color are scarcely present in STEM fields, and getting traction has been slow. BIPOC and women are notoriously known to have low enrollments in STEM degrees including at the faculty level (Li & Koedel, 2017). Studies have shown that it is not a lack of interest that leads to the avoidance, or exit from, STEM fields (Burke, 2007). Some interpersonal reasons for low retention of BIPOC and women in STEM fields include alienation by colleagues, micro-aggressive environments, and the lack of representative mentors or role models to inspire new students/faculty in disadvantaged minorities to pursue these disciplines (Arnim, 2019). Even deeper, systemic threats pervade our academic institutions and steer BIPOC and women away from STEM fields, in forms such as masculine norms/privileging masculine ways of working; unchecked biases in recruitment, hiring, development, evaluation, and promotion processes; unconscious perpetuation of culturally-embedded stereotypes; color-blind racial ideologies; the myth of meritocracy that cripples the ability of historically excluded persons of color from joining STEM fields; and financial and economic barriers related to testing, applications, cost of study, etc. (Block et al., 2019; Dupree & Boykin, 2021; Estrada et al., 2016).

Systemic stereotype threat also leads to the exit from STEM fields (Burke, 2007). "*Systemic stereotype threat* occurs when an individual is in a system that is characterized by racial or gender disparities and the implicit belief about the reason for these disparities is due to stereotypes about deficits of individual group members rather than systemic inequality" (Block et al., 2019, p. 35; see also Steele et al., 2002). BIPOC and women faculty who work in environments characterized by systemic stereotype threat have been seen to cope in different ways (Block et al., 2019; Settles et al., 2019). Block et al. (2019) present empirical research on the coping mechanisms used by women faculty in STEM disciplines to contend with systemic stereotype threats. The three response patterns identified in the article are: (1) Fending off the threat – keeping the threat invisible to self and the association with the stereotype invisible to others; (2) Confronting the threat – making the threat visible to others (3) Sustaining self in the presence of threat – recognizing the threat as a dilemma to be navigated not as a problem to be solved immediately. Settles et al. (2019) studied the response patterns of BIPOC faculty within STEM fields, and have similar results with the use of three invisibility strategies: (1) Strategic invisibility – disengaging with colleagues, while engaging with scholarly activities; (2) Working harder – to prove worth and improve visibility (of work); (3) Disengagement – removed effort from work. While there is much to be understood regarding when and why particular strategies within the response patterns are utilized, it is agreed that the perception of the threat (related to both organizational and individual factors) is a major driver of the response pattern used (Block et al., 2011, 2019; Settles et al., 2019).

3.3 Reflecting on My Own Response Patterns

As a Black, female academic in STEM, I reflect below on how I have utilized different response patterns to navigate my career within the fields of engineering and sustainability science.

3.3.1 To Prove I Can? Or Is It … Who I Am?

Unlike many Black Americans, I was not raised in a society rampant with racial inequity. I was born and raised on a small Caribbean island where most of the population identified as Black. My exposure to the issues of race came from books and television – which perpetuated the belief that this was a 'them, not me' issue. This perception that the threat was not my own, kept it invisible to me when I first moved to the United States.

In my first role as a faculty member, I often felt like an outsider to the faculty body. I reasoned that this could be because of three things: (1) My introverted personality and/or unfamiliarity with me and my culture, (2) My non-tenure-track status, (3) Unfamiliarity with my work or the quality of my research. My response was, therefore, to silently work at building up my research portfolio to make my work (and myself) more visible. I engaged in strategies of 'over-effort' and 'isolation' as discussed by Block et al. (2019). I said yes to everything. I even volunteered for everything. If I proved my worth, my colleagues and department would find me a valuable member of the department. Rather, the result of my over-involvement has been exhaustion, burnout, and an eventual diminishing motivation to continue work that no one ever recognizes. By making the threat and myself invisible – by appearing to take everything in stride – I also made my efforts, and the toll it was taking on me, invisible.

3.3.2 An Attempt to 'Confront the Threat'

I became more cognizant of the issues related to race and stereotypes after volunteering to be on a university committee for faculty women of color. It was in conversations with these women of color, that I realized that our stories were the same. We were unrecognized, overworked, and over-compensating for issues that have been persistent within the academic institution since its genesis. The realization that what I had been experiencing was part of systemic discrimination within the academic institution, left me with deep feelings of hurt, anger, fear, and ultimately, confusion. Not unlike the stages of grief... Having never had discussions or experiences in which I needed to confront discrimination, it left me in a state of anxiety, and an inability to confront a threat that I desperately wanted to address.

There is an assumption that all people of color, particularly those with some standing in the community, or in an established career, should be voicing their frustrations with the discrimination that still pervades our major institutions. If you're not, then you probably are one of those who 'sold out' or are deliberately blind to the injustices of society. I carry deep guilt that I have not been as vocal and active as my female colleagues of color. They have been so deeply embedded and involved in the issues of diversity and inclusion, that I often feel like a 'copout' to my race and gender. It's not that I've cast a blind eye. I've tried to speak up. I've tried to become more aware of the tribulations that black people face in America and the rest of the world. I've even tried to reason with those who unfairly discriminated against me and others. I've given them the benefit of the doubt. But I quickly realized that, especially in an old institution such as academia, those who speak up are labeled as the trouble-makers, the rousers, the 'angry black woman.' This kind of visibility was not what I sought, and so my response adjusted accordingly.

3.3.3 "I'll Just Sit Quietly in the Corner ..."

My natural tendency to avoid confrontation kicked in, not long after. I convinced myself that if I worked *even* harder and created exceptional work, that I would not be seen as a Black person, but rather as an extremely valuable faculty member. This strategy was ultimately unsustainable. With an over-full teaching load (and taking on extra to supplement the low pay), serving as a mentor to many students, sitting on a variety of department, university, and community committees, attempts to carry on research and projects of impact and balancing these with the parenting of two small kids, soon left me battling several chronic illnesses, all tied to high-stress levels.

3.3.4 Sustaining Self in the Presence of Threat

A moment of clarity came after my own body forced me to stop working over a summer break, because of ongoing illness. I reflected on the outcomes of my efforts over the years and realized that I had been supporting department efforts, focusing on building and enhancing programs, preparing students for research and careers, working with communities that needed and appreciated the academic support and others, giving talks and presentations – but that none of these were related to the thing that mattered most in the academic scene – published research. All this work and I had still not attained the one element that was most recognized in my field.

This clarity came at another momentous event in my life. After 3 years with my department, I was assigned a mentor. It was with their help that I was able to define what success meant for me (as opposed to just my department) and learned the important skill of saying "no," even to seemingly great opportunities. I also recognized the need to accept my identity as a Black, female, immigrant scholar, and utilize my unique perspectives and experiences within my work.

Years of feelings of inadequacy, hiding away, keeping quiet, and desperation to prove my worth gave way to an acceptance that the system was indeed flawed, but with an acknowledgment that the system cannot change overnight. I reasoned that continual (maybe even subtle) prodding, increased awareness, and finding the right mentors and support, was a more actionable and strategic way to slowly transform the system from within. I still have days where I toggle between the response patterns, and I think that it is somewhat necessary to shift between response patterns depending on the context. But overall, my definition of success as a scholar, mother, wife, global citizen has transformed the way I work. My drive to be an exceptional scholar is now purely based on my own self-motivation to develop and disrupt my field and make positive changes in the world.

3.3.5 Is There a Right Response?

Despite my having arrived at the 'Sustaining self in the presence of threat' response as being the most appropriate for myself, I have also recognized BIPOC and women colleagues around me using various strategies within the response patterns. In my observations, it appears that for transformation change to occur, we need all three responses simultaneously.

Having stellar outcomes from those engaged in 'Fending off the Threat/Strategic Invisibility/Working Harder' responses, makes the case for the value and appreciation of the work of BIPOC & women scholars. Whilst it is unfair to have this extra burden of proof of self-worth, it helps build the case for transformation. We also need to build awareness of these existing threats within our institutions from those who 'Confront the Threat.' Without awareness, we are perpetuating a cycle of threats that remains visible only to ourselves as BIPOC and women faculty. And finally, we need those who are able to work with and within the current systems, who are 'Sustaining Themselves in the Presence of Threat', to find ways to continuously build awareness, create disruptions, and have non-BIPOC faculty begin to question their role in perpetuating systemic stereotype threats in the academy.

3.4   Moving Forward

As universities continue to push for increasing diversity within their faculty and student bodies, it is important to recognize the factors that affect the *retention* of BIPOC & women. Studies on systemic stereotype threat have found that the presence of such threat can lead to lower confidence, higher turnover (von Hippel et al., 2011 as seen in Block et al., 2019); decreased well-being (von Hippel et al., 2015 as seen in Block et al., 2019); lower beliefs in career advancement (Fassiotto et al., 2016 as seen in Block et al., 2019); and negative physical and psychological health consequences (Juster et al., 2010; Settles et al., 2019). While some BIPOC and women faculty have successfully navigated the academic world, and are valued members of the academic community, many others fall victim to the threats they face. Within this pool lies the untapped, undervalued, and unrealized potential that can contribute tremendously to the university, to the wider discipline, and to better sustainability science (Hunt et al., 2015).

To end, I wish to make a call to universities looking to increase their diversity rates, to consider not just the recruitment of BIPOC and women faculty, but also to recognize the various threats and pressures faced by current BIPOC and women. In doing so we can properly anticipate and prepare for potential barriers that BIPOC and women may face (Block et al., 2011). In particular, in STEM fields where BIPOC & women faculty are represented in disgracefully low numbers, the threats exist throughout our academic institution, from kindergarten to the highest academic achievements. An awareness of these can start the process of dismantling systemic stereotypes that were in-built into our institutions. This is also extremely relevant to environment and sustainability science, where the diversity of perspectives and experiences are necessary assets to sustainability problem-solving (van der Leeuw et al., 2012; Wiek, 2011). This is particularly of great importance, as the negative impacts of sustainability problems are disproportionately felt by poor and vulnerable populations, often which are BIPOC communities (Anguelovski et al., 2019). It is my hope that departments and universities begin to take the necessary steps to create communities where BIPOC and women faculty feel safe, welcomed, equal, and valued.

References

Anguelovski, I., Connolly, J. J., Pearsall, H., Shokry, G., Checker, M., Maantay, J., Gould, K., Lewis, T., Maroko, A., & Roberts, J. T. (2019). Opinion: Why green "climate gentrification" threatens poor and vulnerable populations. *Proceedings of the National Academy of Sciences, 116*(52), 26139–26143. https://doi.org/10.1073/pnas.1920490117

Arnim, E. (2019). *A third of minority students leave STEM majors. Here's why.* EAB.com. Accessed from https://eab.com/insights/daily-briefing/student-success/a-third-of-minority-students-leave-stem-majors-heres-why/

Beech, B. M., Calles-Escandon, J., Hairston, K. G., Langdon, S. E., Latham-Sadler, B. A., & Bell, R. A. (2013). Mentoring programs for underrepresented minority faculty in academic medical centers: A systematic review of the literature. *Academic Medicine, 88*(4), 541–549. https://doi.org/10.1097/ACM.0b013e31828589e3

Block, C. J., Koch, S. M., Liberman, B. E., Merriweather, T. J., & Robertson, L. (2011). Contending with stereotype threat at work: A model of long-term responses. *The Counseling Psychologist, 39*(4), 570–600. https://psycnet.apa.org/doi/10.1177/0011000010382459

Block, C. J., Cruz, M., Bairley, M., Harel-Marian, T., & Roberson, L. (2019). Inside the prism of an invisible threat: Shining a light on the hidden work of contending with systemic stereotype threat in STEM fields. *Journal of Vocational Behavior, 113*, 33–50. https://psycnet.apa.org/doi/10.1016/j.jvb.2018.09.007

Burke, R. (2007). Women and minorities in STEM: A primer. In R. J. Burke & M. C. Mattis (Eds.), *Women and minorities in science, technology, engineering, and mathematics: Upping the numbers* (pp. 3–27). Edward Elgar Publishing.

Dupree, C. H., & Boykin, C. M. (2021). Racial inequality in academia: Systemic origins, modern challenges, and policy recommendations. *Policy Insights From the Behavioral and Brain Sciences, 8*(1), 11–18. https://doi.org/10.1177/2F2372732220984183

Estrada, M., Burnett, M., Campbell, A. G., Campbell, P. B., Denetclaw, W. F., Gutiérrez, C. G., Hurtado, S., John, G. H., Matsui, J., McGee, R., Okpodu, C. M., Robinson, T. J., Summers, M. F., Werner-Washburne, & Zavala, M. (2016). Improving underrepresented minority student persistence in STEM. *CBE Life Sciences Education, 15*(3), es510.1187/cbe.16-01-0038. https://doi.org/10.1187/cbe.16-01-0038

Fassiotto, M., Hamel, E. O., Ku, M., Correll, S., Grewal, D., Lavori, P., Periyakoil, V. J., Reiss, A., Sandborg, C., Walton, G., Winkleby, M., & Valentine, H. (2016). Women in academic medicine: Measuring stereotype threat among junior faculty. *Journal of Women's Health, 25*(3), 292–298. https://doi.org/10.1089/2Fjwh.2015.5380

Hunt, V., Layton, D., & Prince, S. (2015). Diversity matters. *McKinsey & Company, 1*(1), 15–29.

Jimenez, M. F., Laverty, T. M., Bombaci, S. P., Wilkins, K., Bennett, D. E., & Pejchar, L. (2019). Underrepresented faculty play a disproportionate role in advancing diversity and inclusion. *Nature Ecology & Evolution, 3*, 1030–1033. https://doi.org/10.1038/s41559-019-0911-5

Juster, R., McEwen, B. S., & Lupien, S. J. (2010). Allostatic load biomarkers of chronic stress and impact on health and cognition. *Neuroscience & Biobehavioral Reviews, 35*(1), 2–16. https://doi.org/10.1016/j.neubiorev.2009.10.002

Li, D., & Koedel, C. (2017). Representation and salary gaps by race-ethnicity and gender at selective public universities. *Educational Researcher, 46*(7), 343–354. https://doi.org/10.3102/2F0013189X17726535

Settles, I. H., Buchanan, N. T., & Dotson, K. (2019). Scrutinized but not recognized: (In)visibility and hypervisibility experiences of faculty of color. *Journal of Vocational Behavior, 113*, 62–74. https://doi.org/10.1016/j.jvb.2018.06.003

Steele, C. M., Spencer, S. J., & Aronson, J. (2002). Contending with group image: The psychology of stereotype and social identity threat. *Advances in Experimental Social Psychology, 34*, 379–440. https://doi.org/10.1016/S0065-2601(02)80009-0

van der Leeuw, S., Wiek, A., Harlow, J., & Buizer, J. (2012). How much time do we have? Urgency and rhetoric in sustainability science. *Sustainability Science, 7,* 115–120. https://doi.org/10.1007/s11625-011-0153-1

von Hippel, C., Issa, M., Ma, R., & Stokes, A. (2011). Stereotype threat: Antecedents and consequences for working women. *European Journal of Social Psychology, 41*(2), 151–161. https://doi.org/10.1002/ejsp.749

von Hippel, C., Sekaquaptewa, D., & McFarlane, M. (2015). Stereotype threat among women in finance: Negative effects on identity, workplace well-being, and recruiting. *Psychology of Women Quarterly, 39*(3), 405–414. https://doi.org/10.1177/0361684315574501

Wiek, W. (2011). Key competencies in sustainability: A reference framework for academic program development. *Sustainability Science, 6*(2), 203–218. https://doi.org/10.1007/s11625-011-0132-6

Shirley-Ann Augustin-Behravesh is an Assistant Professor and Senior Global Futures scientist at Arizona State University. She received her Ph.D. in Sustainability Science, M.Phil. in Manufacturing Engineering and B.Sc. in Management/Finance. She is from St. Lucia, and was the first in her immediate and extended family to receive a college degree.

Chapter 4
Building Authentic Connections to Science Through Mentorship, Activism, and Community, in Teaching and Practice

Angelica E. Patterson, Tanisha M. Williams, Jorge Ramos, and Suzanne Pierre

4.1 Introduction

Amid the height of the COVID-19 pandemic, which highlighted the disproportionate rates of mortality amongst Black Americans,[1] violent deaths by the hands of the police and white mobs resulted in the deaths of Ahmaud Arbery, Breonna Taylor, and George Floyd.[2] On the same week as George Floyd's murder in the summer of 2020, Christian Cooper, a black birder and prominent member of New York City's Audubon Society was harassed by a white woman in Central Park who weaponized

---

[1] Yancy, C. W. (2020). Covid-19 and African Americans. *JAMA, 323*(19), 1891. https://doi.org/10.1001/jama.2020.6548

[2] Brown, D. N. L. (2021, May 4). *Violent deaths of George Floyd, Breonna Taylor reflect a brutal American legacy*. History. Retrieved December 16, 2021, from https://www.nationalgeographic.com/history/article/history-of-lynching-violent-deaths-reflect-brutal-american-legacy

---

A. E. Patterson (✉)
Black Rock Forest and Miller Worley Center, Mt. Holyoke College, South Hadley, MA, USA
e-mail: apatterson@mtholyoke.edu

T. M. Williams
Biology Department, Bucknell University, Lewisburg, PA, USA

J. Ramos
Jasper Ridge Biological Preserve, Stanford University, Stanford, CA, USA

S. Pierre
Critical Ecology Lab, California Academy of Sciences, San Francisco, CA, USA

© The Author(s) 2023
M. S. Rivera Maulucci et al. (eds.), *Transforming Education for Sustainability*,
Environmental Discourses in Science Education 7,
https://doi.org/10.1007/978-3-031-13536-1_4

a police call after Cooper asked her to leash her dog.[3] The long legacy of violence against Black men and women in the United States not only sparked one of the largest protests in the nation, but also garnered global attention and international support for the Black Lives Matter (BLM)[4] movement. This constant and systematic attack on Black bodies sparked a different kind of movement amongst Black nature enthusiasts and scientists who wanted to increase the visibility of Black members enjoying and participating in the outdoors and within predominantly white spaces. #BlackBirdersWeek,[5] the first social media movement aimed at amplifying Black birders and their experiences was launched on May 31st, 2020. Three days after #BlackBirdersWeek ended, Dr. Tanisha Williams sent out a call *via* Twitter, asking others about sharing the same positivity and awareness of Black people and their love of plants across the botanical community and beyond. Many botanists answered the call and began the work to create the next science-related social media movement. #BlackBotanistsWeek[6] was launched on July 6th, 2020 and gained global participation in their week-long online events that expanded social and professional networks and led to the collaboration amongst the authors of this chapter. Since then, the creation of social media movements in STEM fields, such as #BlackInEnvironWeek,[7] #BlackInAstro,[8] #BlackInMicro,[9] #BlackInChem,[10] #BlackMammalogistsWeek[11] amongst others, have exploded with much support and gratitude for their help in elevating the voices and comradery of members who exist and participate in these predominantly white fields. The systemic economic and cultural barriers in academia and our larger society inhibit racially ethnic and gender-diverse students from fully participating and authentically engaging in the sciences. Building meaningful connections to the field of environmental sciences and sustainability requires integrated approaches that extend "beyond the walls" to

---

[3] Nir, S. M. (2020, June 14). *How 2 lives collided in Central Park, rattling the nation*. The New York Times. Retrieved December 16, 2021, from https://www.nytimes.com/2020/06/14/nyregion/central-park-amy-cooper-christian-racism.html

[4] *Black Lives Matter homepage*. Black Lives Matter. (n.d.). Retrieved December 16, 2021, from https://blacklivesmatter.com/

[5] *#BlackAFinSTEM and #BlackBirdersWeek twitter page*. #BlackAFinSTEM Twitter. (n.d.). Retrieved December 16, 2021, from https://twitter.com/BlackAFinSTEM

[6] *#BlackBotanistsWeek homepage*. #BlackBotanistsWeek. (n.d.). Retrieved December 16, 2021, from https://blackbotanistsweek.weebly.com/

[7] *#BlackInEnvironWeek twitter page*. #BlackInEnvironWeek Twitter. (n.d.). Retrieved December 16, 2021, from https://twitter.com/BlackInEnviron

[8] *#BlackInAstro homepage*. #BlackInAstro. (n.d.). Retrieved December 16, 2021, from https://www.blackinastro.com/

[9] *#BlackInMicro twitter page*. #BlackInMicro Twitter. (n.d.). Retrieved December 16, 2021, from https://twitter.com/BlackInMicro

[10] *#BlackInChem homepage*. #BlackInChem. (n.d.). Retrieved December 16, 2021, from https://blackinchem.org/

[11] *#BlackMammalogistsWeek homepage*. #BlackMammalogistsWeek. (n.d.). Retrieved December 16, 2021, from https://blackmammalogists.com/

encourage, inspire, and motivate students to take an interest in studying and pursuing careers in ecology and the earth sciences. The following conversation centered around critical race theory and intersectionality, features four BIPOC scientists and educators as they discuss the challenges and rewards of helping students navigate science in academia. They also examine how their advocacy for diversity, equity, inclusion, and justice translates into actionable approaches that deconstruct the barriers that have prevented equal representation and empowerment of marginalized groups in the environmental sciences. See also Behravesh (this volume).

Angelica Patterson We are going to get started with formal introductions. Please tell us your name, briefly describe your current position, your current institution, what your job entails, and what is your favorite part about what you do. I will start. My name is Angelica Patterson. I am the Master Science Educator at Black Rock Forest[12] in Cornwall, New York. My job entails being a liaison to about 18 consortium members that are made up of K-12 private and public schools, cultural institutions like the American Museum of Natural History,[13] as well as colleges and universities, both public and private. I am there to connect students to the forest through research and education. I work with educators to create curricula that engage and encourage students to learn about the environment and the ecology. My favorite part about what I do is curriculum development. I really enjoy speaking with teachers about their goals and how we can creatively develop activities students can experience either virtually or hands-on to learn about some complex environmental topics.

Tanisha M. Williams My name is Tanisha M. Williams, and I am the Burpee Postdoctoral Fellow in Botany at Bucknell University[14] in Lewisburg, Pennsylvania. I work with Dr. Chris Martine, a biologist at Bucknell. The overall goal of the Martine lab is to conserve biodiversity, and we do this locally in Pennsylvania and internationally in Australia. In Pennsylvania, we are updating the conservation status of rare plants using genomics methods. We also work with local government organizations like the Western Pennsylvania Conservancy[15] and the Pennsylvania Natural Heritage Program.[16] They are the boots on the ground, and we are the boots in the lab. The research project that really got me excited about coming to a primar-

---

[12] *Black Rock Forest homepage. Black Rock Forest. (2021, June 29). Retrieved December 15, 2021, from https://www.blackrockforest.org/*

[13] *American Museum of Natural History homepage.* American Museum of Natural History. (n.d.). Retrieved December 16, 2021, from https://www.amnh.org/

[14] *Bucknell University homepage.* Bucknell University. (n.d.). Retrieved December 15, 2021, from https://www.bucknell.edu/

[15] *Western Pennsylvania Conservancy homepage.* Western Pennsylvania Conservancy. (2021, December 1). Retrieved December 15, 2021, from https://waterlandlife.org/

[16] *Pennsylvania Natural Heritage Program homepage.* Pennsylvania Natural Heritage Program. (n.d.). Retrieved December 15, 2021, from https://www.naturalheritage.state.pa.us/

ily undergraduate institution is this plants and people project in Australia. We are working with the Martu people. The Martu are Aboriginal people in Western Australia. The Martu are teaching us how they use *Solanum* plants and what they use them for. We are going back to look at the genetics of this plant in order to assess how the Martu people assist in gene flow and how they are conserving and preserving the species by their use of certain plants as a staple food source. This is a long-term collaboration that has taken decades to build. It is nice working with a team that knows the importance of Indigenous knowledge, not only when collecting information, but giving proper credit, authorship, etc. It is nice to know that this is not just a fly-in kind of project. We are doing research that the Martu people are interested in. These are questions that they have about the plants around them. So, that is a really nice part of this work.

My favorite part of the job is working with undergraduate students. I never knew how much I would enjoy working with undergraduate students. I am the postdoc/lab manager, so all of the students' projects come through me, and it is just a joy seeing them. I will tell a very quick story about what a joy it is to work with students. One of my students graduated with an honors thesis on one of the local Pennsylvania rare plant projects I mentioned earlier. They did such great work, and we are currently writing up the manuscript. I cried when they finished their honors thesis presentation. It is wonderful to see undergraduate students' level of research, and it is fulfilling to be a part of their learning process.

Suzanne Pierre   My name is Suzanne Pierre. My pronouns are she and her. My title is Research Scientist at the California Academy of Sciences[17] and the IBSS[18] department, which stands for Institute for Biodiversity, Science and Sustainability. I am also the founder and director of the Critical Ecology Lab,[19] a not-for-profit research, education, and anti-oppression organization. I started the Critical Ecology Lab in early 2020, right around the pandemic's start. The way that I define or describe the organization is meaningfully evolving, and I think that is intentional in some ways. The Critical Ecology Lab focuses on advancing the questions and methods that meaningfully relate the study of power and privilege with the study of biophysical change in response to anthropogenic drivers—applying the kind of critical theory that undergirds our understanding of oppression and extraction as a kind of fundamental, social process and describing or relating those dynamics through observation and experimentation in the biophysical realm.

---

[17] California Academy of Sciences. (n.d.). Retrieved December 15, 2021, from https://www.calacademy.org/

[18] *Institute for Biodiversity Science & Sustainability*. California Academy of Sciences. (n.d.). Retrieved December 15, 2021, from https://www.calacademy.org/scientists

[19] *Critical Ecology Lab. (n.d.). Retrieved December 15, 2021, from https://www.criticalecologylab.org/*

In our lab, because of my background as a biogeochemist and ecosystems ecologist, most of the basic research focuses on the terrestrial environment, nutrient cycling, and plant ecology. But, the deep goal of the organization is to advance how we approach science, how we come up with the questions that we prioritize and who comes up with them and also to influence the ways that we conduct ourselves when we do research. The Critical Ecology Lab is both an academic response to a gap in what we know, but it is also a response to the prevailing culture in academic research and a kind of rebuttal to that, to say we actually can structure our laboratory teams and their culture to orient towards compassion and justice, rather than competition and supremacy. My job entails doing basic research, writing grants, establishing partnerships, and doing this field development work.

What does it look like to find the right collaborators and the right students to advance an idea and then bring that to life through basic research? I do a lot of mentoring. I am talking to undergrads, masters, and Ph.D. students, none of whom I formally advise through university because, as you know, I am at non-traditional institutions. It is a really interesting part of my work to be a mentor, but not in a formal sense. It is also my favorite thing, and I think it is the basis of culture building. My favorite part of my work is being responsible for establishing a culture of compassion and then making sure and holding myself accountable for doing that and living that every day.

Jorge Ramos My name is Jorge Ramos and I am the Associate Director for Environmental Education at Jasper Ridge Biological Preserve[20] at Stanford University.[21] At Jasper Ridge, we are guided by three pillars: research, education, and conservation. I am the Associate Director of the pillar of education. Our Research Director establishes what type of research goes on and the logistics around research. I co-teach a course, and I also guide all of the outreach and education requests, organize the field trips that are requested, and work with all of the community and educational partners that want to visit Jasper Ridge. It is not a recreational space, so every outreach and educational activity has the intent of learning while you are here. Are you interested in something in particular? If it's a bird-oriented activity, we will find you a docent that might focus on birds. The whole program of education ranges from very internal to external. First, we host many Stanford classes, and these can go outside of STEM. For example, if we have language classes, they might request a tour in Spanish, so I can give that one. There are archeology and anthropology classes that focus on history. We have a physics class where they might come to explore some of our projects in physics and engineering. We have a lot of geosciences research projects. I coordinate all of these class visits and make sure they follow safety protocols.

---

[20] Stanford University. (2021, December 8). *Jasper Ridge Biological Preserve homepage.* Jasper Ridge Biological Preserve. Retrieved December 15, 2021, from https://jrbp.stanford.edu/

[21] *Stanford University homepage.* Stanford University. (n.d.). Retrieved December 15, 2021, from https://www.stanford.edu/

What is the culture that we offer here at Jasper Ridge so that the field does not reflect the stereotypical old "macho" white men that used to always be in the field sciences? How do we transform that from the very beginning of their experience as they step into the preserve? We have that other tier that is our educational partners inside of Stanford. I also build relationships with community colleges, K-12 schools, and nonprofits like Latino Outdoors.[22] I am the mentor for Stanford's Society for Advancement of Chicanos, Hispanics, and Native Americans in Science (SACNAS)[23] chapter, as well as the mentor for the Ecological Society of America Strategies for Ecology Education, Diversity, and Sustainability (SEEDS)[24] Stanford chapter. I also build communities outside of Stanford.

Our other big component is our docent community. The course that I co-teach with Professor Rodolfo Dirzo, titled "The Ecology and Natural History of Jasper Ridge," trains Jasper Ridge docents. Since it is not a recreational park, in order to come into Jasper Ridge, you always have to request a tour guide, so this course trains those docents. We have over 100 active docents. They are knowledgeable in natural history, ecology, and many other fields, so they can guide you through the preserve and assist with research projects. My job involves a lot of logistics, but we also write many NSF proposals for funding outreach. Not surprisingly, research at R1 universities such as Stanford is prioritized, so there is always a push for funding research. Still, outreach is where we have to write a bit more to fund more outreach and education activities. So, with the team here, we are always finding ways to collaborate and write proposals. The favorite part of my job is that I get paid to be outside and teach people outside and have that hands-on experience with students or any other learner. The thing that surprises me the most is that no matter the background, anything that you present hands-on for the first time, something that students have not seen, allows me to see that light bulb "ah-ha" moment. But then, when I do this with communities that have not been exposed to outdoor education, the first thing they say is, "Do you get paid to do this?" I am very fortunate that I get to be in an outdoor learning space in nature.

Angelica Patterson Now, that is fantastic! Last Friday, a teacher actually told me, "Your job is awesome." Then they said, "You get to do this every day?" and I said, "I do, thank you." For the next set of questions, I would like to ask, do you consider your job and expertise to be exclusively in science? Do you think that your teaching, your research, or your role in project coordinating is equally relevant to the fields of environmental science, conservation, sustainability, and/or environmental justice? If so, how? You can comment on whether or not your expertise ties into more of an

---

[22] *Latino Outdoors homepage*. Latino Outdoors. (n.d.). Retrieved December 15, 2021, from https://latinooutdoors.org/

[23] *Society for Advancement of Chicanos, Hispanics, and Native Americans in Science homepage*. Society for Advancement of Chicanos, Hispanics, and Native Americans in Science (SACNAS). (n.d.). Retrieved December 15, 2021, from https://www.sacnas.org/

[24] Strategies for Ecology Education, Diversity, and Sustainability homepage. Strategies for Ecology Education, Diversity, and Sustainability (SEEDS). (n.d.). Retrieved December 15, 2021, from https://www.esa.org/seeds/

interdisciplinary focus and, if so, what are those allied fields outside of the science field? Tanisha, do you want to go first?

Tanisha M. Williams I would have to say I like to dabble between fields. I am active in science communication, which is still a part of science. I do some policy work as well, definitely touching on the education realm. The new plants and people project that I talked about is really opening my research into ethnobotany. I am learning a lot from anthropologists, Indigenous groups, and other scientists. I feel like this project is cross-cutting and opening up my research scope. When I did my work for my Ph.D. in South Africa, I was really focused on the plants themselves and only the plants. The genus I worked on has many beneficial plants, but I was not concentrating on those species. I was only focused on, "What is happening to the plants here?" I was looking at this work through an ecological lens. Now, I am broadening that scope of what is happening to the plants and incorporating what is happening to the people using these plants. I really like that link. What I do for my research—is it relevant to environmental science conservation and environmental justice? Most definitely, everything I do ties in some way to those fields. My biggest question is trying to understand how species are responding to climate change. I use plants as my study system because I really love plants. When I talk about climate change and how plants are responding to different audiences, I often bring in examples that make plants relatable and highlight their importance. I touch on how we are using plants and how people connect to plants that they often do not think about—how are you eating? How are you breathing? How are you wearing these clothes? Medicine, all these things. Many people forget that we are here because of the products that we get from plants. As I teach, I like to introduce my field with these points, highlight what we know scientifically, and then try to encourage people to care about the environment. We do not want to lose things here on Earth. We do not know how many pegs (species, ecosystems, etc.,) we are going to knock out of this system until the system is no more.

Suzanne Pierre Do I consider my expertise to be exclusively in STEM? If STEM includes all of the dimensions that Tanisha talked about, like education being a necessary part of STEM, mentorship being a part of STEM, and being a naturalist, then yes. All of those to me are parts of being involved in STEM and I do little bits of all of that. Primarily, I am a scientist, but I am not exclusively a scientist. I think my expertise also lies in noticing dimensions of STEM that are either not being communicated or could be exposed or explained in a new way, which I think is kind of related to science communication (sci-comm). I am not really involved in the broad-scale kind of sci-comm work like a lot of professionals who are really good at targeting a large body of people, really getting messages and facts across. I am kind of in the sci-comm area that is aligned with directives, such as, "Look at this thing that we could explore in a more creative or nuanced way that teaches us something about ourselves." I use science as this series of metaphors or series of processes that really exposes things about society.

Do I think my teaching, research, or role in project coordination is equally relevant to justice, science, conservation, or sustainability? My role involves a little bit of all of those things: teaching, research, and coordination—all of which go towards all of the things that have just been described. In a basic sense, my goal is to continue to do environmental research, but I want to restructure how science is experienced and done. Yes, I am doing environmental science, but I want to do it in a transformative way, and that is the grand experiment of my career. I am probably involved in conservation the least. The other things are more obvious. I am interested in sustainability and environmental justice. It speaks for itself if you look at the work that I have been doing. I think the reason that I am least involved in conservation is that conservation has been racially white-driven. It is driven by white populations and driven by those more resourced, but I hate the word "resourced" because it is this weird, disembodied thing. It is like, they are this group with more resources because they have taken resources, right? So, I think conservation is tied to expropriation in a way that does not interest me. It is really difficult, and maybe you all have feelings about this, but being in love with biology and ecology yet not actively doing conservation feels disingenuous sometimes because it is like, "Shouldn't my work be to protect the thing I love?" But I think the reason that my work is more towards science and justice is that the effort of conservation historically has circumvented how and why we arrived at having [conservation] problems, and who is most harmed by these problems. I trust that there are many, many "well-resourced" white people who will keep doing conservation, but there are so many fewer black women with Ph.D.'s who will do the other stuff. So that is what I do.

Angelica Patterson Yes, that is a great synopsis. Jorge, what are your thoughts?

Jorge Ramos I was trained very "STEM." When you get into this field, and your mentors are all research professors at universities, you are directly and indirectly told you are just going to be a scientist. I was always told not to do outreach because I was not going to be seen with credibility. It was ingrained in my head that I needed to be a pure scientist, where later I found out that I could not leave outreach out of my life. I could not leave mentoring or education out of my life. So, I did not pursue that traditional, expected path from my committee to join a tenure-track faculty job. It was interesting listening to Sue. Everything in my Ph.D. was Environmental Science. My first job was all conservation-oriented, protecting nature for humans. It was a nonprofit world, as with here at Stanford and with the Ecological Society of America (ESA).[25] I do pretty much all education, outreach, and mentoring. I have been a part of the ESA SEEDS program for 17 years and with SACNAS, for 18 years. I have always included authentic DEIJ (diversity, equity, inclusion, and justice) work by interacting with the people in our mentoring programs and talking about science. We did not call it DEIJ back then, we were already doing this work, but it did not have a name.

---

[25] *The Ecological Society of America homepage*. The Ecological Society of America. (n.d.). Retrieved December 15, 2021, from https://www.esa.org/

The weird one is sustainability for me. SEEDS and SACNAS made me realize what I was getting taught at school was all just polarized with whiteness. So SEEDS and SACNAS kept me grounded in a broader reality. I was at Arizona State University (ASU)[26] when the university created the first sustainability school in the USA. It came from an urban perspective and was mostly driven from a very homogenous perspective of wanting to put a name to practices that other people in the world have done, apply a theory behind it, create new terms, which then became money-oriented. Working at Jasper Ridge has made me see conservation as land management instead of the traditional view of conservation of the polar bear or the panda that we used to see as kids. The explorers of National Geographic photographed the snow leopard—that, for me, was conservation growing up. But now, being exposed to more people who have relationships to the land, we can ask how fires are impacting Latinx populations—a population that has the highest risk of fire exposure.[27] Conservation has become about land issues for me, which makes it more relatable to everyone because we all step on land. I do not mean that in a bad way—we walk the land. But sustainability has become a term and a concept that started a long time ago with community-based actions, by communities for communities, and now sometimes in these academic worlds, it has turned into something that is almost impossible to achieve. So sustainability is the one concept I still need to process, especially because Stanford is now creating a new school centered on sustainability, 15 years after ASU founded the first one in 2006. I am sometimes very conflicted with sustainability because I ask, "Is it a theory or is it applied?" The thing that makes me feel better about being here is that earlier this year we created an academic minor on environmental justice.[28] We wanted to make sure the environmental justice curricula were included. I worked with a great group of people led by Dr. Richard Nevle—we just launched it, and I was just notified that the first student declared environmental justice as their minor! We were very happy! This minor in environmental justice is the first one that focuses on history, how to work with communities, and how to translate it into practice.

Angelica Patterson I think you bring up such a great point, and I too feel that it is hard to disentangle our roles in all of these—science, conservation, sustainability, and justice, mostly because of who we are and our marginalized positions in this field. Whether or not I consider my job or expertise exclusively in science? No. Because I work with so many teachers and educators from different types of institutions that impact so many different types of students from different types of places

[26] Arizona State University Technology Office. (n.d.). *Arizona State University homepage*. Arizona State University. Retrieved December 15, 2021, from https://www.asu.edu/

[27] Choi, M. (2021, July 2). *The disproportionate fire risks in Latino communities*. POLITICO. Retrieved December 16, 2021, from https://www.politico.com/newsletters/morning-energy/2021/07/02/the-disproportionate-fire-risks-in-latino-communities-796288

[28] Dulisz, D., & Stanford University Bulletin. (n.d.). *Easys-Min Program: Stanford University Catalog*. EASYS-MIN Program | Stanford University Catalog. Retrieved December 16, 2021, from https://bulletin.stanford.edu/programs/EASYS-MIN

with different types of backgrounds and experiences. How you activate that enthusiasm in young minds comes in different ways. Whether through art, their music, humanities, writing, or purely in the lab or in the field. It is hard for me to see how all those experiences can be siloed. I probably have the least connection as well with sustainability because it seems like such a nebulous term now. What does that really mean, and who is it really for? As we get more people from different backgrounds involved and really evaluate what communities need to survive in this changing climate, we could better understand what that term means for different people in different places.

In my role, I definitely teach environmental science, and like you, Jorge, I understand conservation through land management, because I work in terrestrial forest ecosystems. A big conservation effort that we have is centered around wildlife and habitat connectivity and carbon sequestration in our forests. There is an exploration into whether or not Black Rock Forest could be a place that could be established as a carbon offset in the carbon credit sphere. I do get pulled into conversations through my work as a board member for a land trust. I get pulled into conversations around policy, and I do feel that fields that are deemed to be outside of science have become important and relevant. I do find myself being the one to remind policymakers or scientists to think about bringing in more diverse people into these conversations. Who is on this land? We cannot just think about the green spaces. We also need to think about the urban areas and the communities that are underserved. What grass-root efforts are being made in those communities? A social science part also comes through even though I am not formally trained as a social scientist, but I understand that these things cannot be ignored in these conversations. I definitely see myself as interdisciplinary in that aspect.

How much do you prioritize and value mentoring and teaching students who are preparing for their own careers and lives? Do you prioritize or value this because it is your specific job duty or expectation, or is it also, or instead, a personal advocation or a bit of both? So maybe we will start with you, Sue.

Suzanne Pierre There is no part of my job that formally requires education or mentorship. In my position at the California Academy of Sciences, I am just a research scientist, and I have no service requirements, which is kind of awesome in a way. If I wanted to be more directly involved in teaching and mentorship, I would probably have gone the tenure-track route. There are a number of other reasons why I did not go that kind of faculty direction, but one aspect of it was that I wanted to be more flexible in how I teach, who I teach, and how I mentor. So, I have been able to make that more of a personally directed practice. But that is also very complicated because I prioritize it, and I see teaching and mentoring as deeply embedded in the process of shifting the culture of science and the practice of science. I see it as completely necessary to invest in early-career scientists and students. The other part of it is that I am really trying to create a disciplinary area that I wish existed when I was a student. I think a lot about what kind of resources I would have wanted as a student, and I try to create those and make those available. I also dive into the process of having a nonprofit lab as well as a kind of institutional lab space. But I also think about the academic culture I would have wanted when I was younger, and the ways that I was not served by the academic institutions I attended. I am building a world that I wish existed for me. I am thinking a lot about and talking to students about what they need, constantly reflecting on their feedback about the intellectual space I am creating. It is a weird thing because I am not paid to do that yet. Maybe one day there will be compensation, but I am doing education and mentoring in that way.

Angelica Patterson That is great. Tanisha, what are your thoughts?

Tanisha M. Williams I do prioritize and really value mentoring and teaching students. It is a part of my job as a postdoc to mentor and teach students, and I enjoy it. It is really corny, but the students are our future. When I see them light up about new ideas, listen to them talk to each other about different scientific topics, things like that excite me. It gives me ideas as well, and it gives me hope. Although I am mentoring and teaching, they are also teaching me. It is a back-and-forth relationship; it is not a one-way street. So, I get a lot of energy out of talking to younger folks and hearing more from them. It is just wonderful to see that and to know no matter what form that you are doing it in, whether it is in academia or outside of academia, we are touching not only students' lives, but we are touching people's lives. These are people that we are giving a little bit of us (our knowledge, experience, compassion), a little bit of advice, a little bit of science that hopefully they carry on and pass on to the next person. I did not understand the connection until I really started to do it more on a daily basis. I am also thinking about what my mentors have done and are still doing for me. It makes me reflect on some of the really great mentors and teachers that I've had.

Angelica Patterson That is great. I definitely agree. I highly value and try to prioritize mentoring and teaching students about careers and the future—how to get involved and navigate this field in different ways. I have been prioritizing mentoring since I could officially do so, even before becoming a graduate student when I was working at Barnard College[29] as a research assistant. When students came into the lab, I would always ask them, "How are you feeling today? What challenges are you encountering? This is what I did to overcome feelings of imposter syndrome."

I started to do more peer mentoring as I got into graduate school in an effort to advocate for us as graduate students because we were not getting that mentorship and guidance that we thought we deserved. A cohort of us realized that there was a gap. Resources were not being provided to us as graduate students, and more

---

[29] *Barnard College homepage.* Barnard College in New York City. (n.d.). Retrieved December 15, 2021, from https://barnard.edu/

specifically, as Black, Indigenous, and People of Color (BIPOC) graduate students. That is when we started the Student of Color Alliance[30] at Columbia University, the first [of its kind] graduate student group officially recognized by the university. There was another graduate student group, Women in Science at Columbia,[31] for which I also became an active leader as Co-President. It was all about providing programming to help support us in every career stage of our life up to then and beyond.

Now, my work focuses on working with educators and sometimes with students in the field. I do find opportunities to interact with students one-on-one, and that is usually when I start to give talks about my research or my career journey. When I am invited to speak in classrooms or for seminars, I start getting emails from a student who wants to talk to me about, "How the heck did you graduate from this institution when you almost failed out of undergrad? How did you do that?" I am able to help them, and that is so inspiring because I wish I had that when I was going to college. Any opportunity I can get to mentor someone, is what I do, and what I lean upon. How about you, Jorge?

Jorge Ramos I make it part of my job, so I am actually really lucky because it is one of the things that I love to do. I have all my greetings and thank you cards on my desk. When students start gifting you things as a "Thank you" when you do not even expect it, all you do is cry—I do get teary. The first experience I had with authentic mentoring was when my mentor, who was a Latino geoscientist, Dr. Aaron Velasco, put a bunch of us undergraduate students on a bus to go to a SACNAS conference, from El Paso, Texas to Albuquerque, New Mexico because it was super close to us in distance. After we came back, two of us founded the SACNAS chapter at The University of Texas in El Paso. It was something we thought of in regards to how we can offer what we just experienced during the conference to everyone back on campus. Back then, SACNAS was a small national conference. Now it has grown to thousands of members and participants!

Ever since then, I realized that without mentoring, navigating academia is hard. I realized that during my career, and even now at Stanford, I aim to be a mentor. I do not get to supervise students directly. I do not have a lab, so officially, sort of similar to what Sue was mentioning, I voluntarily mentor within these organizations even though it is not in my job description. I do that because it is fresh in our memories, the lows and the failures that the Principal Investigators (PIs) do have, intentionally or not, because of the system that they are in. I try to fill in those gaps immediately: "You have no idea how to submit your first abstract, so let me help you with that." Angie has seen some of the emails that I have received from some students: "Jorge, no one had mentioned this to me," and they were seniors about to

---

[30] *Columbia Students of Color Alliance homepage.* Columbia students of Color Alliance. (n.d.). Retrieved December 15, 2021, from https://blogs.cuit.columbia.edu/columbiasoca/

[31] *Women in Science at Columbia homepage.* Women in Science at Columbia. (n.d.). Retrieved December 15, 2021, from https://www.womeninscienceatcolumbia.org/

graduate. These are all students of color. I mentor students in these chapters on how to write fellowships and scholarships because I remember how unhelpful and unequal the system was. Some advising can still be brutally exclusive and people that have previous knowledge or support groups are able to succeed. I do that very intentionally, and I love doing it and teaching in informal spaces, like at Jasper Ridge. Also, because very few people in academia have nonprofit career experience in the past, I also mentor students to see different career pathways. I chose not to be in traditional academia for my career path. They are like, "Oh, well, you have an alternative." It is not an alternative. I am happy. I chose this, and this is my way of life. So, I prioritize it. It is a personal mission of mine. I still get mentored on how to mentor by other colleagues that are great mentors. Some students teach me how to mentor. I think mentoring is beautiful, and it makes my days, weeks, months, and years.

Angelica Patterson That is great. How does your research or teaching philosophy intersect with issues in sustainability and/or environmental justice? How have you been engaged in these issues personally or professionally? Using your professional knowledge and lived experiences, have you engaged students or public audiences to take interest and/or action in environmental research or activism, and if so, how? I know it is a loaded question. Maybe, Jorge, you can go ahead?

Jorge Ramos One of the main things I always do, and I have learned through working with Latino Outdoors is how to be a better teacher. Since I teach outdoors, I always have to prioritize that sense of belonging, that sense of comfortability, and that sense of it being safe outdoors. Those are the first ways in which I teach, and have pursued teaching as I strive to improve my teaching. I like to think that the field removes so many other barriers like the walls, the classroom, the power person in front and the sitting down in the back row, the front rows like the overachievers, hand raising—all of that. We have removed that by being outdoors. But then the outdoors brings other aspects like first-timers, or uncertainty, or stereotypes. I actively like to switch that, so I have to be always very conscious. And it is always just listening. Listening also means observing the students—which ones do not even want to go off the trail or which ones are always clenching their arms. You have to observe to understand.

Because our course is a natural history course, my other co-instructor, another Mexican ecologist, and I were very aware of these interactions of the students with each other and with the environment. Maybe it is because we are also outsiders in many ways here in the USA and at Stanford. We bring a lot of the history and the human component to our teaching. We always start with bringing an archaeologist to share the story of Jasper Ridge in the past and its relationship in the present with the Muwekma Ohlone tribe. We always invite our students to attend a Muwekma Ohlone tribal event, so they can see that they are just like us. There is no need to exoticize them. Our archaeologist always says that if we really want to learn from them, we should just attend their public events and participate. They wear tennis shoes, just like us. She is always promoting that we read, "Changing the Narrative about Native Americans: A Guide for Allies,"[32] in order for students to remove any misconceptions, false or harmful content about Native Americans. We try to bring the human element into our course. This leads us to a nice segue to environmental justice. Just bringing the human into nature brings in the ideology of justice because we share it. I think that is one of the ways that we do it in teaching.

Having more diverse organizations, like SEEDS, SACNAS, AISES,[33] Latino Outdoors, the Native American Cultural Center,[34] organize and attend more events here at Jasper Ridge, allows people to see what is inside the Jasper Ridge fence. Some members of these groups had only heard of Jasper Ridge but they thought they could never visit because they saw it as an exclusive location. We had to explain why this is a private preserve. It is all of these concepts of a barrier, the concepts of needing permission to be here, the concept of who is on this side of the fence, do the staff or the docents represent the diversity of the Bay Area? For environmental justice, we try to answer what it means in this space, in the history of this space, and then how do we train the future generation of docents that are going to be explaining to guests: Why is there a fence? Who had land ownership? What is the history here? We train our docents and provide them with materials so that they can share a correct and respectful representation of the history of Jasper Ridge. I do not teach environmental justice courses, and I do not have the theory and deep history knowledge of environmental justice. I do actions here at Jasper Ridge, and maybe I will even take courses in the Environmental Justice minor to learn more about it.

Angelica Patterson Great. Tanisha, do you want to go next?

Tanisha M. Williams Yes, I think I am with Jorge, where I do not teach or research environmental justice or sustainability. I am doing it in a more active way like you were talking about by bringing a human focus to the environment—like, who is currently on the landscape, who was on the landscape, who was taken off the landscape, and how do we teach these different concepts. That is an aspect of environmental and social justice that I have been trying to bring into the classroom as I develop teaching materials for my courses in climate change and ecology. With Jorge, Sue, and Angie, what you all have touched on—I want to explore what I wanted to learn when I was in school, what I wanted to see, and so I am trying to bring a little bit of that into this white academic structure that I am in right now. I want to bring in some of those issues of justice that touch on social justice and bring in the environment.

---

[32] Campisteguy, M. E., Heilbronner, J. M., & Nakamura-Rybak, C. (2018, June). *Changing the narrative about Native Americans: A guide for allies*. Publications. Retrieved December 16, 2021, from https://rnt.firstnations.org/wp-content/uploads/2018/06/MessageGuide-Allies-screen.pdf

[33] *AISES homepage*. AISES. (2021, November 17). Retrieved December 16, 2021, from https://www.aises.org/

[34] *Native American Cultural Center homepage*. Native American Cultural Center. (n.d.). Retrieved December 15, 2021, from https://nacc.stanford.edu/

Our communities already had and have all of this knowledge. We have had knowledge of the land, ecology, biology, medicine, and anything you can think of; we already had these connections. We already had the knowledge, but it was not seen in this western world as knowledge. Because of the dismissal of knowledge systems not known or understood by the western world, I am now digging in archives to bring this historical knowledge to the forefront. I want to make sure that my students are learning this science along with the other things they are required to learn. This inclusion or highlighting of historical and current knowledge systems and science from Indigenous and other underrepresented groups is how I am bringing environmental justice into my research and teaching.

For the last question, I have been doing science communication, science policy, and science outreach since high school. I was in a science camp because I was a nerd all the way back. We would do different outreach activities at the local farmer's markets. I was doing science communication to my great-grandmother, explaining how I built a robot in summer camp and why it is lighting up. I have been doing this work for a while because I think it is so cool and awesome. I want other people to think it is awesome and cool and learn a little bit and have fun with science as well. Where else do you get paid to ask questions and design experiments to answer those questions!? Black Botanists Week has been huge for public awareness of me and my work, although I have been doing it for a while. The organization has helped put the issues I am passionate about and move them to the forefront. I have been able to work as an activist, touch on injustices, highlight unknown people in botany, and highlight people's research that would not get any shine. They are doing incredible work, but they just are not highlighted like their white peers. This is something that Black Botanists Week tries to focus on. We are highlighting Black people in this space. And we also try to reach out to Black, Indigenous, and People of Color (BIPOC). I know we are seen as for "Black people only," but we always do try to have a day to highlight BIPOC members. We want to amplify underrepresented groups in botanical sciences. I know we are moving away from the use of BIPOC, but however we define it, we want to say yes, Black people, Indigenous people, and People of Color have their struggles, and yes we all have our individual struggles (that impact us and our communities differently), but we can come together and say, "Hey, we're struggling together. Let's lift each other up."

Angelica Patterson I agree. I think it is funny that when I wrote this question, I was trying to answer it. Now, I have the terms to describe what is going on, but I feel like it has been a process over the last several years to decolonize my teaching and thinking because we have been trained that way. From the materials that we were taught and knowing that the teaching was not quite right, I had to decolonize the way I thought about science, teaching science, and engaging with my students. I am in the process of re-learning how to diversify the curriculum using my lived experiences. It helps to be able to have conversations with people from my community, who look like me, who have gone through similar training, seeing how they teach and how they do research. That is why I am so glad that all of you are here because you do it in very different and creative ways.

With my job, now, I have the chance to incorporate a lot of creative curricula—not only having teachers teach their students about forest ecology, or botany, or whatever is in the forest but to have them think about an additional layer of the land that we are on and the history of the land. Black Rock Forest is so rich in how the land was used and who occupied the space before European settlers, the Lenape Indians. Thinking of ways to incorporate that type of history into students' experiences and learning about where they are studying is super important. I am hopeful that with this creative programming and curriculum, I can engage students from underserved communities that are right next door to Black Rock Forest and get them enthusiastic about learning about what is in their backyard. I can inspire them to be stewards of the land that they are occupying right now. Hopefully, that will transform into more activism and passion for the larger issues of climate change and other environmental issues that are happening right in their backyard. I think I will end with that. Sue?

Suzanne Pierre Both my research and teaching philosophies or core values are informed by environmental justice, which I think is inherently tied to sustainability. The reason that my research, science, and teaching philosophies are oriented towards environmental justice is that I not only have this deeper feeling of being interested in the environment, alongside having the wonderful experience of nature and loving biology for what it is, but I feel like the environment is really just a reflection of what society is doing and societies are always enacting harm. I was aware of that as an undergraduate, and initially, I was more on the environmental policy and environmental humanities side of things. But I do not really think that the things that we think we know about the environment are fully representative of what is really happening. Look at the people doing the research, and look at who came up with the questions and the methods. I was not a "STEM person" at that time, but I was very scared. Then, I thought, I am going to declare my major to be science-focused. The personal context of it is, if we do not get involved in the process of knowledge-making as non-European and non-white people, then our stories are absent.

Our stories are tied to the justice question of the harm we have been incurring for years. That influences the questions in my research, but it also influences my teaching. My teaching philosophy, with what Jorge said earlier, is the feeling of belonging, and the feeling of comfort, and the sense that you deserve to be somewhere, and that you are at home wherever you go. That is the opposite feeling that I think science has given marginalized people forever, which is that you do not belong here spatially, intellectually, or culturally. My work in environmental justice is not necessarily identifying where contaminants exist in a particular location or measuring some public health parameter. My work with justice is finding out the things that we do not know about global change and global oppression. Justice is knowledge-making and who gets to do it. That is part of my work as well. My professional knowledge is my lived experience, and my lived experience is what connects me to people like you. I never filter out the ways that my personal experience in science has made me who I am, and that just further allows me to engage with students and the public, because it humanizes me. I want to show up in every space as my full self, and my actual full self is simultaneously an activist and researcher.

Jorge Ramos I want to give credit to what you said about the comfortable space and sense of belonging. I said that I learned it through Latino Outdoors, not through academia, which is a community grassroots organization. I am validating your point that back in our day of graduate school, we did not get taught these things, so it was my participation in that group that taught me that, and now I am able to implement that here at Jasper Ridge.

Angelica Patterson This question actually got me thinking also about influencing and engaging students or public audiences to take interest in activism. I have never been one to engage in activism, in the sense of marching in the streets. I have done it in the past, but time escapes me and I feel there are other ways to engage in activism. That got me thinking of the definition of activism: what does it mean to be an activist? Are there different ways you can engage people to make a change without the traditional sense of what I think activism is? It can mean different things to other people. Am I an environmental activist? Do I influence others to become environmental activists? I am not sure because I would have to define what it is. But I do know that I advocate a lot and now speak up in white-led spaces. I make sure to make my voice heard. I make sure to include people who should be stakeholders in these conversations, which are usually from marginalized groups. Whether it is a personal or professional connection, I will be sure to bring somebody to a meeting or will I encourage you to bring someone or encourage someone to bring somebody else who should be sitting at the table. My activism is to inspire Black, Brown, Indigenous, and People of Color to engage in science and to become leaders in science. When "they" want experts in the field, the people we have encouraged can be ready, equipped, and confident to be that voice of "the expert" who can contribute to the conversation.

Tanisha M. Williams As you were talking, I got flashbacks of my life and my activism route, such as what I thought activism was as growing up in Washington, DC. I thought it was that you had to march and be on the street, which we need, but there are so many different roles to being an activist. I went to California State University, Los Angeles,[35] which is a Latino-serving institution, and immediately got involved with Students for Quality Education.[36] This organization was for

---

[35] *California State University, Los Angeles homepage*. California State University, Los Angeles. (2021, December 24). Retrieved December 16, 2021, from https://www.calstatela.edu/

[36] *California State University Students for Quality Education homepage*. California State University Students for Quality Education. (2018, December 18). Retrieved December 16, 2021, from http://csusqe.org/

undocumented students and helped students gain their undergraduate education. We spent most of our time dealing with racist and restrictive legislation but were also marching on the streets and working in the neighborhood. My master's advisor asked if I wanted to change my major. This was a turning point for me. I wanted to finish what I started out to do (get my science education) and also help in the struggle when I could. I feel like my activism has kind of changed, just like you said Angie, from marching on the streets and signing petitions, unless I feel the petition is going to get some traction and will do what it says it is going to do. It has changed to talking to younger people about science, talking to younger people about unlearning certain things and the system, giving them resources about different, diverse scientists, and about different groups of people who have made our society a better place. I see my activism not only to Black, Brown, Indigenous, and People of Color but also to my white students. Wake up, it is not only Darwin! There are real people behind that white face, so I make sure that they are aware of those people, too. Everyone needs to unlearn this system, and everyone needs to be an active participant in making sure we get what all of us are fighting for. Making sure we get environmental justice, social justice, human justice, all of these things are interlinked.

A lot of people of color, myself included, have been saying that we are just scientists, but by being the color we are, from the cultures we are from, and having the experiences we have, we are not just scientists. We are actually doing justice work just by showing up. Someone told me a long time ago that being a Black woman in this career, you are justice, you are showing up and doing what you need to do. Not to say we should just stop there. But what we are doing right now is huge, and then to turn back around and give back to the next person, and not only just one person, but all of these people that we have been mentioning that we are mentoring. We are trying to build up the next generation. I also feel that social media is helping me reach more people. I am not only at my university or within my neighborhood doing activism, but I am talking to people from all over the world. In 2020, the amount of students that I talked to from all over the world, such as in Africa and South America, who reached out and wanted to talk to me about my pathway, different experiences that I have had in science, or just, in general, has been phenomenal. Being able to just talk to people about my story, telling them that this is what I did, but there are so many ways to succeed, and letting them know that whatever they do next, I support them, and they can always use me as a resource.

Angelica Patterson Can you comment on whether you or your students gain a more authentic connection and scientific understanding about research or scientific concepts, if and when a sustainability or environmental justice lens is applied. Jorge, could you start us off?

Jorge Ramos Yes, because it comes more from the heart when they see that you care and you share your story. It is almost like you can do it too and you will not be alone because I did it and I can help you to do it. When we get our feedback on teaching evaluations, the Latinx students say that they never thought they were going to get taught by two Mexicans at Stanford. They are just blown away that they can speak Spanish in the field with us. We give them the advantage of, for example, knowing this scientific name if you are a bilingual Spanish speaker. They have an advantage immediately over someone else, to show that they should be proud of being bilingual and that they are able to learn something faster and better. The students in our Stanford SEEDS chapter developed a whole virtual environmental justice and climate change curriculum to teach high school students enrolled in an alternative/continuation public high school in Redwood City.[37] They used sea-level rise and maps to show how their communities in East Palo Alto were more in danger than other parts of the Bay Area. It is the one area getting pushed by the housing crisis, the one getting pushed by sea-level rise, and it is a community that has almost been forgotten and ignored by many powerful and big institutions and corporations nearby. It was really obvious to see that the SEEDS students from marginalized communities took pride in developing a curriculum for other marginalized communities because it is easier for them to teach something because they care about it, they know it, and they live it.

Angelica Patterson Sue, do you have thoughts on that?

Suzanne Pierre Students come into these environmental science and earth science courses with an underlying knowledge of ecological problems that they are going to be exposed to as well as a latent awareness that these problems do not come from nowhere. They do not solely come from burning fossil fuels. There is more to it than that. Students are not actually presented with that bigger picture, so in a way, I feel like we lose so many students who are not from white or wealthy backgrounds. They feel that others do not have the whole picture because what they are taught does not really address their diasporic, generational, and lived experiences in their communities. It feels incomplete, and without the justice lens or the role of social processes in what we teach in the basic science courses, the students feel that they are being lied to. For first-generation college students or those from communities that do not send a lot of students to college or graduate school, there is a whole bunch of people supporting them in arriving at that university. However, if the work that they are doing does not either directly serve their family or, more generally, serve the community, oftentimes it feels like there is a lot of conflict around what they are doing here and who it is for. I think that one day in some beautiful future,

---

[37] Stanford University. (n.d.). *2021 Jasper Ridge Environmental Education scholar award winners: Sydney Lee Schmitter and Sriram R Narasimhan*. Jasper Ridge Biological Preserve News. Retrieved December 16, 2021, from https://jrbp.stanford.edu/news/2021-jasper-ridge-environmental-education-scholar-award-winners-sydney-lee-schmitter--and-sriram

all people can study whatever they want, regardless of its connection to justice—they can just do it because of their inherent interest. But that is not the reality. The reality is that we are humans, and we are deeply tied to who we are and who we love. The way that we present basic science, research, or concepts, if it is not tied to those realities, we are doing a disservice.

Angelica Patterson That's fantastic, I totally agree. Tanisha, do you have anything to add?

Tanisha M. Williams In the past, it was alright to not even worry about justice and just teach a white-washed version of science. Now and in the future, with the movements that are happening, people are feeling more empowered to advocate for themselves, their communities, their families, and the issues that are important to them. The system is not going to be able to teach environmental science, environmental justice, and sustainability as it has been doing. I am really excited to see this change in how people want to learn, what they want to learn, and how they want to learn the truth and the facts. Students are definitely gaining more of an authentic connection to science because we are actually teaching the truth, and the truth incorporates the environment, the truth incorporates humans, the truth incorporates injustice, inequalities, and disadvantages. It is all connected and people want to learn the truth.

Angelica Patterson Absolutely, and I too feel that it is all about context. If you are being impacted by your family, historically, and thinking about the future, your future family, descendants, and how you will be affected by the climate change crisis, you are going to want to be empowered to do something about it. I was first introduced to the field of environmental justice when I took a class at Cornell University.[38] I read some work by Robert D. Bullard, who is termed the "Father of Environmental Justice." He had written the book, *Dumping in Dixie: Race, Class, and Environmental Quality*[39] about the grassroots "Not in My Backyard" movement around the dumping of chemicals and pollution. People in the community who were getting sick and dying because of these environmental disasters stood up for themselves and spoke up about this injustice and actually made some change. When you can connect those types of experiences, like what is actually happening in other people's backyards and what is happening in your backyard that is affecting you personally, it is hard not to do something about it. So a key motivator is for students to be able to understand how environmental injustice is affecting their experiences so that they can be the ones to do something about it. If they feel they cannot do something about it, they should feel supported and empowered to get the necessary

---

[38] Cornell University Office of Web Communications. (n.d.). *Cornell University homepage*. Cornell University. Retrieved December 16, 2021, from https://www.cornell.edu/

[39] Bullard, R. D. (1994). *Dumping in Dixie: Race, class, and environmental quality*. Westview Press.

education, experience, collaborations, and partnerships to be able to make a change in their community.

You are all active advocates for diversifying the environmental science field while creating spaces that are inclusive, equitable, and accessible for scientists who have traditionally been excluded from this field. In what ways have you made the environmental field diverse, equitable, accessible, and inclusive? What challenges have you encountered along the way and how have you resolved them? Jorge, would you like to begin discussing this question?

Jorge Ramos I think it was Sue who made me think about how unfiltered we are now. I used to be very afraid of being gay and proud, or if I should do outreach or not. I used to be very afraid of telling people that I did not pursue getting a tenure-track position. Did I fail? But now, I am very out and proud about outreach and being gay and Latinx. Showcasing that to any collaborator from the first moment, to students and to faculty, I am not afraid anymore. I give thanks to the communities that supported me, right now: SACNAS and SEEDS. When I came out, some mentors were like "Jorge, we knew. We were just waiting for you." They kept wanting me to write proposals with them, and they kept wanting to hang out with me. By showing that in my space, in my office, in my teaching, and in my research, or just when I talk to you, I hope I will work towards making it more diverse, inclusive, and accessible.

However, the newest challenge I have encountered is the hierarchy in academia. There are many levels in academia. Now, having graduated and being on this side of the university, there are many barriers that as a staff member, make it difficult to change the system where students think we can. I have been called out by some students that have not been on this side who say, "You're already in a position of power. You've failed us." I do not fight back with them, but I do say that I am trying my best, but it is hard as a staff member in a heavily hierarchical institution. There is so much work to be done. It is not by accident that a lot of members from marginalized communities left and ended up in outreach, education, or ended up in staff positions. I do not think it is just by chance or just an accident that we ended up there. Within my power from within my position, I will make it very, very inclusive and diverse because I am not afraid anymore.

Angelica Patterson I love that, yes. Ending in hopefulness, absolutely. Tanisha, do you want to go next?

Tanisha M. Williams What I have actively done is outreach, and just like Jorge, I have been doing outreach for a long time. Because I like to talk and I love new experiences, I find it easy to connect with people. When I got into research, we were extracting information, yet we did not do anything to give back to this community. Yes, our research was being published, but those publications do not help the communities in which we found some of the plants. So, I made sure that with every trip that I took, internationally or locally, I spent time with students and communities to help offer what I could, by mentoring, tutoring, or planting vegetable and fruit gardens. My outreach efforts are not only working with students and working with children, but I really like policy. I am from Washington, D.C., so politics is in my blood. I am up in senators', congressmen's, and congresswomen's offices. I am talking to them and saying, "The National Science Foundation (NSF) needs money and these are the projects from your state that were funded. You are not only helping the scientific community, but these are the people, your constituents, that you are helping." I went to so many people's offices that they started to come to visit me at the University of Connecticut. I love that type of work because it makes me feel like I can change this larger system a little. If I can get policymakers to sign a bill saying that they are going to commit more money to NSF, that is real change.

Moving to some of the challenges that I see, with Black Botanists Week, for example, we have been staying online because of the pandemic. I do not want it to be solely online. I do not want our only connection to be through social media. I want this movement to actually get offline eventually. It is just something that I worry about with some of the movements that are going on. I want to make sure that we get offline and we do things in person. In the way you are saying Jorge, for Latino Outdoors, not only do I want us to do things to uplift each other, like offline and meet one-on-one, take it a little bit further. How can we actually change the system? How does Black Botanists Week, the Critical Ecology Lab, Latino Outdoors, we, with all of our expertise, all of our passions, change this system? Even just a little. I do not have any the answers, or how to resolve it yet, but the first thing is getting offline and talking to people offline and joining forces as well. We are specialized in different fields, but we should come together. I know it does not always work, but there are certain issues that we can all come together on.

Angelica Patterson That's fantastic. And Sue, do you have anything to contribute to this one?

Suzanne Pierre A lot of these questions have centered around the ways that the work we are all doing is about inclusivity and equity. It is so important to me to go back to this idea of the narratives of science and the story that we tell about global change. Supporting that story that is woven into the stories of our diasporas and the processes that shape those diasporas and really explaining those stories through some type of scientific method is the place that I think my advocacy has always been centered. Doing that work of giving examples of basic research, developing tools, and developing methods is my activism. I want to make space for other people to then pick those tools up, pick those frameworks up, and apply them to their work. Activism is not just direct action; it is also practice. It is the way that we make our path in the world. Even if the Critical Ecology Lab is just one organization taking these steps and making these methodological arguments for why justice is inherently a part of global change, anybody who observes it gets to say, "Someone thought about this and someone put effort towards this." That is being a part of a conversation so that is the way I do activism. I think that has natural connections to policy because science informs policy and if we don't have certain answers in science, there will not be a political outcome from it. That is the link that I aim to serve.

I get so many people who reach out and just say, "I'm so delighted to see a brown woman in soil science or in ecology." I am sure you all get those comments as well. It starts with relationships, and just by being present in a very white space, I can form relationships with people who need that connection. Activism, in my expansive view of it, is also about transforming ourselves and you all have touched on that. But one of the ways that I think my self-transformation needs to happen and continues to happen is related to the presence of so much hurt around these topics. There is so much alienation, there is so much pain. Even though we are advancing in our careers, we are becoming mentors and leaders, and we still carry that. Building relationships with young people and other non-white scientists is a sort of therapy. That is world-building and that is transformation. Relationships that all of you have touched on have been a form of activism for me.

Angelica Patterson   Absolutely, and I love that you stated relationships and world-building, especially doing work for ourselves. As we all mentioned, us being here is transformative for young people who see somebody like them represented in this field. However, the challenge for us and for me, in particular, is community building. I am so glad that we are on this call and talking about it, and we have yet to meet in person, but at least we have the space to be able to communicate, bounce ideas, resonate our thoughts, conflicting ideas, and concepts of whatever comes into mind with each other. After this phone call, I am going to go back to my place of work and where I live, and it is predominantly white. It is a white space, and I am the only one. I have been the only one, since grade school, growing up in northeastern Pennsylvania. I think the challenge lies in how we, as early-career scientists of color, build our community to have it be stronger. There are groups, such as Black in Geosciences,[40] SEEDS, and SACNAS. But when it comes to postgraduate students, I do not see much of that community. If we can create a stronger support network for all of us and uplift each other along the way, we could really make some changes. With that, I am going to end this, and I want to say thank you all for being part of this conversation. I am so honored to be able to talk to you and meet you and have you be a part of this project.

Acknowledgments   The authors would like to thank Taspia Rahman for their help in transcribing the conversation into text and for editing this work.

---

[40] *Black in Geosciences homepage*. Black in Geoscience. (n.d.). Retrieved December 16, 2021, from https://blackingeoscience.org/

References

AISES homepage. AISES. (2021, November 17). Retrieved December 16, 2021, from https://www.aises.org/

American Museum of Natural History homepage. American Museum of Natural History. (n.d.). Retrieved December 16, 2021, from https://www.amnh.org/

Arizona State University Technology Office. (n.d.). *Arizona State University homepage.* Arizona State University. Retrieved December 15, 2021, from https://www.asu.edu/

Barnard College homepage. Barnard College in New York City. (n.d.). Retrieved December 15, 2021, from https://barnard.edu/

Behravesh, S. (this volume). Diversity in academia and sustainability science: The STEM blindspot. In M. S. Rivera Maulucci, S. Pfirman, & H. S. Callahan (Eds.), *Education for sustainability: Discourses on authenticity, inclusion, and justice.* Springer.

#BlackAFinSTEM and #BlackBirdersWeek twitter page. #BlackAFinSTEM Twitter. (n.d.). Retrieved December 16, 2021, from https://twitter.com/BlackAFinSTEM

#BlackBotanistsWeek homepage. #BlackBotanistsWeek. (n.d.). Retrieved December 16, 2021, from https://blackbotanistsweek.weebly.com/

#BlackInAstro homepage. #BlackInAstro. (n.d.). Retrieved December 16, 2021, from https://www.blackinastro.com/

#BlackInChem homepage. #BlackInChem. (n.d.). Retrieved December 16, 2021, from https://blackinchem.org/

#BlackInEnvironWeek twitter page. #BlackInEnvironWeek Twitter. (n.d.). Retrieved December 16, 2021, from https://twitter.com/BlackInEnviron

#BlackInMicro twitter page. #BlackInMicro Twitter. (n.d.). Retrieved December 16, 2021, from https://twitter.com/BlackInMicro

#BlackMammalogistsWeek homepage. #BlackMammalogistsWeek. (n.d.). Retrieved December 16, 2021, from https://blackmammalogists.com/

Black in Geosciences homepage. Black in Geoscience. (n.d.). Retrieved December 16, 2021, from https://blackingeoscience.org/

Black Lives Matter homepage. Black Lives Matter. (n.d.). Retrieved December 16, 2021, from https://blacklivesmatter.com/

Black Rock Forest homepage. Black Rock Forest. (2021, June 29). Retrieved December 15, 2021, from https://www.blackrockforest.org/

Brown, D. N. L. (2021, May 4). *Violent deaths of George Floyd, Breonna Taylor reflect a brutal American legacy.* History. Retrieved December 16, 2021, from https://www.nationalgeographic.com/history/article/history-of-lynching-violent-deaths-reflect-brutal-american-legacy

Bucknell University homepage. Bucknell University. (n.d.). Retrieved December 15, 2021, from https://www.bucknell.edu/

Bullard, R. D. (1994). *Dumping in Dixie: Race, class, and environmental quality.* Westview Press.

California Academy of Sciences. (n.d.). Retrieved December 15, 2021, from https://www.calacademy.org/

California State University Students for Quality Education homepage. California State University Students for Quality Education. (2018, December 18). Retrieved December 16, 2021, from http://csusqe.org/

California State University, Los Angeles homepage. California State University, Los Angeles. (2021, December 24). Retrieved December 16, 2021, from https://www.calstatela.edu/

Campisteguy, M. E., Heilbronner, J. M., & Nakamura-Rybak, C. (2018, June). *Changing the narrative about Native Americans: A guide for allies.* Publications. Retrieved December 16,

2021, from https://rnt.firstnations.org/wp-content/uploads/2018/06/MessageGuide-Allies-screen.pdf

Choi, M. (2021, July 2). *The disproportionate fire risks in Latino communities*. POLITICO. Retrieved December 16, 2021, from https://www.politico.com/newsletters/morning-energy/2021/07/02/the-disproportionate-fire-risks-in-latino-communities-796288

Columbia Students of Color Alliance homepage. Columbia students of Color Alliance. (n.d.). Retrieved December 15, 2021, from https://blogs.cuit.columbia.edu/columbiasoca/

Cornell University Office of Web Communications. (n.d.). *Cornell University homepage*. Cornell University. Retrieved December 16, 2021, from https://www.cornell.edu/

Critical Ecology Lab homepage. Critical Ecology Lab. (n.d.). Retrieved December 15, 2021, from https://www.criticalecologylab.org/

Dulisz, D., & Stanford University Bulletin. (n.d.). *Easys-Min Program: Stanford University Catalog*. EASYS-MIN Program | Stanford University Catalog. Retrieved December 16, 2021, from https://bulletin.stanford.edu/programs/EASYS-MIN

The Ecological Society of America homepage. The Ecological Society of America. (n.d.). Retrieved December 15, 2021, from https://www.esa.org/

Institute for Biodiversity Science & Sustainability. California Academy of Sciences. (n.d.). Retrieved December 15, 2021, from https://www.calacademy.org/scientists

Latino Outdoors homepage. Latino Outdoors. (n.d.). Retrieved December 15, 2021, from https://latinooutdoors.org/

Native American Cultural Center homepage. Native American Cultural Center. (n.d.). Retrieved December 15, 2021, from https://nacc.stanford.edu/

Nir, S. M. (2020, June 14). *How 2 lives collided in Central Park, rattling the nation*. The New York Times. Retrieved December 16, 2021, from https://www.nytimes.com/2020/06/14/nyregion/central-park-amy-cooper-christian-racism.html

Pennsylvania Natural Heritage Program homepage. Pennsylvania Natural Heritage Program. (n.d.). Retrieved December 15, 2021, from https://www.naturalheritage.state.pa.us/

Society for Advancement of Chicanos, Hispanics, and Native Americans in Science homepage. Society for Advancement of Chicanos, Hispanics, and Native Americans in Science (SACNAS). (n.d.). Retrieved December 15, 2021, from https://www.sacnas.org/

Stanford University homepage. Stanford University. (n.d.). Retrieved December 15, 2021, from https://www.stanford.edu/

Stanford University. (2021, December 8). *Jasper Ridge Biological Preserve homepage*. Jasper Ridge Biological Preserve. Retrieved December 15, 2021, from https://jrbp.stanford.edu/

Stanford University. (n.d.). *2021 Jasper Ridge Environmental Education scholar award winners: Sydney Lee Schmitter and Sriram R Narasimhan*. Jasper Ridge Biological Preserve News. Retrieved December 16, 2021, from https://jrbp.stanford.edu/news/2021-jasper-ridge-environmental-education-scholar-award-winners-sydney-lee-schmitter-and-sriram

Strategies for Ecology Education, Diversity, and Sustainability homepage. Strategies for Ecology Education, Diversity, and Sustainability (SEEDS). (n.d.). Retrieved December 15, 2021, from https://www.esa.org/seeds/

Western Pennsylvania Conservancy homepage. Western Pennsylvania Conservancy. (2021, December 1). Retrieved December 15, 2021, from https://waterlandlife.org/

Women in Science at Columbia homepage. Women in Science at Columbia. (n.d.). Retrieved December 15, 2021, from https://www.womeninscienceatcolumbia.org/

Yancy, C. W. (2020). Covid-19 and African Americans. *Journal of the AMA, 323*(19), 1891. https://doi.org/10.1001/jama.2020.6548

Dr. Angelica E. Patterson is the Curator of Education and Outreach in the Miller Worley Center for the Environment at Mount Holyoke College and was formerly the Master Science Educator at Black Rock Forest. She received her B.S. degree in Natural Resources from Cornell University and her M.A., M.Phil., and Ph.D. degrees from Columbia University, where she studied the physiological drivers behind climate-induced tree migration. Angelica develops innovative curricula that emphasize inclusive teaching practices. She is a strong advocate for diversity, equity, and inclusion in the environmental sciences.

Dr. Tanisha M. Williams completed her Ph.D. in Ecology and Evolutionary Biology at the University of Connecticut and is the Burpee Postdoctoral Fellow in Botany at Bucknell University. Her dissertation research examined the impacts of climate change on plant species throughout South Africa. Her postdoctoral research elucidates the role Aboriginal peoples have on the movement and maintenance of plant species in Australia. She is the founder of Black Botanists Week.

Dr. Jorge Ramos is the Executive Director of Jasper Ridge Biological Preserve at Stanford University where he oversees research, education, and conservation programs. In his former position as the Associate Director for Environmental Education, he co-taught an outdoor course on ecology and natural history and managed educational programs that encounter up to 8000 educational visits annually. Jorge earned a B.S. degree at The University of Texas at El Paso, an M.S. degree at the University of Washington, and a Ph.D. at Arizona State University.

Dr. Suzanne Pierre is a biogeochemist and ecosystems ecologist focused on the role of oppression as a driver of global change. She received her doctorate from Cornell University, completed a President's Postdoctoral Fellowship at the University of California, Berkeley, and is currently a research scientist at the Institute of Biodiversity and Sustainability Science at the California Academy of Sciences. Dr. Pierre is also the founder and lead investigator of the Critical Ecology Lab.

Chapter 5
A Commons for Whom? Racism and the Environmental Movement

Ralph Ghoche and Unyimeabasi Udoh

On June 1, 2017, much of the world watched, disappointed—but not especially surprised—as President Donald J. Trump announced his decision to pull the United States out of the Paris climate agreement. The landmark environmental accord, ratified in 2015, had been supported by all but two of the world's independent states, and now it was lacking the backing of the world's biggest polluter as well. The United States, superpower, defender of democracy, and champion of the West, effectively announced that being the 'Leader of the Free World' did not apply to the fight against climate change.

The other two nations—Nicaragua and Syria—that stood outside the Paris climate accord, did so for vastly different reasons. Nicaragua, one of the four nations most endangered by climate change, found that the agreement was too lenient. Syria was uninvited and unable to attend (Taylor, 2017). Only President Trump rejected the accord out of disbelief in the facts of, and human responsibility for, global warming. The credit for this radical turn was given, primarily, to two men: Scott Pruitt, the then-head of the Environmental Protection Agency, a man who was so hostile to his role that he had sued that self-same agency in the interest of polluting industries before being appointed to run it; and Steve Bannon, perhaps Trump's most notorious former advisor (Baram, 2017).

Bannon, then executive chairman of far-right "news" site Breitbart and noted white nationalist (Politi, 2017), opposed the Paris agreement both for its climate science and its "globalist" agenda. Bannon is credited with the line of reasoning that Trump invoked in his announcement about the agreement: that the Paris accord was "cheating" America economically, putting its coal miners, who supported Trump

R. Ghoche (✉)
Department of Architecture, Barnard College, New York, NY, USA
e-mail: rghoche@barnard.edu

U. Udoh
School of the Art Institute of Chicago, Chicago, IL, USA

© The Author(s) 2023
M. S. Rivera Maulucci et al. (eds.), *Transforming Education for Sustainability*,
Environmental Discourses in Science Education 7,
https://doi.org/10.1007/978-3-031-13536-1_5

overwhelmingly in the 2016 election, at a particularly gross disadvantage. Paying for other, poorer countries' continued survival in the face of potential environmental disaster, as the agreement encouraged the United States to do, runs counter to nearly everything Bannon, Breitbart, and the far-right internet apparatus purport to stand for. It gives a "handout" to those less fortunate, who *clearly* should pull themselves up by their bootstraps and save themselves from the literal rising tide of America's creation. It considers the interests of nations who are not the United States and whose citizens, not coincidentally, are not white. It is capitulating to the requests of the "socialist" and "globalist" European Union and United Nations, both boogey-men to those who would prefer that America comes first and stands alone in all things.

Under Steve Bannon's guidance, President Trump announced his decision to withdraw from the Paris Climate accord. Thus, concern for the continuation of a myth of America had superseded care for the globe and its community. White nationalism had trumped (pun intended) environmentalism. This, nowadays, would appear to be an unfortunate but natural progression: reactionary conservatism and environmentalism are thought to be antithetical to each other. The media's look into Steve Bannon's history regarding the latter, however, threw things into confusion. Could it be that such a renowned nationalist (had) cared for the environment? Had Bannon merely had a sour change of heart—or could it be that such nationalism and ecological stewardship were not only *not* mutually exclusive but even had some cause for overlap?

The progressive, tech-savvy news sites of the left-leaning internet are littered with articles linking three things: Steve Bannon, a project called Biosphere 2, and a general sense of bewilderment. The explanation is as follows: in 1991, eight people sealed themselves into a complex reminiscent of the geodesic domes of Buckminister Fuller, with plans not to reemerge for another 2 years. The complex, Biosphere 2, and the eponymous mission it supported was the pet project of Edward P. Bass, a Texas oil scion-turned-environmentalist. At the sprawling, 3.1-acre compound, rep-lica ecosystems—including marshland, rainforest, desert, savannah, and a miniature 'ocean'—were enclosed, along with the human subjects, in what was meant to be a sealed, self-sustaining system. If the Earth itself was Biosphere 1, then Biosphere 2 was the 1970s' fascination with closed-system ecology made concrete: a miniature Earth that could be studied. Perhaps, as its participants hoped, it would give valu-able insight into the workings of the Earth's (changing) climate. Furthermore, if the project proved successful, they could apply the closed ecosystems to other endeav-ors such as Mars colonization.

Unfortunately, from its outset, Biosphere 2 was not strictly successful. After a series of mishaps—financial, procedural, structural, and public-relational—Bass hired a Wall Street banker named Steve Bannon to keep the project from leaking money. Bannon, as *Wired's* Eric Niiler (2016) puts it, was "all business." And despite the retrospective incongruity of his participation in the project, it does seem that Bannon's involvement with Biosphere 2 was strictly professional. Of all the outlets taken with the apparent oxymoron of Biosphere Bannon, the *New Republic* gives the most thorough take on the subject. Through interviews with scientists once involved with Biosphere 2 and analysis of Breitbart's output under Bannon, journalist Emily Atkin (2017) pieces together a portrait of a man who perhaps does not believe much of *anything* when it comes to climate change.

As Atkin notes, statements on the subject from Bannon himself are in relatively short supply. Though Bannon gave a platform for climate change denialists on Breitbart, some staffers doubted how invested he was in the subject, beyond irritation at the amount of public energy expended on the "manufactured crisis" (Atkin, 2017). Bruno Marino, onetime scientific director at Biosphere 2, remembers Bannon as a man who kept books on climate science on his desk, who promoted the project's research potential in the press, and who now "could be willfully ignoring his knowledge of climate science" (Atkin, 2017). Bannon once produced a movie called *The Steam Experiment*, a B-film set in a sauna whose aim was to "prove that humanity will go crazy under the pressures of global warming" (Atkin, 2017). To be fair, it is difficult to read much intellectual depth or intent into a film of this quality. In the movie, the scientist traps the attractive protagonist in the steam bath who is depicted as mad, risking human lives to prove his 'crackpot' environmental theory. Whether or not it can be said to prove the professor's theory, the captives *do* suffer a fair deal.

On a more conspiratorial note, Atkin suggests that "perhaps Bannon believes that a warming world could cause chaos—and even welcomes it. After all, he is an alleged Leninist who wants to deconstruct the administrative state" (Atkin, 2017). Perhaps, in addition to finding the fight against climate change too extraneous to and globalist for his agenda, Bannon *does* believe in the scientific consensus on the subject. Maybe he yearns for the brave new world that would be ushered in by a climate apocalypse fueled by Breitbart's propaganda. Until he addresses his past and present relationship to the specifics of climate science, though, all this is simply speculation. However, should Bannon prove to have an ideological stake in such environmental matters, he would not be the first American white nationalist to do so. Arguably, he would not (until recently) even be the most famous. If ecological stress, such as that brought about by anthropogenic climate change, would destroy the current world order, then there are those for whom the defense of the planet and the white race go hand in hand.

Unfortunately, as with Bannon's involvement with Biosphere 2, the connection between white supremacy and environmentalism is not always readily apparent. Buzzwords like 'overpopulation' are used regularly, even in progressive climate discourse. But these discourses have their roots in theories of ecology that are disdainful of—if not outright hostile to—those who contribute the least, yet are most exposed to, the effects of climate change: the world's poor and non-white. The racist and misanthropic underpinnings of the modern environmental movement, left unexamined, can undermine post-Paris attempts to move towards a more sustainable way of life. On a shared planet, no climate action can be effective if it is not equitable.

5.1   A Commons for Some: Garrett Hardin's Enclosures Act

The question of whether we must 'save the world,' in an ecological sense, has in its wake another, more insidious question: for whom? Although this second question may not be asked outright, it often gets answered. For a glaring example, one needs to look no farther than the work of the late Garrett Hardin: a man who, troublingly, is much more famous for his contributions to the field of ecology than he is for the intense racism that motivated that work. To engage with ecology, in general, and environmentalism, in particular, as fields of knowledge and practice that benefit the *whole* world, it is vital that we understand how its foundations run counter to those aims.

Hardin's "The Tragedy of the Commons" is a foundational text in ecology. The 1968 article, which further stretched the idea of Earth's 'carrying capacity,' is a staple in classrooms to this day. Hardin paints a picture of a world doomed by human numbers and greed. The tragedy of the commons, according to Garrett Hardin, is that people cannot—and *should not*—share. Imagining a pasture open to all, Hardin then introduces a herdsman who, "as a rational being… seeks to maximize his gain. He concludes that the only sensible course for him to pursue is to add another animal to his herd. And another; and another…." (p. 1244). The pasture becomes overburdened, and each member of the increased population suffers.

Already there is something odd, something sinister about Hardin's logic. For these hypothetical herders, *more* is always better, even with stiffer competition for the resources needed to support increased numbers. This approach is not a particularly rational course of action, despite Hardin's claims. The individual herders are entirely unable to consider themselves members of a community that might act in the interest of shared goals. For Hardin, what prevented the despoiling of the commons in the past was not the success of interpersonal communication or the awareness of a common humanity. Instead, "tribal wars, poaching, and disease" (p. 1244) kept the population too low for the 'fact' that man's interests run directly counter to that of society at large to become an obstacle.

For this grave problem, Hardin offers a brutal solution. Neither education nor community development will solve the tragedy of the commons, where a Hobbesian species of man proliferates in great numbers. Instead of questioning the narrow individualism on which he has built his argument, Garrett Hardin challenges the right of certain individuals to exist. "The only way we can preserve and nurture other and more precious freedoms is by relinquishing the freedom to breed, and that very soon" (p. 1248), he writes, else "freedom to breed will bring ruin to us all" (p. 1248). *Coercion*, a buzzword at the time, is the specific term that he uses to describe the means for curbing what he viewed as devastating overpopulation: no longer can we fill the Earth with so much human waste.

Hardin implicitly includes human beings with contaminants, listing 'breeding'— already an exceptionally dehumanizing term—right after toxic waste dumping and air pollution. The rather antisocial belief in humanity as a 'parasite' on the planet was not uncommon in the heyday of twentieth-century environmentalism (more on this later). However, Hardin's take on the matter is perhaps even more vicious. Instead of pitting the desires of the human race against that of the planet, Hardin sets the comfort of the few against the existence of the many. To him, a positive ecological future is more a question of *which* race. He includes a few dog-whistles in "The Tragedy of the Commons" that hint at who he does not include in this specific few. He writes, "The most rapidly growing populations are (in general) the most miserable" (p. 1244). He continues with, "because "our society is deeply committed to the welfare state" (p. 1246), the poor have less chance of suffering the negative consequences of "overbreeding"—that is, a rank inability to feed their children. Intelligent, temperate people simply don't reproduce at the same rates, and their attitudes are drowned out in the genetic pool. Where might these poor, miserable, and intemperate people be found? Well, predominantly in the Global South, where population growth rates have outpaced those of the West for quite some time. Is it fair to ask that these brown people exist in fewer numbers? Hardin writes, "Injustice is preferable to total ruin" (p. 1247).

It does not take much sleuthing to discover that Garrett Hardin, one of the leading intellectual voices of the ecological movement, was a strident white nationalist. The Southern Poverty Law Center describes his contributions to both movements as such:

> Hardin used his status as a famous scientist and environmentalist to provide a veneer of intellectual and moral legitimacy for his underlying nativist agenda, serving on the board of directors of both the anti-immigrant Federation for American Immigration Reform and the white-nationalist Social Contract Press. He also co-founded the anti-immigrant Californians for Population Stabilization and The Environmental Fund, which primarily served to lobby Congress for nativist and isolationist policies. (SPLC, n.d.)

It is, unfortunately, not difficult to see why these two causes might converge. If brown people are bad, and an overpopulation of humans is bad, then it makes good racist sense that a preponderance of this already-unsavory type would be a particular problem. It is difficult to argue against racist logic, based as it is on irrational hatred. The introduction of racist logic into ecological thought brings into the discipline a strain of wrongheadedness that undermines even more than the idea that a healthy planet is (a) good for *all* people.

5.2 "A White, Racist Plot"

The source of Hardin's arguments stems from the post-Second World War revival of interest in population theories advanced by the English political economist Thomas Robert Malthus. In *An Essay on the Principle of Population,* first published in 1798, Malthus drew attention to the relationship between the human population and the productive capacity of the soil. Malthus argued that increased food production resulting from new mechanical instruments introduced during the first industrial revolution would inevitably lead to unsustainable growth in the human population.

Two books, published months apart from each other, are generally credited for the revival of Malthus' population scare in the twentieth century: Fairfield Osborn's *Our Plundered Planet* (1948) and William Vogt's *Road to Survival* (1948). These works were critical to the emergence of the modern environmental movement (Robertson, 2012), introducing a new style and sense of urgency to environmental writing that was blunt, alarmist, and highly contagious. They had a tremendous impact on succeeding generations of environmentalists, especially on the best-sellers of the 1960s, such as Rachel Carson's *Silent Spring* (1962) and Paul Ehrlich's *The Population Bomb* (1968). Despite their extreme popularity during the 1950s and 1960s, only recently have these works elicited the attention of environmental historians. The oversight is due, in part, to the unwillingness to complicate our views of the environmental movement.

Many of the problematic aspects of Hardin's worldview were derived from Osborn and Vogt. For example, Vogt castigated Indians for "breeding with the irresponsibility of codfish" (Desrochier & Hoffbauer, 2009, p. 46). He argued that a high death rate should be seen as "one of the greatest national assets" of the developing world and called for foreign food aid to poor countries be cut so as not to "subsidize … unchecked spawning" (Desrochier & Hoffbauer, 2009, p. 46). Paul Ehrlich seemed to echo these sentiments in the first lines of *The Population Bomb*, which vividly described an overcrowded market as "one stinking hot night in Delhi" (p. 1).

The sense of disdain for the Global South was recast by Hardin in an article published in 1974 with the disturbing title, "Lifeboat Ethics: The Case against Helping the Poor." In it, Hardin attacked Christians and Marxists, and all those who "feel guilty about their good luck" living in the United States for their "justice-based" moral conscience (p. 39). Lifeboat ethics, by contrast, were "reality" based; less-affluent nations needed to "learn the hard way" (p. 40). In short, Hardin recommended that the UN's World Food Program, which facilitated emergency aid to famine-stricken nations, and a similar domestic program, Food for Peace, be permanently closed and all food aid terminated. Hardin warned against programs that didn't export food but empowered communities to develop more robust agricultural practices. He also called for an immediate halt to all immigration into the United States to "save at least some parts of the world from environmental ruin" (p. 43).

The idea that the poor and the nonwhite cause more ecological damage than the predominantly white developed world is simply false. Ehrlich came to understand and promote this stance. He realized that a good deal of ecology centered on population planning (he estimated one-third) was "a white racist plot" and that "fundamentally the rich of the world are still stealing from the poor" (Robertson, 2012, p. 174). Racist ecology is self-defeating. It is an oxymoron because the attitudes behind the former occlude the realities of the latter. It is predominantly the rich who are eating the world. Shunting the blame for environmental degradation onto disadvantaged minorities does almost nothing to solve the real problems at hand.

In addition to being scapegoated, the poor experience greater environmental pollution than other socio-economic classes, such as the disposal of toxic or radioactive waste on African American or tribal lands in the United States or the smog in Asia generated by the production of commodities mainly for Western markets. As feared by Garrett Hardin and others, overpopulation is only an existential threat if we all consume at the selfish and excessive rates of the herders in his dark parable—the rates of the 1%. Even overpopulation does not have the same culprit that Hardin identifies. The biggest driver of both population growth and ecological strain is not the wanton behavior of the poor and brown but the brutal churning of global capitalism.

While Hardin borrowed key arguments from the Neo-Malthusians, he ignored or underemphasized notable elements from the work of Osborn and Vogt in his analysis. Among these was their criticism of American consumption. As Thomas Robertson (2012) has argued, the population debates of the immediate post-war era are best understood in reaction to the popularity of ideas put forth by the economist John Maynard Keynes and the "new obsession with growth" overtaking the American imagination (p. 30). Osborn and Vogt were fiercely critical of the spread of American consumerism across the globe. They condemned the free enterprise system, which they understood to be primarily responsible for the rapid destruction of wilderness. Therefore, the unjust focus on the poverty of developing economies in their work was balanced with a critique of the significant American contribution to that dynamic. By comparison, Hardin's call to reduce the world's population–he set the limit at 100 million people–was not paired with a significant reduction in consumption (Miele, 2002, p. 267). Instead, he imagined that, were there fewer people on earth, they would be "living a hell of a good life" (Miele, 2002, p. 267), and their standard of living would undergo a dramatic rise as "clean beaches, unspoiled forests and solitude" (Hardin, 1974, p. 41) would be available to all. In addition, Hardin presented private property as the antidote to a failed system of shared environmental commons. "Under a system of private property," Hardin explained, "the men who own property recognize their responsibility to care for it" (Hardin, 1974, p. 40). The historical precedent of the commons in England, the very commons to which Hardin was alluding and which were privatized by the passage of the Enclosures Acts beginning in the seventeenth century, would seem to show the inverse. The enclosure of the commons precipitated a massive population shift from rural settings to the swelling industrial cities and increased misery and poverty tenfold.

There are several problems with Hardin's views, not least that they conceal the social inequalities that cause the dramatic imbalances in the distribution of resources. Resources, after all, have tended to flow from developing nations to rich ones. This oversight points to a broader difficulty at the core of much Neo-Malthusian thinking on the environment: its tendency to subsume humankind under the larger category of living beings. For Hardin, as for Vogt, Osborn, and the early work of Ehrlich, the human impulse to reproduce and over-populate an ecosystem was seen as scarcely different from that of deer, fish, or locusts, and the consequences could be just as dire. Their omission of the social and political vectors that drive resource depletion was due, in large part, to their desire to apply lessons from the animal world directly to the human one. Indeed, humanity's fate was continually compared to the lesson of the Kaibab deer incident, an episode in Northern Arizona that saw the deer population plummet precipitously upon exceeding the carrying capacity of its habitat (Robertson, 2012, p. 73).

The Neo-Malthusians were not alone in their inability to contend with human agency and differentiate between humanity and the rest of the animal kingdom. Indeed, generations of environmentalists have found an intractable problem in the human/nature divide. Despite its outright racist subtext, "Tragedy of the Commons" has come to be reprinted in over 100 anthologies since its publication in 1968, most providing no critical discussion of the motivations of its author. Its racism comes out of a profound process of dehumanization endemic to more than a few environmental movements.

Hardin, of course, was not the first to inject nativist ideas into the discipline. Indeed, as a more or less holistic system of thought, ecology has long been co-opted towards troubling ends. The problem, as Joel Kovel (2003) has rightly pointed out, lies in the temptation to "naturalize" human culture in a reductive way, in other words, to assume that nature or biology determines all facets of human conduct. The German concept of *Lebensraum* (translated loosely as "living space"), weaponized by the Third Reich, provides the most explicit cautionary tale. Friedrich Ratzel, a pupil of Ernst Haeckel, the noted nineteenth-century zoologist who coined the term "ecology" in 1866, developed the notion. For Ratzel, the German people were an organism whose development could only be assured if given the proper room to grow in the environment for which it was adapted. Nazi Germany reconfigured these ideas into racial policies that denied the political boundaries of the German state in favor of expansionist practices that sought to provide the "Aryan" race the territory deemed necessary for it to flourish.

If naturalizing human conduct is a problem, so is the inverse of that fallacy, *denaturing* humanity. Both dynamics were clearly at work in the Neo-Malthusian environmental movements of the twentieth century. For example, Hardin, Ehrlich, Vogt, and Osborn all tended to blame humanity—often in the form of newborn babies—for the despoiling of the environment, not specific practices, lifestyle choices, or belief systems. Indeed, one of the most popular political lapel pins of the 1970s featured the simple words: "People Pollute," as though harmful pollution was endemic to human behavior (Chase, 1980, p. 366).

5.2.1 Misanthropic Ecologies

A similar process of naturalizing and denaturing human beings underpinned environmentalists' efforts at the other end of the spectrum, such as the wilderness preservation group Earth First!. With Earth First!, the rhetoric about population growth shifted from focusing on the fear of war, famine, and resource depletion—effects that would harm human beings—to emphasizing threats to the well-being and biodiversity of non-human species. Earth First! traced its roots back to the nature worship of Henry David Thoreau and John Muir and to such mid-twentieth-century ecologists as Aldo Leopold (1949), whose criticism of conservation efforts based on human and "economic self-interest" (p. 213) left a lasting imprint on succeeding generations. The group's views also were shaped by the Deep Ecology movement of the 1970s and its challenge to the prevailing anthropocentric system of values. Deep Ecologists, such as Arne Naess, argued that nature's well-being was an intrinsic value of its own; "the earth does not belong to humans," Naess often insisted. Earth First! radicalized these principles into an eco-centrism that not only privileged the diversity and expansion of the non-human world but demonstrated a veritable antipathy to humankind.

Earth First! participants were militant and uncompromising, and their activism promoted direct action in contrast to the lobbying efforts and legal mechanisms employed by mainstream environmental organizations. They saw their group as the "action wing" of the Deep Ecology movement, and their mission was to create as many wilderness expanses as possible (Chase, 1991, p. 8). For some in the group, the ultimate end lay in jolting humanity to revert to its nomadic and pre-agricultural roots, and "return to the Pleistocene" was a familiar rallying cry at Earth First! events (Chase, 1991, p. 21). Participants in Earth First! employed methods that were controversial, such as "ecotage" tactics—from nail spiking old-growth trees to the sabotage of earth-damaging machinery—in order to defend the environment from, what they termed, "parasitic" humanity.

In addition to tactics that posed a danger to human welfare, the viewpoints of a faction of participants in Earth First! reveal the depths of the group's misanthropic, if not apocalyptic, streak. For example, an article published by a certain Miss Ann Thropy (the *nom de plume* of the environmental activist Christopher Manes) in the May 1987 issue of *Earth First! Journal* argued that AIDS was "nature's way of protecting the planet from an excess population" and that the disease was not a problem, "but a necessary solution" (Manes, 1987, p. 32). Like the characterizations of the Neo-Malthusians, humankind was presented as an undifferentiated mass, one subjected to the full brunt of nature's population control mechanisms. Humanity was treated no differently from other forms of life; either it was futile, or it stood against the course of nature to limit exposure to the deadly virus or remedy its effects.

Indeed, some participants in Earth First! also held views that steered undeniably close to the nativist policies championed by Garrett Hardin. Dave Foreman, the group's most prominent participant, argued against foreign aid to quell the famine in Ethiopia in 1986, "the best thing we can do is let nature seek its balance, to let the people there just starve" (Scarce, 2016, p. 92). On the issue of immigration to the United States, Foreman advocated for a moratorium in the 1980s and, even as he retracted some of his most dogmatic opinions in the following decades, his promotion of deeply restrictive immigration policies remained long after his participation with Earth First! had concluded.

5.3 The Return of the Commons

Hardin and Foreman's disturbing views are essential reminders that even well-intentioned causes, such as the preservation of the environment, can be redirected to entrench the very attitudes that impoverish citizens of the Global South. An ecological movement that, purposefully or not, implicitly or explicitly castigates or excludes entire swaths of the population is antithetical to its aims. The views of Hardin and Foreman also demonstrate the danger of attempts to naturalize and denature human conduct. Efforts to naturalize away human agency, denature human actions, and thus remove humans from the rest of the web of life are not constructive. Both approaches fail to acknowledge and treat the entirety of the situation in which we find ourselves.

These lessons are particularly important in pedagogical settings, such as college classrooms, where anxieties about environmental degradation can lead to misanthropic or discriminatory reasoning. We have seen two problematic attitudes resurface in classroom discussions about ecological crises, which bear mentioning. The first tends to subsume human action under nature's own set of priorities or prescribed sense of agency. Human conduct is perceived as an extension of nature's inner purpose. The human capacity for self-destruction (and some would say humanity's drive towards self-extinction) is often explained away as the direct result of nature's protective mechanism. It is not hard to recognize the profound difficulties this kind of perspective poses if one is serious about mitigating human harm to the environment. Humanity, for one, cannot at once be a passive *object* of nature's directives and a constructive *subject* engaged in environmental rehabilitation. The second characterization that surfaces in the classroom is equally prominent in the attitudes of Neo-Malthusians and sees humans as "parasites" existing outside of nature and acting on it in a wholly disruptive way. Here, the familiar binaries—nature/culture, reason/instinct—reappear in such a way as to make their reconciliation an impossibility.

More nuanced understandings of humankind's relationship to the natural environment are essential, especially in the classroom, if antisocial and misanthropic ecologies are to be dispelled. In the 1980s, the most vociferous challenges to Neo-Malthusian and Deep Ecology perspectives were issued from advocates of Social Ecology, who argued that humankind's relation to nature could be explored only with a simultaneous critique of social structures. Social Ecologists, such as Murray Bookchin, contended that the domination of nature could not be overcome until hierarchy in all its forms—gender, racial, economic, and political—was eliminated from society. Bookchin's ecological philosophy highlighted both the continuities and the ruptures between human and natural worlds. While not merely reducible to the instincts and impulses in "biotic nature," human society, for Bookchin, was, at its point of origin, profoundly related to the evolutionary processes which were understood as mutualistic practices of diversification, differentiation, and flexibility. Bookchin (1993) advanced "an ethic of complementarity" in which human beings "must play a supportive role in perpetuating the integrity of the biosphere" (p. 369).

Over the past two decades, philosophers such as Bruno Latour and Timothy Morton have invited us to move beyond subject/object ecologies and towards a recognition of the complex imbrications between humankind and the environment. Latour (1998), for example, has criticized Green movements in Europe, arguing that political ecology ought to trigger a more sweeping re-evaluation of governing structures. Political representation, in his imagined system of governance, would be extended to incorporate the relational webs that "entangle" human and non-human nature into what he terms "a parliament of things" (Latour, 2004). Morton also demonstrates considerable antipathy towards the notion of a stand-alone nature which he sees as a primary source of humankind's shunting of ecological responsibility. For Morton (2007), the concept of nature suggests a passive object to be used up and polluted for human gain. Nature, he writes, is "the *away* to which things are flushed" (Morton, 2012). A more constructive approach, in his view, entails disclosing the matrix of "affiliations" between human and non-human objects.

In the last few years, environmentalists have become more fully aware of the racism and misanthropy that once resided in elements of the environmental movement and are quick to recognize the enduring nature of some of these biases today. Indeed, the term "Anthropocene," coined in 2000 by the atmospheric chemist Paul J. Crutzen to denote the measurable geological stratum in the natural landscape that is produced by mankind, has raised alarms for fear of resuscitating the problematic tropes that all human beings share equal blame for the climate emergency and that human existence is in itself irreconcilable with a healthy natural environment. As the environmental historian Jason W. Moore has recently explained, "Capitalocene" may be a more appropriate term if one hopes to pinpoint the source most responsible for planetary degradation in the last two centuries (Moore, 2015).

These contemporary perspectives in ecological thought come when a new environmental crisis, the rapid warming of the globe due to unchecked carbon emissions, has replaced the population scare of the previous generation. And though, motivated by population growth, Hardin and Foreman arrived at the troubling conclusion that the United States should be saved while the rest of the world should be left to its own devices, climate change today knows no geographical or political boundaries. Indeed, the analogy of the earth as a spaceship, or a closed system, an analogy famously rejected by Hardin due to its inference of a shared commons, has become all the more relevant in a world that is beginning to experience the effects of increased greenhouse gases. In the face of this new hazard, the approach of Neo-Malthusians, which assigned blame to the very existence of people and not to specific carbon emissions-producing practices, is simply untenable.

The current situation presents possibilities that can move us past the impasse occasioned by the population scare of the 1950s and 1960s, and away from the divisive rhetoric of Hardin and Earth First!. Most importantly, climate change compels us to focus on the global atmosphere as a continuous and shared territory, uninterrupted by geographic features or national boundaries. Indeed, environmentalists today are rethinking the global atmosphere in ways that draw more rigorous attention to the nature and source of contemporary pollution. Recent approaches to curbing the impacts of climate change, such as The Green New Deal, proposed by

Congresswoman Alexandria Ocasio-Cortez and Senator Ed Markey, reflect an awareness of the environmental movement's past biases by deliberately conjoining social equity and justice with economic and environmental rehabilitation. The attention has helped shift the discussion of the environmental crisis from blaming people for pollution to understanding the outsized role that business and industry play in perpetuating it. The figure of the commons today has reappeared, not as a tragedy to be avoided, but as a tool for better understanding global responsibility and sustainability.

References

Atkin, E. (2017, March 7). *Steve Bannon used to believe in science. Now he's America's top climate villain.* The New Republic. Retrieved from https://newrepublic.com/article/141102/steve-bannon-used-believe-science-now-hes-americas-top-climate-villain

Baram, M. (2017, June 2). *Trump's climate decision reeks of Steve Bannon's influence.* Fast Company. Retrieved 20 December 2021 from https://www.fastcompany.com/40426884/trumps-paris-decision-reflects-the-unexpected-resurgence-of-steve-bannon-for-now

Bookchin, M. (1993). What is social ecology? In M. E. Zimmerman (Ed.), *Environmental philosophy: From animal rights to radical ecology* (pp. 354–373). Prentice Hall.

Carson, R. (1962). *Silent spring.* Houghton Mifflin.

Chase, A. (1980). *The legacy of Malthus: The social costs of the new scientific racism.* University of Illinois Press.

Chase, S. (1991). *Defending the earth: A dialogue between Murray Bookchin and Dave Foreman.* South End Press.

Crutzen, P. J., & Stoermer, E. F. (2000). The "Anthropocene". *Global Change Newsletter, 41,* 17. http://www.igbp.net/download/18.316f18321323470177580001401/1376383088452/NL41.pdf

Desrochier, P., & Hoffbauer, C. (2009). The post war intellectual roots of the population bomb. Fairfeld Osborn's 'Our plundered planet' and William Vogt's 'Road to survival' in retrospect. *Electronic Journal of Sustainable Development, 1*(3), 38–61. Retrieved from http://www.ejsd.org

Ehrlich, P. R. (1968). *The population bomb.* Ballantine Books.

Hardin, G. (1968). The tragedy of the commons. *Science, 162*(3859), 1243–1248. https://doi.org/10.1126/science.162.3859.1243

Hardin, G. (1974). Lifeboat ethics: The case against helping the poor. *Psychology Today, 8,* 38–43.

Kovel, J. (2003). Racism and ecology. *Socialism and Democracy, 17*(1), 99–107. https://doi.org/10.1080/08854300308428343

Latour, B. (1998). To modernize or to ecologize? That's the question. In N. Castree & B. Willems-Braun (Eds.), *Remaking reality: Nature at the millenium* (pp. 221–242). Routledge.

Latour, B. (2004). *Politics of nature: How to bring the sciences into democracy.* Harvard University Press.

Leopold, A. (1949). *A Sand County almanac, and sketches here and there.* Oxford University Press.

Malthus, T. R. (1798). *An essay on the principle of population, as it affects the future improvement of society. With remarks on the speculations of Mr. Godwin, M. Condorcet, and other writers.* J. Johnson.

Manes, C. (Thropy, Miss Ann.) (1987). AIDS and population. *Earth First! Journal,* 32–33.

Miele, F. (2002). Living within limits—An interview with Garrett Hardin. In *The battlegrounds of bio-science: Cross-examining the experts on: Evolutionary psychology, race, intelligence, & genetics, population, environment, & cloning* (pp. 259–273). 1st Books Library.

Moore, J. A. (2015). *Capitalism in the web of life: Ecology and the accumulation of capital.* Verso Press.

Morton, T. (2007). *Ecology without nature: Rethinking environmental aesthetics.* Harvard University Press.

Morton, T. (2012). Architecture without nature. In *Tarp architecture manual.* Pratt Institute.

Niiler, E. (2016, December 7). *Trump's chief strategist Steve Bannon ran a massive climate experiment.* Wired. Retrieved from https://www.wired.com/2016/12/trumps-chief-strategist-ran-massive-climate-experiment

Osborn, F. (1948). *Our plundered planet.* Little, Brown.

Politi, D. (2017, January 29). *Trump gives Steve Bannon, champion of white nationalism, key national security seat.* Slate. Retrieved from http://www.slate.com/blogs/the_slatest/2017/01/29/trump_gives_steve_bannon_champion_of_white_nationalism_key_seat_on_national.html

Robertson, T. (2012). *The Malthusian moment: Global population growth and the birth of American environmentalism.* Rutgers University Press.

Scarce, R. (2016). *Eco-warriors: Understanding the radical environmental movement.* Routledge.

Southern Poverty Law Center. (n.d.). *Extremist files: Garrett Hardin.* Retrieved from https://www.splcenter.org/fighting-hate/extremist-files/individual/garrett-hardin

Taylor, A. (2017, May 31). Why Nicaragua and Syria didn't join the Paris Climate Accord. *The Washington Post.* Retrieved from https://www.washingtonpost.com/news/worldviews/wp/2017/05/31/Poliwhy-nicaragua-and-syria-didnt-join-the-paris-climate-accord/

Vogt, W. (1948). *Road to survival.* W. Sloane Associates.

Ralph Ghoche is an Assistant Professor in the Department of Architecture at Barnard College. He teaches on a wide range of topics, including environmental and landscape histories. Ghoche's scholarship centers on French architecture and urbanism in the nineteenth century, particularly in colonial Algeria.

Unyimeabasi Udoh graduated with a BA in Architecture from Columbia University, and holds an MFA in Visual Communication Design from the School of the Art Institute of Chicago. Their work as an artist and graphic designer explores systems of knowledge and communication.

Chapter 6
Pathways to Sustainability: Examples from Science Teacher Education

María S. Rivera Maulucci

6.1 Why Do We Need Education for Sustainability?

Education for sustainability is rooted in environmental education. In 1977, the first intergovernmental environmental education conference was convened in Tbilisi, Georgia, of the former Soviet Republic. The conference attendees included 265 delegates and 65 representatives and observers. Participants came from 66 states, 2 non-member states, eight United Nations (UN) agencies and programs, 20 international nongovernmental organizations, and three intergovernmental organizations. Organized by the United Nations Education, Scientific, and Cultural Organization (UNESCO) in collaboration with the U.N. Environment Programme (UNEP), the conference declaration stated: "Environmental education, properly understood, should constitute a comprehensive lifelong education, one responsive to changes in a rapidly changing world" (UNEP, 1978, p. 24).

Now, more than 40 years later, it is easy to see how rapidly the world has changed. We can cite the rise of the internet, widespread home and office computers, smartphones that bring the world to your fingertips, subscriptions for everything, increased consumption, ubiquitous streaming services, more people living in cities, and more people overall. Despite these changes, a proper understanding of environmental education – much less sustainability education – has yet to take root in many societies. It is fair to say that too many children do not attain a *comprehensive* understanding of the relationship between the environment, society, and our economy. Instead, environmental and sustainability education remain at the margins of most K-12 curricula.

Many educators see the responsibility for environmental and sustainability education as belonging to science teachers. For example, environmental science topics

M. S. Rivera Maulucci (✉)
Education Program, Barnard College, New York, NY, USA
e-mail: mriveram@barnard.edu

© The Author(s) 2023
M. S. Rivera Maulucci et al. (eds.), *Transforming Education for Sustainability*,
Environmental Discourses in Science Education 7,
https://doi.org/10.1007/978-3-031-13536-1_6

are typically found in state science standards (cf. NYSED, n.d.-a), and in New York State, one of the High School Regents exams is called Living Environment (NYSED, n.d.-b). Yet, science education is often marginalized at the elementary level to emphasize tested subjects, English language arts, and mathematics (Rivera Maulucci, 2010). At the middle and high school levels, science teachers often have curricula with a depth and breadth that affords limited opportunities for addressing environmental education or sustainability topics. As a high school Biology student, since we did not have time to cover environmental science topics in class, I had to read the chapters on my own to prepare for the New York State Regents examination. As a middle school Biology teacher, environmental topics competed with an extensive unit on human biology. I proposed two science electives, *Zoology* and *Sea and Shore*, during which students could explore concepts such as habitats, endangered species, water quality, pollution, and conservation. However, since these courses were electives, only students that took these electives benefitted from them.

When environmental issues are raised in a community there are often calls for education as part of the solution to the problem. For example, as Director for Urban Forestry of the Environmental Action Coalition (EAC), I was part of a team of environmental educators and activists that developed an *Action Plan for New York City's Urban Forest* (Pirani et al., 1998). The report calculated the extent of NYC's urban forest, including the number of trees (5.2 million) and the percent of tree cover (16.6%). Other sections extolled the benefits of the urban forest, provided suggestions for trees to plant, and called for four immediate actions (enhance growing conditions, diversify the urban forest, improve service delivery, and manage and restore public woodlands), and four long term actions (establish an Urban Forestry Board, inventory and develop a management plan, increase the total tree population by one million trees, and *foster an engaged and knowledgeable public*). Another EAC project was called Trees for East Harlem. Working with local community activists, we identified places where trees could be planted, then contacted the NYC Parks Department to request a street tree. The Parks Department would survey the site, and then plant a tree if the site was suitable. We worked with the community activists to pass out flyers about the benefits of the trees and engage in honest conversations with local business owners and community members about how to care for the new trees. Environmental education was central to both initiatives.

In 2007, the NYC parks department announced Mayor Bloomberg's MillionTreesNYC initiative (MillionTreesNYC, n.d.). Eight years later, in November of 2015, they planted the millionth tree (Coneybeare, 2015) and the third decadal census of NYC street trees in 2015 which engaged many volunteers and emphasized citizen stewardship placed the number of street trees at 666,134 (NYC Parks, 2016). Today, the efforts are noticeable. Every time I pass a new street tree planting or a green swale, I smile. Yet, I would argue that progress with fostering an engaged and knowledgeable public in New York City has lagged far behind. How many New Yorkers walking by a street tree or a tree in one of the city's parks do so with full knowledge of the benefits the trees provide in the urban environment, including improving air quality and mitigating the urban heat island effect? How many New Yorkers can articulate the policy issues that affect the quality of the air they breathe, the water they drink, and the food they consume? How many New Yorkers commit, beyond compliance, to doing what they can to reduce, reuse, and recycle? Are educators providing students with a comprehensive education that helps them understand environmental and sustainability issues, sift through the misinformation that abounds, and make informed, sustainable decisions in their everyday life? To be clear, I do not blame teachers. Instead, I contend that like the lack of traction around environmental and school reform issues, we need a more comprehensive approach.

This chapter will put forward four key arguments. First, education for sustainability (EfS) needs to be the responsibility of all educators in all subjects, especially in the sciences. Second, educators need a framework and pathways for thinking about integrating sustainability into the curriculum. Third, EfS can help us answer the timeless question, "Why are we learning this?" Finally, educators, researchers, and policymakers must take up the call of EfS seriously so that it becomes impossible for students to graduate from high school without a comprehensive education that also provides a basis for lifelong decisions and choices that foster sustainability.

6.2 What Is Education for Sustainability?

Both environmental education and EfS have their roots in early studies of nature. Concern for the effects of development and industrialization on the natural environment emphasized conservation, and concern for human health and well-being emphasized social or environmental justice (McCrea, 2006). For example, Aldo Leopold's *Sand County Almanac* (1949) helped establish the land ethic of the conservation movement, and Rachel Carson's, *Silent Spring* (1962), galvanized the modern environmental movement as she drew a direct link between pesticide use and the loss of songbirds. The Tbilisi conference report (UNEP, 1978) advocated a holistic, interdisciplinary approach to environmental education. According to the authors,

> [A]dopting a holistic approach, rooted in a broad interdisciplinary base…recreates an overall perspective which acknowledges the fact that the natural environment and man-made environment are profoundly interdependent. It helps reveal the enduring continuity which links the acts of today to the consequences for tomorrow. It demonstrates the interdependencies among national communities and the need for solidarity among all mankind (UNEP, 1978, p. 24).

Key ideas that we can draw from the Tbilisi report, are the "interdisciplinary" nature of environmental education, as well as the interdependence of the "natural and man-made environment" and "national communities." Finally, the description calls for education that "links the acts of today to the consequences for tomorrow," and recognizes the "need for solidarity." At the same time, this early description of environmental education does not question or problematize individualism (emphasizing freedom of action for individuals) or a rapidly changing world, and it seems to equate natural and man-made environments as if both are equally "good."

More recently, the United Nations (2002) declared 2004–2015, The Decade of Education for Sustainable Development (DESD) as part of their Education for Sustainable Development (ESD) initiative. UNESCO (2017, based on Wiek et al., 2011) also set out a bold vision for student learning outcomes. For example, they identify eight competencies that students should develop as part of their education for sustainable development goals: systems thinking, anticipatory, normative, strategic, collaboration, critical thinking, self-awareness, and integrated problem-solving. These competencies are described in Table 6.1.

UNESCO (2017) identifies specific learning objectives for each of the Sustainable Development Goals (See Fig. 6.1) that span the cognitive domain (what students should know), socio-emotional domain (what social interaction, communication, and self-reflection skills students should have), and behavioral domain (what students should be able to do).

While at the surface level, the UNESCO competencies and learning objectives seem laudable, there are some concerns and critiques regarding approaches to developing and measuring global competencies. Neoliberal approaches have been critiqued for focusing on global economic and labor concerns; global consciousness

Table 6.1 UNESCO education for sustainable development competencies

Competency	Description
Systems thinking competency	The abilities to recognize and understand relationships; to analyze complex systems; to think of how systems are embedded within different domains and different scales; and to deal with uncertainty.
Anticipatory competency	The abilities to understand and evaluate multiple futures – possible, probable and desirable; to create one's own visions for the future; to apply the precautionary principle; to assess the consequences of actions; and to deal with risks and changes.
Normative competency	The abilities to understand and reflect on the norms and values that underlie one's actions; and to negotiate sustainability values, principles, goals, and targets, in a context of conflicts of interests and trade-offs, uncertain knowledge and contradictions.
Strategic competency	The abilities to collectively develop and implement innovative actions that further sustainability at the local level and further afield.
Collaboration competency	The abilities to learn from others; to understand and respect the needs, perspectives and actions of others (empathy); to understand, relate to and be sensitive to others (empathic leadership); to deal with conflicts in a group; and to facilitate collaborative and participatory problem solving.
Critical thinking competency	The ability to question norms, practices and opinions; to reflect on one's values, perceptions and actions; and to take a position in the sustainability discourse.
Self-awareness competency	The ability to reflect on one's own role in the local community and (global) society; to continually evaluate and further motivate one's actions; and to deal with one's feelings and desires.
Integrated problem-solving competency	The overarching ability to apply different problem-solving frameworks to complex sustainability problems and develop viable, inclusive and equitable solution options that promote sustainable development, integrating the above-mentioned competences.

Fig. 6.1 United Nations sustainable development goals. (From SDG Academy and Creative Commons – Creative Commons)

approaches have been critiqued for furthering universalistic values that do not account for or challenge imbalances of power; critical approaches are largely theoretical and untested but they raise the concerns that mainstream approaches tend to be transmissive rather than transformative and privilege members of higher socioeconomic groups; and finally, advocacy approaches focus on rights and responsibilities towards other groups or nature but learning from advocacy is hard to measure (Connolly et al., 2019). The authors note that there are regional differences with Europe and Canada favoring global consciousness approaches and the United States and Asia favoring neoliberal approaches. In addition, there are tensions between competency-based approaches, which can focus attention on technical skills, and virtue-based approaches, which are not necessarily universal, that also need to be resolved (Petersen, 2022).

At its heart, ESD and EfS recognize that we need to make changes in how we live to ensure a more viable future for humans and all living things. While many changes need to occur at the macro-level of governments and nations, sustainability education embraces the hope that individuals can participate in that change while simultaneously making a difference at the micro-level. EfS also grapples with the tension between individual and collective action, and struggles to integrate not just human-centered approaches but other aspects of the environment. At the same time, EfS has an array of benefits for students, including enhancing student engagement and motivation, boosting student learning and achievement, improving student behavior and attendance, significantly increasing students' perceptions of self-efficacy, encouraging students to connect with local and global systems, and helping students gain an appreciation for civic engagement and the democratic process (Cloud, 2014). For teachers, EfS supports "both new and veteran teachers in achieving strong academic outcomes from their students" (Cloud, 2014, p. 4).

Given the wide scope of knowledge, skills, understanding, and competencies educators need to foster across cognitive, socio-emotional, and behavioral domains, it is clear that these goals can and should be developed across the curriculum. While the early association of EfS with environmental science meant that it was mostly relegated to science teaching, EfS can and should appear across the curriculum, and the interdisciplinary nature of sustainability allows it to align well with all subject areas, not just science. Furthermore, students need repeated exposure to EfS to fully understand its importance and what they can do to make a difference. EfS is greatly enhanced when taught from the multiple perspectives that different curriculum areas bring. Finally, EfS should afford opportunities for students to engage in different modes of representation, such as visual, written, spoken, and numerical, and provide students with various ways to integrate their learning with local and global action.

To achieve the goals and benefits of EfS, we must move from rhetoric to reality. As Maria Hofman (2015, p. 213) explains, "If one wants to change the society and education, one of the cornerstones to start with is the education and training of already qualified teachers and teacher educators." Although I argue that sustainability can and should be taught across the curriculum, in the following sections I focus on the challenges and possibilities within science education, specifically.

6.3   What Are Some of the Challenges of Science Education for Sustainability?

Education for sustainability presents many challenges to science educators that teacher education, professional development, and schools must address. A central challenge for sustainability in science education is that sustainability is an ill-defined and ill-structured concept (Wals & Dillon, 2015). First, there is no consensus about the definition of sustainability, let alone education for sustainability. Most definitions include the idea that we must "meet the needs of the present without compromising the ability of future generations to meet their own needs," (Brundtland, 1987) and emphasize the three pillars – economic, social, and ecological. Other definitions of sustainability include a political dimension with nations that are "secure, thriving and peaceful" (cf. Cortese & Rowe, 2016) or a cultural dimension (United Cities and Local Governments (UCLG), 2010, p. 3), which "[a]ffirms that culture in all its diversity is needed to respond to the current challenges of humankind." Furthermore, calls for inclusion of Indigenous perspectives on sustainability invoke attention to gratitude instead of entitlement, and reciprocity instead of consumption (Kimmerer, 2015) For example, Kimmerer notes, "our role as human people is not just to take from the earth, and the role of the earth is not just to provide for our single species" (In Tippett, 2016).

According to Neil Taylor et al. (2015), definitions of sustainability can range from "weak" to "strong." Weak sustainability definitions focus on anthropocentric (human) needs, economic development, and technological solutions. Strong sustainability definitions focus on ecocentric (natural) needs, preserving ecological integrity, and properly valuing ecosystem services. Without a consensus regarding how to define sustainability, teachers lack a clear vision for sustainability, let alone how they might integrate it into the curriculum. Consequently, education from these different perspectives can take the form of education for sustainable development (ESD), environmental education (EE), or education for sustainability (EfS) (Taylor et al., 2015).

For example, according to UNESCO (2022), ESD "gives learners of all ages the knowledge, skills, values and agency to address interconnected global challenges including climate change, loss of biodiversity, unsustainable use of resources, and inequality." A definition of EE from the US EPA (2012) states,

> Environmental education is a process that allows individuals to explore environmental issues, engage in problem solving, and take action to improve the environment. As a result, individuals develop a deeper understanding of environmental issues and have the skills to make informed and responsible decisions.

Finally, Cloud (2014) defines EfS as,

> …a transformative learning process that equips students, teachers, schools, and informal educators with the knowledge and ways of thinking that society needs to achieve economic prosperity and responsible citizenship while restoring the health of the living systems upon which our lives depend.

All three of these definitions recognize the role of problem-solving and decision-making in solving local or global issues or challenges. However, not all explicitly recognize the roles of values (UNESCO, 2022) or ways of thinking (Cloud, 2014) that are required, and not all emphasize the integral importance of "restoring the health of the living systems upon which our lives depend" (Cloud, 2014).

A second difficulty with sustainability education is that sustainability, as a concept, is neither neutral nor value-free (Rogers, 2016). Judy Rogers (p. 217–18) explains that "sustainability is not a thing or an endpoint but rather a way of thinking, talking, writing, and acting in response to an identified and urgent need for transformation and change at the global, regional, and local level." Rogers explains that imagined futures and proposed changes might be "deeply contested" or even contradictory. For example, individuals, stakeholders, or communities might have different, and even incompatible, ideas about economic prosperity, responsible citizenship, or plans for restoring the health of living systems noted above in Cloud's (2014) description of education for sustainability. Further evidence of this difficulty can be found in a review of US science standards, completed by Kim A. Kastens and Margaret Turrin in 2006. Kastens and Turrrin (2006) reviewed 49 state science standards and found:

> There is little support among state standards developers for the notion that science lessons or science teachers should proactively encourage students to change the nature of their own interactions with the environment, or to seek to bring about such changes in their own family, school or local community (p. 431).

A more recent review of the Next Generation Science Standards (NGSS Lead States, 2013), which have been adopted by 20 states and used by 24 states to develop their science standards (*NGSS Hub*, n.d.), found that the standards emphasize learning *about* sustainability from a scientific point of view rather than *for* sustainability, or the types of decisions and actions students could or should take to promote sustainability (Egger et al., 2017).

Given the lack of guidance from state standards, teachers may be concerned that when education for sustainability advocates particular values or solutions to environmental issues, it can walk the line of indoctrination.

For example, the idea that reusing, reducing, and recycling are all actions that in most communities can be viewed as good for individuals, society, and the environment, thus, advocating for these behaviors might not be seen as indoctrination. However, some issues, such as climate change, have become entangled with social and political ideologies that may make it risky for teachers to address these topics in the classroom (see Pfirman & Winckler, this volume, Chap. 19). Other issues, such as the Flint Water Crisis, expose environmental racism, and it is difficult, if not impossible, to ignore the necessary critique of the racist ideologies that allowed the crisis to occur. Although teachers must steer clear of indoctrination in the classroom, or teaching from an uncritical, partisan, or ideological standpoint, dialogic approaches that make it seem like any position is valid might also be untenable (Wals, 2010). Arjen E. Wals describes the paradox encountered by educators who delve into the social aspects of sustainability issues:

> On the one hand, there is a deep concern about the state of the planet and a sense of urgency that demands a break with existing unsustainable systems, lifestyles and routines, while on the other, there is a conviction that it is wrong to persuade, influence or even educate people towards pre- and expert-determined ways of thinking and acting.

Wals (2010) is speaking to the ways in which topics like climate change and sustainability have become politicized with "experts" on both sides weighing in on what should be taught or what should be done about it. In this context, educators need to recognize pluralism (multiple ideas about sustainability, none of which can be prescribed), relativism (multiple perspectives on sustainability--the learners, educators, families, communities, local to global, multicultural, multispecies, multidisciplinary approaches), and the role of democracy, agency and self-determination, or "a collaborative search for and engagement in sustainability that is not limited to small elites in society but accessible to all stakeholders, including those who are or feel marginalized" (Wals, 2010, p. 147). Absent clear guidelines for democratic, participatory, agentic approaches that foster good for the entire ecosystem, not just humans, teachers may shy away from teaching about sustainability to avoid controversy in their classroom or school communities.

The third challenge of education for sustainability is its interdisciplinary nature. As noted above, in addition to the three widely recognized pillars of sustainability, ecological, economic, and social dimensions, some educators also recognize a fourth pillar, a political dimension (Taylor et al., 2015) or a cultural dimension (UCLG, 2010). Although the science curriculum deals with ecological systems, many science educators may feel that the economic, social, and political dimensions are beyond the scope of their mandated science curricula. Teachers in the humanities may consider sustainability's scientific or economic aspects to be beyond their responsibility and areas of expertise. Thus, many teachers may feel ill-prepared to teach the interdisciplinary content sustainability education requires (see also Pfirman and Winckler) and may lack the flexibility to teach interdisciplinary curricula given accountability regimes, such as strict curriculum mapping or high-stakes tests.

Fourth, there is a level of indeterminacy to sustainability. What we see as sustainable now might not be viewed as sustainable in the future, and what we see as sustainable in one context might not be sustainable in another (Wals, 2010). Thus, the complexity and urgency of sustainability education impedes the development of a universal vision for both content and process that other content areas may have. As a result, modes of integrating sustainability topics into the science curriculum can vary. For example, educators can teach *about* sustainability (about recycling), *for* sustainability (why recycle and how to recycle), or *sustainable education,* living and learning more sustainably (engaging in recycling audits and school or community-based recycling programs) (See Sterling, 2003, examples added).

Fifth, as with justice-oriented curricula, teachers may find instilling a sense of hope challenging when environmental, racial, and other equity issues seem intractable (Davis & Shaeffer, 2019). Taylor et al. (2015, p. 6) explain, "All too often, sustainability issues are presented and discussed solely as a series of problems, and while it is important to acknowledge that problems do exist, an overemphasis on them can leave children feeling severely disempowered." On the other hand, "[a]ction-oriented projects allow [children] to see that change at a community level is possible and are vital to keeping young people engaged with sustainability issues and positive about the future" (Taylor et al., 2015, p. 6).

Finally, teachers may struggle to find the best way to integrate sustainability with the curriculum, whether with science or another content area. For example, curriculum integration efforts for multicultural education have been critiqued for focusing on "heroes and holidays" in ways that tokenize or essentialize the contributions of other racial, ethnic, or cultural groups" (Lee et al., 2002). Drawing on Banks' (1995) Typology for Curriculum Integration developed for multicultural education, I propose five pathways for the integration of sustainability (see Fig. 6.2).

Given the range of curricula and goals and purposes of education, I purposely do not attach levels that might imply that some approaches are better or more preferred than others. In some courses, content specificity and limited time may mean that educators can only *model* sustainability. Some ways to model sustainability include going paperless, bringing a compost bin to class, making sure the class knows how to recycle properly, or turning lights off when everyone leaves the room.

On the other hand, course content might allow the entire curriculum to be *transformed* by nesting it within a meaningful sustainability context (See Snow, this volume, Chap. 8). Some units or lessons might benefit from *adding* resources (i.e., texts, images, video clips) about sustainability. Other units or lessons might benefit from including the special *contributions* of Indigenous scholars, traditional

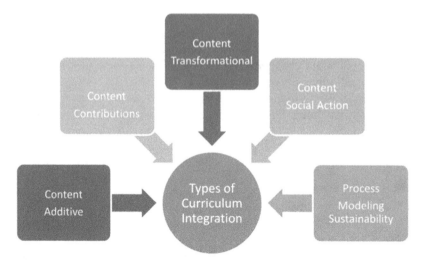

Fig. 6.2 Five approaches to integrating sustainability

ecological knowledge, or local case studies (see Heskel & Mergenthal, this volume, Chap. 15. For example, case studies of the Flint Water Crisis (Davis & Shaeffer, 2019) or the Vieques National Wildlife Reserve (Withers, 2013) can provide a more personal context or sense of urgency for sustainability. While the Flint Water crisis has received international attention, the struggles for sustainability and environmental justice in Vieques, Puerto Rico, a former United States military site, are lesser-known. Vieques is an island off the coast of Puerto Rico with "high ecological values and a toxic legacy including unexploded ordnances as a result of the area's former use as a naval bombing range" (Guzman et al., 2020, p. 13). Finally, in some cases, students can go out into the real world and engage directly with sustainability issues through *social action*, such as planting a school garden, conducting a BioBlitz (see O'Donnell & Brundage, this volume, Chap. 11) or engaging in community-activism to address local air quality concerns (Morales-Doyle, 2017).

6.4   What Are Some of the Possibilities of Science Teacher Education for Sustainability?

Given calls for integrating sustainability, science teachers require education and professional development to understand the economic, political, social, moral, and technical dimensions of sustainability and how to incorporate them into the science curriculum (Stratton et al., 2015). As noted by Stratton and co-authors (2015), quality teacher education for sustainability requires: (1) a holistic approach; (2) place-based methods; (3) connections between the natural and built world; (4) conceptual frameworks for educating for sustainability; (5) development of sustainability

Table 6.2 Quality sustainability teacher education themes

Theme	Description
A holistic approach	Taking an inclusive and multifaceted approach to teaching and learning about sustainability that spans ecological, natural and built, economic, social, cultural, and political components
Place-based education methodology	Learning that is rooted in local issues or the history, environment, culture, economy, literature, and art of a particular place
Connections between the natural and built world	Challenging learners to become more aware of how their individual actions are intimately linked to the environment and are vitally important in light of sustainability initiatives
Conceptual frameworks for EfS	Explicitly defining conceptual frameworks for the study of EfS
Development of sustainability literacy	Including ways to plan for and assess sustainability literacy: Education for a sustainable life
Transformative pedagogical approaches	Developing new forms of pedagogy to educate [students] in sustainability practices
Links to climate change	Explicitly linking to climate change and related issues of social equity

Note: Adapted from Stratton et al. (2015)

literacy; (6) transformative pedagogical approaches; and (7) links to climate change. Table 6.2 provides descriptions of each of these themes.

The Next Generation Science Standards (NGSS Lead States, 2013), a United States science education reform document that provides a consensus vision for the development of disciplinary core ideas, science and engineering practices, and cross-cutting themes across grades K-12, includes a subcategory called "human sustainability (see NGSS Lead States, 2013, p. 24).

The performance standards associated with this topic, which falls under Earth and Space sciences, include the following:

- create a computational simulation to illustrate the relationships among management of natural resources, the sustainability of human populations, and biodiversity,
- evaluate competing design solutions for developing, managing, and utilizing energy and mineral resources based on cost-benefit ratios (NGSS Lead States, 2013, p. 287).

Egger et al. (2017) note that many teachers may not be prepared through the undergraduate curriculum to teach the sustainability content and Earth systems thinking required by the NGSS standards.

Feinstein and Kirchgasler (2015) provide an in-depth review and critique of the explicit and implicit ways that the NGSS Standards address sustainability. For example, they note that the standards mainly provide a "technology-centered perspective" that advances the view that technological solutions can address most environmental problems rather than a more in-depth exploration of the ethical and political dimensions of sustainability. In particular, they raise the concern that:

...science education will advance an oversimplified idea of sustainability that diminishes its social and ethical dimensions, exaggerating the role of technology and the importance of technical expertise at the expense of non-STEM (science, technology, engineering, and mathematics) disciplines and nontechnical expertise. Rather than supporting a generation of students to engage with science in realistic and productive ways as they address sustainability challenges, this approach might lead students to systematically misinterpret and underestimate the challenges that confront their local, regional, and global communities (Feinstein & Kirchgasler., 2015, p. 123).

The authors identify three troubling themes: (1) universalism (seeing humanity as a single indivisible element), (2) scientism (seeing science as the type of knowledge that is most relevant to sustainability), and (3) technocentrism (seeing technical knowledge and technological solutions as the best way to address sustainability issues). Together, these troubling approaches can lead to students developing a "technology-centered perspective called *ecological modernization* that defines sustainability as a set of global problems affecting all humans equally and solvable through the application of science and technology" (p. 121).

6.5 What Are Some Examples of Teacher Education for Sustainability?

The following sections describe two examples of science teacher education for sustainability. The first example explores an in-class set of activities that demonstrate how I introduce the EfS Curriculum Standards (articulated below) to students and help them explore how and why to integrate sustainability into science and engineering curricula. The second example describes a case study from the Summer STEM Teaching Experiences for Undergraduates (TEU) program that incorporated sustainability as a central element of the science curriculum that TEU interns designed for high school youth attending the program.

6.5.1 Introducing and Justifying the Need for EfS Standards

I teach two science pedagogical methods courses incorporating the EfS Standards and the Cloud Institute EfS Framework. In both classes, students do a science or engineering activity, and then they redo the activity through a sustainability pathway of their choice. For example, in one class that focuses on integrating science and engineering, students are presented with a pile of Keva planks, a toy car, and an engineering challenge to build a bridge that spans a 20-cm distance. While students are working in groups on this challenge, I notice them looking at other groups to see how they are solving the problem. There is a competitive edge and a focus on finding a solution, not why the problem might be important or to whom. Students find the activity fun, challenging, collaborative, inquiry-based, constructivist, and creative. Yet, they struggle to see clear connections to specific science content. As they reflect on the extent to which the activity was inclusive, culturally relevant, or meaningful, they realize that without clear instructor intent, these aspects might happen naturally, but they might not.

Next, we shift to discussing some of the struggles with integrating engineering in K-12 science classrooms. I present several challenges, including standardized science curricula and balancing teaching science deeply alongside engineering design. We also discuss student diversity and the different needs teachers must address for students of varying grade level, age, gender, ethnicity, language, culture, and ability. We discuss the difficulties of finding good engineering curricula that develop strong mathematics or science content knowledge and practices and the challenge of having adequate resources. Some schools may have plentiful resources for integrating engineering and can offer robotics classes, while others may lack resources altogether. Finally, I note that with the emphasis on STEM, some students may have had extensive experiences with engineering, and some may have had little to none. Then, I ask the class to turn and talk to a partner to identify additional challenges. After their shared talk, we cogenerate a list of additional challenges for students and teachers when integrating engineering with the science curriculum (see Table 6.3). While the challenges the participants identify are nested within the context of integrating engineering, many are relevant to integrating sustainability. For example, lack of repeated exposure, standardized curricula, student diversity, inadequate resources, and student experience levels can all present challenges for integrating sustainability. An emphasis on English language arts, test prep mandates, and lack of testing incentives for sustainability can also impact the time available for teaching for sustainability.

At the end of this discussion, I tell students I see a bigger problem. To me, *the purpose of engineering* presents a more significant problem. Why do we want students to learn about engineering in the first place? To have fun? To do something

Table 6.3 Additional challenges for integrating engineering and science education

Role	Challenge
Students	Lack of repeated exposure to engineering design
	Not enough resources to repeat the activity
	Developing social skills, collaboration, teamwork
	Learning how to disagree
	Gendered nature of engineering, females might be intimidated by the process
	Student misconceptions about engineering
Teachers	How to cover the different types of engineering
	Emphasis on English language arts in every subject
	Design projects are time-consuming
	Working with students with special needs, fine motor skills
	Engaging students who are just not interested
	Test-prep mandates
	Lack of incentive if students are not tested
	Helping teachers to develop creative formative assessments in STEM
	Determining what metrics to use to assess design project

hands-on? For them to be introduced to lucrative careers? To contribute to a consumer culture? To solve problems? To help people? To reduce poverty? To build sustainable cities? To restore habitats and ecosystems? To save the planet? I raise these questions and then propose the EfS Standards as the reason why.

The EfS Standards from the United States Partnership for Education for Sustainable Development (USPESD, 2009) specify three goals. Students should be able to:

1. understand and apply "the basic concepts of sustainability (i.e., meeting present needs without compromising the ability of future generations to meet their needs)" (EfS Standard 1, USPESD, 2009, p. 3).
2. ...recognize the concept of sustainability as a dynamic condition characterized by the interdependency among ecological, economic, and social systems and how these interconnected systems affect individual and societal well-being. They develop an understanding of the human connection to and interdependence with the natural world (EfS Standard 2, USPESD, 2009, p. 3).
3. ...develop a multidisciplinary approach to learning the knowledge, skills, and attitudes necessary to continuously improve the health and well-being of present and future generations, via both personal and collective decisions and actions. They are able to envision a world that is sustainable, along with the primary changes that would need to be made by individuals, local communities, and countries in order to achieve this (EfS Standard 3, USPESD, 2009, p. 3).

After presenting these three goals, I share the nine core content standards of the Cloud Institute (Cloud, 2012). Table 6.4 provides a brief description of each of the core content standards.

Students look at the titles and descriptions and tell me where they see science, mathematics, and other curriculum areas. Then, I send them back to their groups to build another bridge. They have the same constraint; it must span a 20-cm divide. However, now they must discuss the pathways with their groups, choose a particular sustainability pathway, and build a new bridge that addresses their sustainability context. The room buzzes with excitement and creativity. Students do not look around at other groups because everyone is doing something different.

When the students finish building, they share their sustainability context and show their bridges, and what a difference! Before, they were just constructing a bridge. Now, students care about who the bridge is for, how the bridge will be used, and what it connects. Now, they care about the sturdiness of the bridge, and they try to use fewer resources to build it. There are parks with swings, slides, and monkey bars, bike lanes, pedestrian paths, safety rails, signs to highlight a sense of place, and windmills to produce the energy to light the bridges (see Table 6.5).

In our discussion, I ask students who is the judge of the success of their bridge-building. They realize that they are now the judges. They recognize that the criteria for success now depend on the sustainability pathway they choose. They see that although the groups chose different sustainability pathways, all the groups converged on the idea of a multi-purpose bridge for the community. All of them had bike paths!

Table 6.4 Cloud Institute core content standards/pathways and performance indicators

Standard/ pathway	Description	Performance indicators (students will…)
Cultural preservation and transformation	The preservation of cultural histories and heritages, and the transformation of cultural identities and practices contribute to sustainable communities.	Develop the ability to discern with others what to preserve and what to change in order for future generations to thrive.
Responsible local/global citizenship	The rights, responsibilities, and actions associated with leadership and participation toward healthy and sustainable communities.	Know and understand these rights and responsibilities and assume their roles of leadership and participation.
The dynamics of systems & change	A system is made up of two or more parts in a dynamic relationship that forms a whole whose elements "hang together" and change because they continually affect each other over time.	Know and understand the dynamic nature of complex systems and change over time. Apply the tools and concepts of system dynamics and systems thinking in their present lives and to inform the choices that will affect our future.
Sustainable economics	The evolving theories and practices of economics and the shift towards integrating our economic, natural, and social systems, to support and maintain life on the planet.	Know and understand twenty-first-century economic practices and will produce and consume in ways that contribute to the health of the financial, social, and natural capital.
Healthy commons	Healthy commons are what we all depend upon and for which we are all responsible (i.e., air, biodiversity, climate regulation, our collective future, water, libraries, public health, heritage sites, topsoil, etc.).	Be able to recognize and value the vital importance of the commons in our lives and for our future. Assume the rights, responsibilities, and actions to care for the commons.
Natural Laws and Ecological principles	The laws of nature and science principles of sustainability.	See themselves as interdependent with each other, all living things and natural systems. Put their knowledge and understanding to use in the service of their lives, communities, and places in which they live.
Inventing and affecting the future	The vital role of vision, imagination, and intention in creating the desired future.	Design, implement and assess actions in the service of their individual and collective visions.
Multiple perspectives	The perspectives, life experiences, and cultures of others, as well as our own.	Know, understand, value, and draw from multiple perspectives to co-create with diverse stakeholders shared and evolving visions and actions in the service of a healthy and sustainable future locally and globally.
Strong sense of place	The strong connection to the place in which one lives.	Recognize and value the interrelationships between the social, economic, ecological, and architectural history of that place and contribute to its continuous health.

Note: From: The Cloud Institute for Sustainability Education (Cloud, 2012, p. 3)

Table 6.5 Sustainable bridge building projects

Group	Sustainability pathway	Bridge description
1	Multiple perspectives	Our original bridge was just for cars and there was no way for people to walk on it, so we built some ramps. We also have some stairs, and we added some bike lanes
2	Inventing and affecting the future	We built a two-lane bridge with a bike path to encourage people to bike more.
3	Responsible local and global citizenship	We were trying to build a bridge that was exclusively for bikes to help people to be better community members. We tried to use the minimal amount of resources. We thought about there being a watershed across the bridge and access to clean water.
4	Healthy commons	We put a divider in the middle with one side strictly for bikes and pedestrians. We added a park with slides, see saws, and monkey bars, two windmills to power the lighting on the bridge, and a plaque/sign for a sense of place.
5	Responsible local and global citizenship	We are connecting two communities to become more culturally diverse. We did this because our first bridge was not very good. We have un-level tables, so it still did not end up great but it is better than what we had before. That actually works with what we were trying to do—Connect two different places—And they might not be even. They might be uneven. So the communities have to figure out how to make it work.
6	Healthy commons	We have a bike lane on one side. This is the car lane in the center, and this is for the pedestrians on the other side because we have seen problems in other bridges with pedestrians and bikers on the same path. We put them on different levels so there is no risk of overlap.
7	Responsible local and global citizenship	We made bike lanes and walking paths and automobiles go down the middle. We think that the bike and walking paths really support healthy and sustainable communities.

Students see the value of adding a sustainability pathway to the engineering activity. One student noted that the sustainability pathway provided a compelling answer for the question, "Why are we doing this?" Students also see that sustainability pathways can expand their ideas about the meaning and relevance of building bridges. For example, a student noted:

> When we were just building a bridge, we just focused on the bridge itself. Once we had the freedom to look at all the different sustainability pathways, we could decide what the bridge was for and what else to include in our bridge area, which really changed up what we were doing. At first, we were just trying to build a good bridge. Now, we were all trying to put a purpose to this bridgework, and we started to think about what's on the other side of it, what are we actually using this bridge for.

Another student explained that the pathways allow students to bring in cultural relevance. For example, some students might care about the local context, but others might care more about the global aspects of sustainability.

By providing sustainability pathways, each group's vision for bridge-building goes beyond just spanning a 20-cm divide. Students have nine different ways to engage and connect with the content or activity set before them (see Table 6.4). Students have multiple ways to bring their knowledge, interests, and experiences into the classroom. The curriculum can also foster communication around issues that have significance beyond the main course of study. In addition, presenting the sustainability pathways attends to four of the seven components of quality teacher education for sustainability identified by Stratton et al. (2015) (see Table 6.2). The pathways incorporate a holistic approach to sustainability and provide a clear conceptual framework for EfS. Students also consider place-based concerns and draw explicit connections between the natural and built world. The possibilities and connections afforded by the pathways are rich and meaningful and this was just one activity.

Teacher educators are responsible for presenting EfS Standards to teacher candidates, but simply presenting them is not sufficient. Pre-service teachers also need opportunities to develop and implement curricula drawing on EfS Standards to solidify their ability to teach for sustainability. The following section describes a project where undergraduates develop and teach sustainability curricula in science. I ask the following questions: What does it mean to engage preservice teachers in learning about sustainability? How do they envision planning and teaching about sustainability? What connections do they make between sustainability and the science curriculum?

6.5.2 Summer STEM Teaching Experiences for Undergraduates Program

The Summer STEM Teaching Experiences for Undergraduates from Liberal Arts Institutions (TEU) program is developing and testing a model program that provides undergraduate STEM majors with an immersive summer experience in secondary mathematics (Math TEU) or science teaching (Science TEU). The model integrates a high-quality, discipline-specific pedagogy course with an urban teaching practicum and mentoring from experienced science teachers. A college in the northeast hosted the Science TEU Program in Summer 2017. The science pedagogy course included science education standards, science lesson planning, place-based education, sustainability standards, and aspects of classroom management. The Science TEU interns (Science TEUs) taught two cohorts of rising tenth-grade high school students from a local magnet school who took summer enrichment courses in science and writing. Attendance was required for the high school students, and they had to complete final presentations to obtain credit. The summer science enrichment course focused on river ecology, water quality, and sustainability. The high school students visited and sampled a local river and brought the samples back to the college laboratories where they analyzed them. Alongside the river curriculum, as part of the approach advocated by the instructor of the pedagogy course, the Science TEUs developed additional lessons to augment students' understanding of sustainability. High school students presented their findings by the end of the session. This chapter describes a two-week sustainability curriculum unit developed and implemented by one team of Science TEUs. The analysis shows the challenges and possibilities of developing and implementing a high-quality sustainability curriculum.

The course begins with a river study. The high school students put on waders to explore a local brook in their community. They gather samples to investigate macroinvertebrates, turbidity, and coliforms (see Figs. 6.3 and 6.4). The students complete their analyses and then prepare group presentations for the final day of their two-week session. The Science TEUs are responsible for leading students in these investigations and developing lessons to support students' understanding of river ecology, sustainability, and scientific communication. After teaching the first cohort of students, the Science TEUs have the opportunity to revise and reteach their curriculum to the second cohort of students.

During Summer 2017, I spent 2 days on-site visiting the Science TEU program, observing lessons, interviewing TEU interns and their mentors, and collecting other qualitative data. In addition, the TEU interns shared their curriculum development folders and samples of students' work. Thus, data include direct observations of teaching sessions, lesson plans, samples of student work, and other artifacts, notes from debriefing sessions with mentor teachers and team meetings with the science teacher educator, as well as interviews with TEUs, mentor teachers, college science faculty, and the director of the Science TEU program.

I used a comparative case study approach (Yin, 2008) to develop a detailed understanding of the impacts of the seminar on the development of the TEU interns.

Fig. 6.3 High school students collecting macroinvertebrate samples

Fig. 6.4 A TEU intern helps students collect macroinvertebrates

I compiled data for each case study team into a single case record (Yin, 2008). The data were then analyzed using open coding methods, noting distinct themes or categories in the data, followed by axial coding to determine the representativeness of the themes and possible relationships between them (Huberman & Miles, 1994). I looked for within-case patterns and themes and then compared patterns and themes across cases (Yin, 2008). I also probed the data for the seven themes outlined by Stratton et al. (2015) (see Table 6.2). For this chapter, I selected a case that best represented some of the tensions in developing sustainability curricula, especially when the curriculum developers are: (1) novice teachers, (2) they are new to place-based education, issues of sustainability, and sustainability standards; and (3) their goal was to integrate sustainability with an existing curriculum.

   The team consisted of three STEM undergraduates. Anna (pseudonym)[1] was a Senior majoring in Environmental Analysis and Chemistry. She planned to teach in the Fall and applied to the program out of a deep commitment to science, justice, and a desire to work for science literacy. Anna had prior experience working with high school youth through community gardening and college access programs. Rina was a Junior majoring in Computer Science. Rina was exploring teaching as a meaningful and purposeful career. She looked forward to focusing on theory (through the pedagogy course) and experiential learning (through the practicum). Rina had prior experience with tutoring students. Lee was a sophomore majoring in Biology. He was attracted to the TEU program to improve his ability to design

---

[1] Pseudonyms are used for all TEU interns.

lessons, manage laboratory sessions, and incorporate relevant pedagogical methods. He had prior experience as a Teaching Assistant for an undergraduate Chemistry Lab.

6.5.3 How Do the TEU Interns Envision Planning and Teaching About Sustainability?

The Science TEU team took up the challenge of developing a two-week sustainability curriculum and river study. Furthermore, between Session I and Session II, opportunities to revise the unit plans provided avenues for improving instruction, content delivery, questioning, activity structure, and smoother transitions between activities. Yet, in analyzing their curriculum materials, I noticed they seemed to have two parallel streams of activities loosely connected by an overarching umbrella of sustainability. Thus, the curriculum lacked some thick connections that would have helped create a more cohesive sustainability curriculum. The first stream, shown in Fig. 6.5, focused on the river curriculum. Students were introduced to the river, they learned about the Healthy Rivers Program and macroinvertebrate, turbidity, and coliform tests for water quality. They also collected field specimens, conducted the tests in the college labs, analyzed the data, prepared presentation slides, and gave final presentations. Since the high school students were taking the course for credit, these aspects of the curriculum were required, especially the final presentations.

The second curriculum stream, shown in Fig. 6.6, focused on sustainability topics. The TEU interns prepared lessons to help students explore their relationship to nature, including how Indigenous cultures relate to the land. Students engaged in a role play about greenhouse gases and their effects on the atmosphere and learned about global warming and its causes. In subsequent lessons, students watched the Story of Stuff (Leonard et al., 2007), learned about the Amazon rainforest,

Fig. 6.5 The river curriculum

Fig. 6.6 The sustainability curriculum

conducted an endangered species simulation, learned about the Flint, Michigan Water Crisis, discussed redlining in the local community and environmental justice, followed by a power mapping activity where the HS students identified a goal, capacities, stakeholders, target, and tactics for a chosen environmental issue in their community.

As I mapped out the curriculum, I could see that for the high school students, the two streams might seem rather disjunct. Furthermore, the sustainability curriculum seemed to provide a crash course in various environmental topics rather than a cohesive curriculum that could support students in their analysis of the river samples or deepen their understanding of the implications for water quality, and the important role of sustainability in water resource management. At the same time, when I coded the activities for the seven themes identified by Stratton et al. (2015), I noticed that the two curricular streams collectively addressed all seven themes (see Table 6.6).

Table 6.6 Mapping the river and sustainability curriculum to sustainability education themes

Theme	River curriculum	Sustainability curriculum
A holistic approach	Ecological: Trout brook eutrophication, river study and analysis Natural and built: Trout brook study, CT River flooding	Ecological: Amazon rainforest and endangered species simulation Economic: The story of stuff, redlining Social: The story of stuff Cultural: Indigenous People's views of the land Political: Flint water, environmental justice, power mapping
Place-based education methodology	Trout brook lecture CT River history CT River ecology, local sampling and analysis	Redlining in Hartford
Connections between the natural and built world	River water sampling and analysis	The story of stuff
Conceptual frameworks for EfS		Sustainability defined
Development of sustainability literacy	Final presentations	Global warming comics Power mapping
Transformative pedagogical approaches	Place-based pedagogy: Trout brook	Sustainability role plays Simulations Power mapping Video clip analysis
Links to climate change		Global warming Endangered species simulation Video clip: Dear future generation: Sorry

The River Curriculum addressed five out of the seven themes but lacked a clear conceptual framework for EfS and a connection to climate change. Since having a clear sustainability pathway can help justify why students are learning about something, identifying these gaps can help improve the curriculum. Although the sustainability curriculum had activities that addressed all seven themes, as noted earlier, the connections between many of them were thin, which lessened the overall cohesiveness of the curriculum for the high school students.

A particular strength of the river curriculum was the place-based river study. According to Sobel, (2004, p. 6) place-based curricula "emphasiz[e] hands-on, real-world learning experiences, [and] this approach to education increases academic achievement, helps students develop stronger ties to their community, enhances students' appreciation for the natural world, and creates a heightened commitment to serving as active, contributing citizens." However, one of the challenges to implementing place-based curricula in the program was that the TEU interns came from all over the country. They had to develop a quick understanding of the local area's cultural, historical, and ecological history, including science content knowledge about the river, the ecosystem, local plants, history, management, human uses, and wildlife macroinvertebrates, microbiology. A strength of the sustainability curriculum was that the TEU interns had a strong grounding in sustainability concepts, whether from prior college study or through the pedagogy course. At the same time, connections between the place-based study of the river and the sustainability curriculum could have been stronger and made more explicit.

6.5.4   What Are Some Recommendations for Improving the Sustainability Curriculum?

In analyzing this case study and others, several recommendations to improve EfS in the program focus on better integrating sustainability with the river study, drawing more strategically on the TEU interns' science backgrounds, and enhancing student engagement. First, the TEU interns' initial visit to the brook occurred with the high school students. Bringing the TEU interns to the brook during the first 2 weeks of the pedagogy course before they went with the high school students would allow them to learn about the physics, chemistry, and biology of the river and build their local knowledge of the river. The TEU interns would be able to explore the sampling techniques and relevant sustainability connections. The inclusion of two investigation sites, the brook and a river in another community that the high school students come from would engage more of the students in studying their local community (a way to increase relevance and engagement) and allow the HS students to engage in comparative analysis. Finally, the TEU interns should develop a sustainability curriculum that features the river study as a central component, but then expands students' learning to the physics, chemistry, and biology concepts underlying sustainability of river ecosystems, as well as the social, historical, and political issues of land and water use and urban development. The TEU interns did not feel that their backgrounds in their major, particularly those with Physics backgrounds, had strong connections to the curriculum they were teaching. Some Physics majors even thought that if they knew this in advance, they might not have applied to the program because they lacked the expertise to teach about river ecosystems or sustainability. However, the physics of rivers can be studied in the context of fluid dynamics or hydrostatic pressure to name a few examples, and they could make authentic connections to sustainability with these topics.

The bulk of the pedagogy I observed and noted in the sustainability unit for this case-study team was discourse-based, discussing sustainability topics such as redlining, ocean ecosystems, and rain forests, or watching video clips and analyzing them. One way to improve student engagement would be to center the NGSS science and engineering practices (NGSS Lead States, 2013) to convey a clear sense of the nature of science and engineering to the high school students. The science and engineering practices include: (1) asking questions and defining problems; (2) developing and using models; (3) planning and carrying out investigations; (4) analyzing and interpreting data; (5) using mathematics and computational thinking; (6) constructing explanations and designing solutions; (7) engaging in argument from evidence; (8) obtaining, evaluating and communicating information. The TEU interns could build the curriculum around EfS as the goal and the scientific and engineering practices as the method to achieve EfS. This approach would enable the TEU interns to enact a central element of NGSS reforms and develop a curriculum that addresses the full set of scientific and engineering practices and sustainability in a more coherent way.

6.6 Concluding Thoughts on Education for Sustainability

As noted earlier, a key challenge for EfS includes integrating sustainability with the existing curriculum, whether it be science or another content area. Just as the balance between science and engineering can be uneven, the balance between science and sustainability can also be uneven. Furthermore, the siloed nature of subjects particularly in the upper grades often creates barriers to the types of interdisciplinary learning that sustainability concepts require. One way to support preservice and inservice teachers in developing a more cohesive approach would be to construct a checklist drawing on the seven themes developed by Stratton et al. (2015) (see Table 6.6) to identify and assess the components of sustainability in curriculum units. However, addressing all seven sustainability themes is not necessarily the goal. Instead, the goal is to clarify which themes they address, how and why they address them, and their rationale for not including other themes.

Education for sustainability is complex, draws on multiple disciplines and perspectives, and helps foster an array of highly desirable competencies for students that are urgently needed to address the myriad local and global sustainability issues that surround us. EfS helps bring meaning and connection and the answer to the timeless question of, "Why are we learning this?" Given its broad scope, EfS needs to be the responsibility of all educators in all subjects, but especially in the sciences. At the same time educators need clear frameworks and pathways for thinking about how to integrate sustainability into the curriculum. We need to move beyond rhetoric and fragmented action to serious and concerted efforts on the part of educators, researchers, and policymakers so that it becomes impossible for students to graduate from high school or college without a comprehensive education that also provides a basis for lifelong decisions and choices that foster sustainability locally and globally.

Acknowledgements The Summer STEM Teaching Experiences for Undergraduates program was supported by a 5-year award from the Improving Undergraduate STEM Education (IUSE) program of the National Science Foundation, Grant No. 1525691. Any opinions, findings, and conclusions or recommendations expressed in this material are those of the author and do not necessarily reflect the views of the National Science Foundation.

References

Brundtland, G. H. (1987). *Our common future: report of the world commission on environment and development.* Geneva, UN-Document A/42/427. https://digitallibrary.un.org/record/139811?ln=en

Carson, R. (1962). *Silent spring.* Houghton Mifflin.

Cloud, J. P. (2012). *Education for sustainability (EfS) standards and corresponding performance indicators.* The Cloud Institute for Sustainability Education. Retrieved July 17, 2017, from https://cloudinstitute.org/cloud-efs-standards

Cloud, J. P. (2014). The essential elements of education for sustainability (EfS) editorial introduction from the guest editor. *Journal of Sustainability Education, 6.* Retrieved July 29, 2021, from http://www.jsedimensions.org/wordpress/wp-content/uploads/2014/05/Cloud-Jaimie-JSE-May-2014-PDF-Ready2.pdf

Coneybeare, M. (2015). Million Trees NYC just planted 1,000,000th tree in Joyce Kilmer Park. *Viewing NYC.* Retrieved December 15, 2021, from https://viewing.nyc/million-trees-nyc-just-planted-1-000-000th-tree-in-joyce-kilmer-park

Conolly, J. Lehtomäki, E., & Scheunpflug, A. (2019). *Measuring global competencies: A critical assessment. Academic Network on Global Education and Learning.* Retrieved May 31, 2022 from https://static1.squarespace.com/static/5f6decace4ff425352eddb4a/t/5fc8eeb922891c6875c8a465/1637594418818/measuring-global-competencies.pdf

Cortese & Rowe. (2016). *In Haverford Sustainability Plan Draft.pdf.* https://www.haverford.edu/sites/default/files/HaverfordSustainabilityPlanAprilDraft.pdf

Davis, N. R., & Schaeffer, J. (2019). Troubling troubled waters in elementary science education: Politics, ethics & black children's conceptions of water [justice] in the era of Flint. *Cognition and Instruction, 37*(3), 367–389. https://doi.org/10.1080/07370008.2019.1624548

Egger, A. E., Kastens, K. A., & Turrin, M. K. (2017). Sustainability, the next generation science standards, and the education of future teachers. *Journal of Geoscience Education, 65*(2), 168–184. https://doi.org/10.5408/16-174.1

Feinstein, N. W., & Kirchgasler, K. L. (2015). Sustainability in Science Education? How the Next Generation Science Standards Approach Sustainability, and Why It Matters. *Science Education, 99*(1), 121–144. https://doi.org/10.1002/sce.21137

Guzman, A., Heinen, J. T., & Sah, J. P. (2020). Evaluating the conservation attitudes, awareness and knowledge of residents towards Vieques National Wildlife Refuge, Puerto Rico. *Conservation and Society, 18*(1), 13–24. https://doi.org/10.4103/cs.cs_19_46

Hofman, M. (2015). What is an education for sustainable development supposed to achieve—A question of what, how and why. *Journal of Education for Sustainable Development, 9*, 213–228. https://doi.org/10.1177/0973408215588255

Huberman, A. M., & Miles, M. B. (1994). *Data management and analysis methods.* In *Handbook of qualitative research* (pp. 428–444). Sage Publications, Inc.

Kastens, K. A., & Turrin, M. (2006). To what extent should human/environment interactions be included in science education? *Journal of Geoscience Education, 54*(3), 422–436. https://doi.org/10.5408/1089-9995-54.3.422

Kimmerer, R. W. (2015). *Braiding sweetgrass.* Milkweed Editions.

Lee, E., Menkart, D., & Okazawa-Rey, M. (2002). *Beyond heroes and holidays: A practical guide to K-12 anti-racist, multicultural education and staff development.* Teaching for Change.

Leonard, A., Fox, L. & Sachs, J. (2007). *The story of stuff.* Free Range Studios. https://www.storyofstuff.org/movies/story-of-stuff/

Leopold, A. (1949). *A Sand County almanac: And sketches here and there.* Oxford University Press.

McCrea, E. J. (2006). *The roots of environmental education: How the past supports the future.* Retrieved October 6, 2018, from: https://files.eric.ed.gov/fulltext/ED491084.pdf

MillionTreesNYC (n.d.). *NYC Mayor Michale Bloomberg and Bette Midler plant tree one of one million.* Retrieved December 15, 2021, from https://www.milliontreesnyc.org/html/newsroom/pr_milliontreesnyc_launch.shtml

Morales-Doyle, D. (2017). Justice-centered science pedagogy: A catalyst for academic achievement and social transformation. *Science Education, 101*(6), 1034–1060. https://doi.org/10.1002/sce.21305

NGSS Hub. (n.d.). *About the NGSS science standards.* Retrieved December 19, 2021, from https://ngss.nsta.org/about.aspx

NGSS Lead States. (2013). *Next generation science standards: For states, by states.* The National Academies Press.

NYC Parks. (2016). *NYC Street Tree Census – 2015 Report.* http://media.nycgovparks.org/images/web/TreesCount/Index.html

NYSED. (n.d.-a). *Science Learning Standards | New York State Education Department.* Retrieved December 21, 2021, from http://www.nysed.gov/curriculum-instruction/science-learning-standards

NYSED. (n.d.-b). *Living Environment: Science Regents Examinations: OSA:P-12: NYSED.* Retrieved December 21, 2021, from https://www.nysedregents.org/livingenvironment/

Petersen, K. B. (2022). Global citizenship education for (unknown) futures of education: Reflections on skills-and competency-based versus virtue-based education. *Futures of Education, Culture, and Nature – Learning to Become, 1*, 89–101.

Pirani, R., Berizzi, P., Rivera Maulucci, M. S., & Wolf, N. (1998). *Keeping the green promise: An action plan for new York City's urban forest.* Regional Plan Association and Environmental Action Coalition.

Rivera Maulucci, M. S. (2010). Resisting the marginalization of science in an urban school: Coactivating social, cultural, materials, and strategic resources. *Journal of Research in Science Teaching, 47*(7), 840–860. https://doi.org/10.1002/tea.20381

Rogers, J. (2016). Sustainability and performativity. In J. P. Davim & W. Leal Filho (Eds.), *Challenges in Higher Education for Sustainability* (pp. 217–229). Springer International Publishing. https://doi.org/10.1007/978-3-319-23705-3_10

Sobel, D. (2004). *Place-based education: Connecting classrooms and communities.* The Orion Society.

Sterling, S. (2003). *Whole systems thinking as a basis for paradigm change in education: Explorations in the context of sustainability* [Ph.D. thesis]. University of Bath.

Stratton, S. K., Hagevik, R., Feldman, A., & Bloom, M. (2015). Toward a sustainable future: The practice of science teacher education for sustainability. In S. K. Stratton, R. Hagevik, A. Feldman, & M. Bloom (Eds.), *Educating science teachers for sustainability* (pp. 445–457). Springer. https://doi.org/10.1007/978-3-319-16411-3_23

Taylor, N., Quinn, F., & Eames, C. (2015). *Educating for sustainability in primary schools: Teaching for the future.* Sense Publishers.

Tippett, K. (2016). *Robin Wall Kimmerer—The Intelligence of Plants.* Retrieved December 18, 2021, from https://onbeing.org/programs/robin-wall-kimmerer-the-intelligence-of-plants/

United Cities and Local Governments (UCLG). (2010). *Culture: The Fourth Pillar of Sustainability.* https://www.agenda21culture.net/documents/culture-the-fourth-pillar-of-sustainability

United Nations Environment Programme (UNEP). (1978). *Intergovernmental conference on environmental education, Tbilisi, USSR, 14–26 October 1977.* (p. 101). UNESCO. Retrieved December 5, 2021, from https://unesdoc.unesco.org/ark:/48223/pf0000032763

UNESCO. (2017). *Education for sustainable development learning objectives.* Retrieved December 18, 2021, from https://en.unesco.org/themes/education-sustainable-development

UNESCO. (2022). *What you need to know about education for sustainable development.* https://www.unesco.org/en/education/sustainable-development/need-know

United Nations. (2002). *Resolution adopted by the General Assembly 57/254. United Nations Decade of Education for Sustainable Development.* UN Documents. Retrieved December 19, 2021., from http://www.un-documents.net/a57r254.htm

US EPA. (2012). *What is environmental education?* [Overviews and Factsheets]. Retrieved December 18, 2021, from https://www.epa.gov/education/what-environmental-education

US Partnership for Education for Sustainable Development. (USPESD). (2009). *National Education for Sustainability K-12, Student Learning Standards, Version 3.* http://k12.uspartnership.org/standar/national-education-sustainability-k-12-student-learning-standards-version-3

Wals, A. E. J. (2010). Between knowing what is right and knowing that is it wrong to tell others what is right: On relativism, uncertainty and democracy in environmental and sustainability education. *Environmental Education Research, 16*(1), 143–151. https://doi.org/10.1080/13504620903504099

Wals, A. E., & Dillon, J. (2015). Foreword. In Stratton, S. K., Hagevik, R., Feldman, A., & Bloom, M. (Eds.). *Educating science teachers for sustainability* (pp. v-viii). Springer International Publishing. https://doi.org/10.1007/978-3-319-16411-3

Wiek, A., Withycombe, L., & Redman, C. L. (2011). Key competencies in sustainability: A reference framework for academic program development. *Sustainability Science, 6*(2), 203–218. https://doi.org/10.1007/s11625-011-0132-6

Withers, G. (2013). *Vieques, A target in the sun.* Washington Office on Latin America. Retrieved December 15, 2021, from https://www.wola.org/sites/default/files/downloadable/Regional%20Security/Vieques%20a%20Target%20in%20the%20Sun%205.1.13.pdf

Yin, R. K. (2008). *Case study research: Design and methods* (4th ed.). SAGE Publications, Inc.

María S. Rivera Maulucci is the Ann Whitney Olin Professor of Education at Barnard College. Her research focuses on the role of language, identity, and emotions in learning to teach science for social justice and sustainability. She teaches courses in science and mathematics pedagogy and teacher education for social justice.

Part II
Sustainability and Ecological Perspectives on Biodiversity

Chapter 7
When a Titan Arum Blooms During Quarantine

Nicholas Gershberg

7.1 The Arthur Ross Greenhouse at Barnard

The Arthur Ross Greenhouse has been an architecturally striking presence at Barnard College, atop Milbank Hall for nearly 25 years. It appears in almost any view of campus, from the ground, other buildings, or the air. For more than nine decades, research-active greenhouses have occupied Milbank's roof, with two previous facilities preceding today's. Having a campus greenhouse is emblematic of a commitment to botany and science, operationally and symbolically supporting the education of women and STEM research projects by women. Today's facility enables a variety of research projects while also housing fantastic biodiversity, with more than a thousand plants from nearly 500 species, representing plant biodiversity from almost every continent and biome on the planet. This living collection supports core STEM disciplines like biology, chemistry, and environmental science, plus anthropology and ethnobotany, political science and biogeography, economic botany, education, the arts, literature, and history.

Being so intent on the greenhouse as central to Barnard's identity and mission raises questions about the practicality of any such facility. Extensive labor and energy resources are required for greenhouse maintenance and operations. In turn, living collections and other resources must therefore align, at least partially, with the curricula in many courses about science or sustainability. Instructors and students should be able to devise and complete class activities, projects and other

From *When a Titan Arum Blooms During Quarantine: (aka, Making a Stink Online)*, by N. Gershberg. 2021. Plant Science Bulletin, 67(1): 34–37. (https://issuu.com/botanicalsocietyofamerica/docs/psb67_1_2021). Reprinted with permission.

N. Gershberg (✉)
Arthur Ross Greenhouse, Barnard College, New York, NY, USA
e-mail: ngershberg@barnard.edu

© The Author(s) 2023
M. S. Rivera Maulucci et al. (eds.), *Transforming Education for Sustainability*,
Environmental Discourses in Science Education 7,
https://doi.org/10.1007/978-3-031-13536-1_7

research with the support of the Ross Greenhouse. And, our greenhouse is a visible and effective locus for STEM outreach to the college's alumni, area schools and many other neighbors in northern Manhattan.

These myriad purposes are conveyed in this essay by full-time Greenhouse Administrator, Nicholas Gershberg. A skilled botanical horticulturist, he has long prioritized the sustainability of horticultural practices, aiming to minimize or mitigate negative impacts of greenhouse practice while also optimizing the care of the hundreds of plant species living inside. Gershberg also focuses on the well-being of the workers who provide that care and shares his sustainability and outreach mindset through an extensive network of plant-loving humans throughout New York City and beyond. A perennial part of Gershberg's job is spotlighting and documenting the "added value" of the Ross Greenhouse to campus life. Beyond achieving specific research objectives and educational outcomes, why is it worthwhile? As his narrative explains, sustainable practice and caregiving can add to its routine support of research and teaching by delivering intangible rewards, truly "green" and beautiful gifts to inspire and uplift the entire campus community.

7.2   Of All the Times …

In April through June of 2020, during the early COVID-19 quarantines and the height of the virus outbreak in New York City, our collection's *Amorphophallus titanum* (aka, Titan Arum, Corpse Plant) decided it was going to send up its first-ever bloom (Fig. 7.1). Following in the footsteps of the Brooklyn Botanic Garden's

Fig. 7.1 'Berani' (Indonesian for "Bold"), appx 10–12 years old, blooming for the first time, May/ June 2020

(BBG) *Amorphophallus titanum* specimen "Baby", which sent up the first recorded bloom in NYC in nearly 70 years, we had cultivated ours with great anticipation. That 2006 event was met with great public celebration and media coverage throughout the city and brought a widespread fascination with the plant into public consciousness. A number of years later, BBG decided to develop a new crop of *titanums*. Ours came from that crop, as an inter-conservatory gift of goodwill. Upon arrival at our greenhouse, it was a humble, young, 3-inch corm weighing about a pound. For nearly a decade, we have been cultivating it, with great affection, patience, and aspirations. Only a lucky few ever get to see these plants bloom in the native wilds of central Sumatra; the rest of us botanists and plant lovers must bring about a bloom *ex-situ* if we are ever to bear witness to its singular presence. For an institutional greenhouse, blooming events such as these are prime opportunities to engage a wide public audience and bring attention to a host of relevant issues. We hoped that we could hold our own festive on-site event to inspire others, as the BBG event had inspired us. We were looking forward to the opportunity to enthrall both our campus and the local community in the wonder of the Titan's bloom.

As some readers are likely to already know, the blooming of *A. titanum* in a conservatory setting is no longer the uncommon event it once was. Today, it is not as momentous an occasion as when Britain's Kew Gardens bloomed the first in cultivation in 1889, or when The New York Botanical Garden bloomed their first in 1937. As seed availability and horticultural knowledge have proliferated, mature specimens and their inflorescences have become staples in conservatories. Nonetheless, it is still a superlative event in the botanical world, and the blooming at Barnard was still an important and long-anticipated event for our community.

Any team that has successfully brought an *A. titanum* to bloom, or achieved a similarly rare botanical event, understands this sentiment. Beyond the sheer excitement of the occasion, a corpse plant blooming is evidence that a greenhouse is doing a solid job of providing care 365 days of the year. Beyond being a point of pride (clearly), it also comes with a certain sense of relief, which may feel familiar to growers reading along. On the long road to a bloom, any number of things can happen (e.g., one cold night, a few days of too much or too little moisture, a random hapless accident) and years of cultivation can be voided before one has even realized it has happened. So for many of us, being able to enjoy the tangible product of a capable team of horticulturists is a welcome occasion.

Each time one of these plants blooms, it is an opportunity for a greenhouse to put its best foot forward, and make the most of the outreach potential of these "megaflora" events. At the very center of any conservatory's mission lies the message of stewardship. While the plant effortlessly does the work of drawing public interest (among the beetles and flies), the story of its lengthy cultivation inherently reinforces the notion that stewardship takes more than just good intentions and wholesome aspirations. It demands hard work, dedication, knowledge, and patience.

But, nature is on its own timeline. Despite our best-intended interventions as growers, it more often than not shows little concern for our ends. And, famously, these titans tend to keep a particularly unpredictable schedule. It was only on the final day of our quarantine preparation, just before the "stay at home" orders were issued in New York State, that our plant began to emerge from a long dormancy period. As much as we had hoped for a bloom for so many years, we would have been content with just another majestic non-reproductive leaf this time around. Weeks later, when there was no doubt that our plant had begun to produce not a leaf but a bloom in the midst of the nationwide quarantine, it was frustrating, and a bit of a disappointment. There was no way we could safely receive visitors at our facility. Not even our student workers, who had been instrumental in the plant's daily cultivation, were allowed on campus at that stage.

As it happens, all was not lost. It turns out folks are quite fond of the internet. And despite the setback, we were intent on sharing the bloom with as much of our community and beyond as we possibly could.

First, in a matter of a few days, we prepared a livestream of the plant, which was posted on YouTube. Often just a supplement to the in-person experience, for this bloom a live stream feed would be our community's primary window into the greenhouse. Initially, I was surprised at how much technical know-how was involved in setting up the various elements of the stream. But, largely thanks to the help of an impromptu, remote collaboration of our college's informational technology (IT) and digital media staff, we were soon able to broadcast 24/7. In addition, using the greenhouse's existing Instagram account (@barnardgreenhouse), we staged a public naming contest (as is the custom), culminating in the decision to name our bloom Berani, a word meaning "bold" in Indonesian. We updated and informed our audience with regular posts, and used "Instagram Live" as an impromptu platform for interacting directly with the public. Finally, a few days after the inflorescence waned, we created a beautiful time-lapse of the 48 h of opening and closing, which has proven a cherished memento. In retrospect, thanks to the staggering connectivity of the internet, it is possible that we reached a much larger, more diverse audience than if we only had an on-site event. Without a doubt, it was an object lesson in the potential that digital and social media offer for science and education outreach.

7.3   Making the Best, of the Best

As a conservatory, fostering science literacy and educational outreach is a large part of our mission. When visitors come to the greenhouse, it is our job to elicit the "spark" of interest in the plants we house and to create the momentum that will generate further interest and desire for more knowledge. We try our best to provide our guests with clear, factual, thoughtfully contextualized, and relevant information. Greenhouse curators routinely rely on an array of charismatic plants to facilitate conversations with visitors. It is important to remember that we are speaking to audiences with a diversity of interests. Some of our guests are of course already very interested in plants, and connoisseurs in their own right. Others are less so, and some are not initially interested at all. When young children come to the

Fig. 7.2 Detail within the spathe

greenhouse, after an initial introduction, we usually begin by asking whether they like ice cream. Even if they are not particularly interested in the plants, they are definitely interested in chocolate and vanilla! So we show them our vanilla orchid vine and our cacao trees. When burgeoning scientists come, we point out the fascinating potential for biotechnology that plants such as cacao possess, and discuss how our particular plants were grown from tissue culture for potential commercial use. And if a plant eats insects, moves rapidly—or both—it is sure to engage a crowd.

It has long been known that the use of charismatic flora and fauna is a highly effective tool in science outreach about biodiversity (e.g., Lindemann-Matthies, 2011; Shah & Parsons, 2014), and provides great opportunities to cleverly introduce more sophistication into the scientific vernacular. For example, a key point we emphasized from the start of our coverage is that what people were witnessing was not *the world's largest flower*, but instead *the world's largest unbranched inflorescence*, or single structure containing many dozens or hundreds of individual flowers (Figs. 7.2, 7.3 and 7.4). Throughout the 2 to 3 weeks of the ongoing event, we were able to introduce many more concepts, of greater nuance and sophistication, from topics including tropical biogeography, systematics and taxonomy, ethnobotany, morphology, pollination and reproductive ecology, and conservation biology. We were also able to bring attention to the history of often exploitive colonialist aspects of "discovering" any tropical plant taxa and to affirm the importance of local guides and Indigenous knowledge, both historical, and current (Adams et al. 2014; Yudaputra et al. 2021). [HC4] Over the years, we have come to appreciate that there are myriad cultural connections with almost all of our taxa, making them interesting subjects for classes in sociology, history, literature, etc. In that sense, we have promoted the greenhouse collection as a resource for most, if not all of the other departments at the college, on some level. Here is an example of one of the interdisciplinary connections we made during the corpse plant blooming (Fig. 7.5).

Fig. 7.3 From the
ground up

Fig. 7.4 Overhead view

Fig. 7.5 Interdisciplinary connections: "Who wore it best?"

7.4 Improvise, Adapt, Overcome

In this digital age, online media is now the mainstream, and certainly the norm among our student body. The Titan Arum blooming event affirmed that as educators, we need to embrace this online trend. Moreover, that means not just using the preferred platforms of our student audience, but also adapting to the pace they set. That is not necessarily an easy proposition, as education and advocacy have to keep up and maintain relevancy in the midst of nearly constant stimulation. We must continually strive to provide high-quality content, in an eloquent manner, often in a way that is interdisciplinary, and cross cultural. Given the urgency to start an engaged awareness of the natural world early, it is our goal to reach as wide an audience as possible. By establishing these connections using modern platforms, we can fulfill this job of raising awareness, encouraging scholarship, fostering a committed sense of stewardship, and promoting informed action. Hopefully, if we can connect on common ground with broad appeal, we will help create a truly open and robust entry point for the STEM pipeline, available to the widest possible range of individuals.

There were some hurdles to clear in creating our online event. The main lesson we learned is that it is of vital importance to have a working familiarity with the setup and gear involved, and to practice to gain proficiency. To my surprise, no one person or department on our campus was familiar with the soup-to-nuts process of setting up a live stream video for public viewing. I had to enlist the help of about a dozen individuals, across several departments to make it work, and under quite unusual circumstances since almost nobody was actually on campus. It may have seemed like a fairly basic operation, given the ubiquity of live streaming lately. But

Fig. 7.6 Screenshot from YouTube

in actual fact, it was seldom needed, so the process took some effort to bring to bear. It is advisable to build these skills and interdepartmental relationships early, so that all of the components are effectively ready to go (Fig. 7.6).

Fortunately, the fact that we had already cultivated a social media follower base meant that, when we had exciting, ephemeral content, there was someone to share it with. While we did of course use the college's existing traditional media outlets, they were somewhat static by comparison, and lacked the innately interactive quality of social media. Just as with the live stream hardware setup though, online platforms take time to develop well. As a relative social media novice myself, learning to craft content that is at the standard of our in-person visitor experience, and that has a comparable "voice", has taken time. A big thanks goes to a number of our greenhouse student worker staff, who put time in over the course of the last few years to help cultivate an engaging and worthwhile platform. Creating more immersive informational hubs like websites can be extremely time-consuming as well, and are subject to a host of technical issues. Our own site has been a long work in progress, and is a high priority as the management and purposeful utilization of data continues to pave new paths in science. So it makes sense to view not only developing content, but also the means to disseminate it, as an ongoing process.

In the last year, this concept of a hybrid model for outreach and institutional engagement has, by sheer necessity, taken a huge leap forward. If anything though, the process has just been accelerated by circumstance, as there has long been a need to close the gap between "real world" and online scholarly resources. In cultivating on-site and online content synergistically, ideally, each one serves as a follow-up resource for the other. Whether you are a small institutional greenhouse with a modest collection like ours at Barnard College, or a large, world-class institution, there exists great potential to enhance the user experience, and provide access to more diverse audiences, with nearly instant pathways for cultivating meaningful content.

7.5 Looking Back

I would like to take a moment to acknowledge that in addition to happening during the quarantine lockdowns, this bloom event converged with a time of unprecedented social upheaval. The nation, and the entire Barnard community, were outraged and heartbroken over the murder of George Floyd, Jr., and were activated in the ubiquitous decrying of systemic violence against Black Americans. As the protests and national dialogue emerged as the events of true importance at that moment, how we proceeded was intently focused on not overshadowing them. Furthermore, our own community was already in a state of coping with the recent devastating death of a student. So we must remember that whenever these celebrated institutional events occur, they occur necessarily in the human context of everything else going on at the time. We hope, at the very least, that our event was uplifting in some small way, in the face of these truly serious, important, and deeply personal matters.

Now that our plant has returned to dormancy, we hope that the traditionally mounted pressing of the full inflorescence we have successfully made will be a tangible memento that students can marvel at in person upon their return. But we look forward to pursuing our outreach role online as well, as best we can, with all the latest tools at our disposal. Looking back at 2020, a year of highs and lows, I am ever more grateful our plant provided us an opportunity to focus on something positive, that brought our community together for a short while in mutual admiration. If the study of plants teaches us anything, it is that nature persists, and finds ways to adapt even in the most unlikely of environments. Hopefully, we can continue to reflect on this lesson, into 2021, and beyond.

References

Adams, M. S., Carpenter, J., et al. (2014). Toward increased engagement between academic and indigenous community partners in ecological research. *Ecology and Society, 19*(3), 5. https://doi.org/10.5751/ES-06569-190305

Gershberg, N. (2021). When a Titan Arum blooms during quarantine: (aka, making a stink online). *Plant Science Bulletin, 67*(1), 34–37. https://issuu.com/botanicalsocietyofamerica/docs/psb67_1_2021

Lindemann-Matthies, P. (2011). 'Lovable' mammals and 'lifeless' plants: How children's interest in common local organisms can be enhanced through observation in nature. *International Journal of Science Education, 27*(6), 655–677. https://doi.org/10.1080/09500690500038116

On the importance of local guides and indigenous knowledge.

On the roles of charismatic flora and fauna in science outreach about biodiversity.

Shah, A., & Parsons, E. C. M. (2014). Lower public concern for biodiversity than for wilderness, natural places, charismatic megafauna and/or habitats. *Applied Environmental Education & Communication, 18*(1), 79–90. https://doi.org/10.1080/1533015X.2018.1434025

Yudaputra, A., Fijridiyanto, I. A., Witoni, J. R., & Astuti, I. P. (2021). The plant expedition of an endangered giant flower *Amorphophallus titanum* in Sumatra. *Warta Kebun Raya 19*(1), 23–29. https://publikasikr.lipi.go.id/index.php/warta/article/view/735/603. Accessed 18 Dec 2021.

Nicholas Gershberg, administrator of Barnard's Arthur Ross Greenhouse, curates and maintains its worldwide collections, assists researchers, and leads education activities for visitors of all ages. He also teaches at the New York Botanical Garden. He is trained in propagation, greenhouse management, and bonsai technique.

Chapter 8
What Does Cell Biology Have to Do with Saving Pollinators?

Jonathan Snow

8.1 Introduction

To explain why I work on honey bees, I first want to tell you a little bit about my journey as a biologist. Biology is often portrayed as a continuum made up of different levels of organization, from cells to tissues, to organs and organ systems, to individual organisms, and to populations, communities, and ecosystems. Wikipedia defines cell biology as the "branch of biology that studies the different structures and functions of the cell and focuses mainly on the idea of the cell as the basic unit of life" (Wikipedia contributors, 2018). As an undergraduate majoring in Biology at Williams College, I became fascinated with the cellular and molecular aspects of biology. As part of a continuum, the concept of cell biology in isolation is relatively abstract without attempts to integrate it with broader biology, chemistry, and physics. Especially when thinking about multicellular organisms, cell biology is most relevant when it can be applied to understanding higher levels of organization. The cell biology definition in Wikipedia continues on to suggest that "knowing the components of cells and how cells work is fundamental to all biological sciences; it is also essential for research in biomedical fields such as cancer, and other diseases" (Wikipedia contributors, 2018). True to form, I was thinking about how biology at this level contributed to disease in humans. After graduation, I studied mammalian blood and immune development and function. I was involved in looking at the cell biology of how we make our red blood cells, our white blood cells, and our platelets, often in the context of blood disorders and malignancies. Primarily focusing on mouse models and cell lines, I examined the role of signaling pathways and transcriptional regulators affecting these processes in both my Ph.D. and post-doctoral work. The relevance of my studies to disease in humans was always in the backdrop.

J. Snow (✉)
Department of Biology, Barnard College, New York, NY, USA
e-mail: jsnow@barnard.edu

M. S. Rivera Maulucci et al. (eds.), *Transforming Education for Sustainability*,
Environmental Discourses in Science Education 7,
https://doi.org/10.1007/978-3-031-13536-1_8

During my postdoctoral training, I had the opportunity to teach a course on the American food system with a longtime friend. This served as the beginning of a continuing interest in the systems, and especially the organisms, involved in making our food. As I looked towards the future, I hoped to use the techniques and ways of thinking I had learned in my graduate and postdoctoral training with a new model organism that might bring together these interests. With bees, I felt perhaps I had found the perfect model. Bees are fascinating and they are currently experiencing a crisis, but we do not know much about them at the cellular and molecular level. Luckily, however, their genome is sequenced and we can use knowledge gleaned from other model insects to inform our questions. Perhaps most importantly, a cell biological understanding of biology is highly transferable between animal species, so I began to use my skills in this new area.

While honey is one of the most commonly thought of benefits supplied by these insects, the most important benefit is the pollination services they offer (reviewed in Snow & Rivera Maulucci, 2017). The Western Honey Bee, *Apis mellifera,* provides pollination services of critical importance to humans in both agricultural and eco-logical settings (Goulson et al., 2015). When bees fly from flower to flower foraging for nutrients, they transfer pollen between flowers of the same species allowing for fertilization of the ovule and completion of the plant's reproductive cycle (see Fig. 8.1).

The co-evolution of pollinators and plants over millions of years has made bees and other animals a critical part of many ecosystems (Potts et al., 2016). In addition to their importance in natural ecosystems, this pollination service is absolutely required for the growth of many agricultural crops important to humans.

Plants like almonds absolutely require insect pollination. As our agricultural sys-tems have become more homogeneous, and monocultures have become the norm, honey bees have become the only pollinator with the ability to effectively pollinate

Fig. 8.1 Honey bee pollinating a cherry tree on Riverside

many of the crops in our modern agricultural system. Some examples include apples, cranberries, melons, and broccoli with some crops like almonds, blueberries, and cherries being greater than 90% dependent on honey bee pollination. Honey bees are easy to move, provide a high number of individuals, and they are generalists in the plants they will pollinate. To satisfy the pollination needs of an industrialized agriculture system, "industrial beekeeping" practices are now employed. According to a 2015 survey, over 85% of the country's 2.6 million colonies were in commercial migratory beekeeping operations with an average of >4000 hives (Lee et al., 2015). Recently, meeting the pollination demand has become even more challenging as honey bee colonies have suffered from increased mortality, most likely due to a complex set of interacting stresses (Goulson et al., 2015). Key stresses thought to be involved include nutritional stress due to loss of appropriate forage, chemical poisoning from pesticides, changes to normal living conditions brought about through large-scale beekeeping practices, and infection by arthropod parasites and pathogenic microbes (see Fig. 8.2).

The challenges facing pollinators mirror those facing diverse organisms in both natural and managed environments. These issues will have far ranging impacts on the health of individuals and communities and the negative effects will be felt disproportionately by marginalized and disadvantaged groups. For example, decreased abundance and diversity of animal pollinators like bees (Ollerton, 2017) is predicted to have potentially dire consequences for food security (Smith et al., 2015) that will be felt especially acutely in areas that are 'food deserts' where little food stuffs are produced, such as urban centers (Lawson, 2016).

As no single cause for the recent increase in honey bee disease is evident, there is enhanced focus on the impact of interactions between various stressors. In seeking to understand how stresses might synergize to impact honey bee health, we have undertaken efforts to more completely define common cellular processes and cell stress pathways that are impacted by multiple stressors. My lab studies cell stress and infectious disease in honey bees. Undergraduates are incorporated into every aspect of this research throughout the year, with particular emphasis on designing and performing experiments as well as analyzing data. Figure 8.3 shows undergraduate students working with the bees.

Together, we are interested in two big questions. First, how do bees respond to stress at the cellular level? Second, how does one of their common parasites, *Nosema*

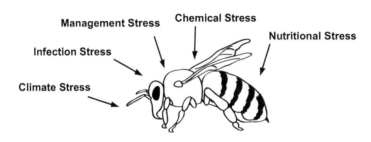

Fig. 8.2 Key stressors impacting honey bee health

Fig. 8.3 Snow lab members installing a 'package' of bees

ceranae, respond to stress at the cellular level? We hope to use the answers to these two questions to improve honey bee health.

We are specifically focused on the process of proteostasis, which refers to the homeostasis of protein synthesis, folding, function, and degradation (reviewed in Taylor et al., 2014). In multiple organisms, a number of normal and pathologic conditions shown in Fig. 8.2 can lead to disruption of proteostasis. Environmental factors, as shown in Fig. 8.4. can modify proteostasis by acting through these cellular and organismal pathways, including microbial infection, aging, and nutrition, all of which are implicated in honey bee health.

Work in other organisms has defined a number of cellular pathways that are involved in recovering from proteotoxic stress, including the Unfolded Protein Response (UPR) centered in the endoplasmic reticulum (Walter & Ron, 2011) and the Heat Shock Response (HSR) centered in the cytoplasm (Vabulas et al., 2010). In our previous work, a number of very talented undergraduates have worked to characterize these two proteostasis pathways (Johnston et al., 2016; McKinstry et al., 2017; Shih et al., 2020, 2021; Adames et al., 2020; Bach et al., 2021; Flores et al., 2021). These studies are ongoing with new cohorts of students helping to define unique aspects of honey bee cell stress pathways and the stress pathways of their pathogen, *N. ceranae* (McNamara-Bordewick et al., 2019).

In the context of a research laboratory, it has been relatively easy to discuss the plight of the honey bee with students and to draw broad links between our work on

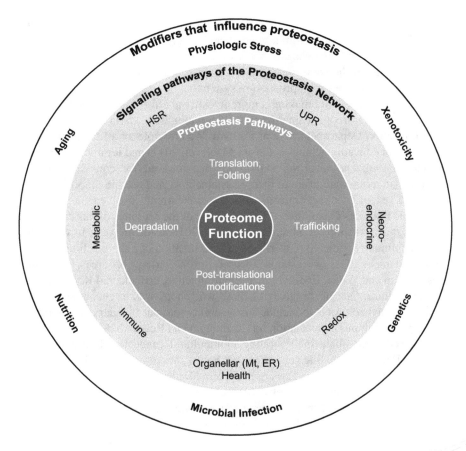

Fig. 8.4 Proteostasis is regulated at many levels of biology. (Adapted from Powers & Balch. 2013)

cell biology and larger issues of sustainability. This has been, in part, because research students have direct interaction with bees; helping to tend the hives, assisting in extracting honey, collecting and caring for bees in caged experiments, and dissecting specimens to obtain samples (see Fig. 8.3). In the next section, I will talk about my efforts to bring sustainability in thought and action into my classroom, with specific focus on my course, *Laboratory in Cell Biology*.

8.2 Sustainability in the Classroom

As a faculty member at Barnard, I have been fortunate to teach exceptional students. I have drawn on my experiences as an undergraduate at Williams College, where I was nurtured in both intellectual creativity and curiosity as well as challenged to develop sharp critical-thinking skills. I was hired as a cell biologist and I am primarily responsible for teaching three classes that deal with this level of biological organization, *Laboratory in Cell Biology, Cell Biology*, and *Introduction to Cell and Molecular Biology*. I come from a biomedical background and am still quite interested in questions of human health, but as described above, my big picture view has shifted to focus on the environment and sustainability, especially as they connect with agriculture and our food production system.

My first thoughts about making a more intentional effort to incorporate sustainability into my classrooms came from a collaboration with María Rivera Maulucci in the Education Department here at Barnard. We co-wrote an article about using honey bees in education (Snow & Rivera Maulucci, 2017) and it turned out to be a fun, educational, and inspiring experience that allowed for considerable self-reflection and a renewed desire to merge sustainability and teaching more purposefully. This experience synergized with my concurrent involvement with the Sustainable Practices Committee at Barnard. One of the activities of this committee was to hold a workshop series during the 2017–2018 academic year, in which the college community gathered to work towards the creation of a shared climate action vision statement. The workshops focused on a variety of areas, including Consumption and Waste, Energy, Local Environment, Curricula and Research, and Campus Culture. One of the goals of these workshops was to encourage faculty members to envision ways to include sustainability into our broader curriculum.

During the campus conversation on Curriculum and Research, María introduced us to Banks' ideas about incorporating multicultural concepts into a curriculum (Banks & Banks, 2016). She suggested that Banks' framework could be applied to introduce sustainability into a classroom (Rivera Maulucci, Pathways, this volume). The strategies proposed by Banks, and adapted by María, allow for different amounts of integration of this new perspective, in this case sustainability, with the 'core' material of the class. The 'levels of integration' proposed by Banks include: The Contributions Approach (Level 1), The Additive Approach (Level 2), The Transformation Approach (Level 3), and The Social Action Approach (Level 4). According to Banks, Level 1 would see inclusion of a small number of discrete exposures to the new perspective without substantial connection to the existing course material. Level 2 would incorporate the concepts associated with the new perspective into the whole course without a substantial change to the course structure. Level 3 involves a major restructuring of the curriculum with the intent to fully integrate new material. Level 4 encourages students to apply the new perspective to real-world issues and pursue activism to support social change shaped and supported by what they have learned. Thus, the intensity of sustainability inclusion is dependent on the flexibility of the original material, the teacher, and the students involved in the class.

The collaboration with María and the work as part of the Sustainable Practices Committee helped me to imagine incorporating sustainability more fully into one class I teach, *Laboratory in Cell Biology*. In the case of this class, the broader context of the material is quite flexible, so I could employ the more intensive Transformation Approach (Level 3) and restructure the curriculum to fully integrate sustainability. In addition, although the students are Biology and Neuroscience majors primarily interested in health-related fields, they are some of the best and the brightest STEM students at Barnard, and I knew they would be ready to meet any challenge. To describe the proposed changes, I first want to start by describing the class and giving some context on how Barnard's Biology Department envisions laboratory classes.

Research on science pedagogy has revealed that involvement with the theory, practice, and culture of science as undergraduate researchers increases long-term persistence in STEM (Linn et al., 2015). Providing such experiences is a major goal of a research program at a small liberal arts college. In my research lab, student researchers play a significant role in working to elucidate cellular stress responses of honey bees. However, as the number of students that can be included in a research program at a small college is necessarily limited, another approach to exposing a broader set of students to the 'authentic' theory, practice, and culture of science is through the use of upper-level labs. Authentic learning is an instructional approach that allows students to apply their knowledge to real-world problems while modeling the theory, practice, and culture of professionals in the field. Important aspects of authentic learning include open-ended inquiry, learning in groups, and the use of independent project work (Rule, 2006).

In our department, there has been a major push to implement these authentic learning principles in lab classes, mirroring a trend that has been gaining popularity in science education for some time nationally. One model for achieving these goals in known as Course-Based Undergraduate Research Experiences (CURE) (Auchincloss et al., 2014) and is defined as a course in which "students work on a research project with an unknown answer that is broadly relevant to people outside of the course" (Cooper et al., 2017). While not a CURE class in the strictest sense, the lab class I teach, *Laboratory in Cell Biology*, incorporates many of its principles, pursuing cell biology through weekly lab experiments that allow about 16 upper-level students to gain experience in the theoretical and practical aspects of how cell biologists think about and perform experiments, collect data, and present scientific findings. The incorporation of original research questions interconnected with those pursued in my research lab represents a key aspect of this class that increases the engagement of students by incorporating "authenticity" into laboratory class. The inclusion of an independent project for the last third of the class is especially valuable in helping meet the goals of authentic learning. In addition to providing an authentic research experience, classes based on the CURE model have been shown to help increase the participation and retention of underrepresented groups in STEM education (Rodenbusch et al., 2016). Thus, CURE-type classes can be a major force for increasing inclusion in the science education and hopefully in the scientific endeavor more broadly.

When I first developed the class, I decided to focus on a small part of cell biology, specifically choosing an area that was directly related to my research; proteostasis and the cellular responses to disruption of this process. As mentioned above, proteostasis is maintained by the responses of the proteostatic network within individual cells. These pathways sense proteostatic stress through the detection of a build-up of unfolded proteins and trigger a suite of responses designed to return the cell to homeostasis. Three highly conserved pathways are responsible for sensing

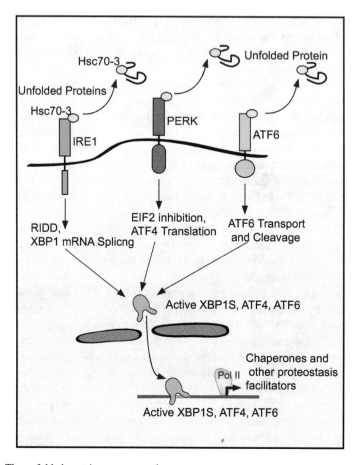

Fig. 8.5 The unfolded protein response pathways

proteotoxic stress in the endoplasmic reticulum and trigger the coordinated response known as the UPR (Fig. 8.5). These pathways are characterized by unique trans-membrane receptors and signal transduction machinery which activate UPR target genes to return the cell to homeostasis (see Fig. 8.5).

For the class, we use a fruit fly (*Drosophila melanogaster*) cell line (Schneider, 1972), because they are easy to grow and many protocols exist for experimenting with them. But as a fruit fly-derived cell line, they are far removed from bees and pollinators. For framing and readings, I have focused on the cell biology and human diseases associated with the dysfunction of the processes we study, such as Alzheimer's (Hetz et al., 2020). Because of the flexibility of the cell line model, we can easily look at the process of proteostasis and the response to it by measuring various steps in the process including signaling, changes in gene expression, and cellular outcomes (Samali et al., 2010). Based on my assessments and student evaluations, this class has been quite successful in its aims of introducing the theoretical and practical aspects of how cell biologists consider their field while increasing student enthusiasm and engagement. However, most students consider the broader biological context of the class to be about human disease, with approximately 40% of biology majors entering MD programs (for years 2010–2015). As biology is a continuum made up of different levels of organization and cell biology is highly transferable between metazoan species, it should be relatively straightforward to change the context to one involving themes of sustainability.

I devised five strategies to help link cell biology to problems impacting honey bees (see Table 8.1). First, I framed the class in terms of pollinators and bee health. Instead of beginning with human diseases, I started with pollination, honey bee biology, and honey bee health. Readings now reflect this change in framing (e.g., inclusion of Potts et al., 2016). I still make connections with medicine, but by altering the structure to move medicine to later dates in the course, I can shift the emphasis to sustainability. Second, I established this new emphasis by bringing live honey bees into class on the first day (with trips to see the apiary on the roof when weather permits). Third, I incorporated original research from my lab into the readings. Previous work examining proteostasis pathways in honey bees have resulted in publications with student coauthors that can be used in the class, including those cited above (Johnston et al., 2016) and (Adames et al., 2020). Fourth, we started with cell biology experiments using honey bees before focusing on cells. By getting to know the organisms before thinking about cells, I hoped to move through the levels of biological organization from organisms to tissues to ultimately zero-in on cells and the molecular component involved in their function. Finally, I encouraged students to explore links between cellular biological organization and sustainability problems that affect pollinators in their final projects.

After implementing the majority of these changes for one semester in the fall of 2018, I found the project was overall a success with some key additions to be undertaken in subsequent years. Anecdotally, I found that students were excited by the inclusion of sustainability concepts in the course and that students felt like the structural and content changes introduced them to new ideas and connections between cell biology and sustainability that they would not otherwise have been exposed to. However, I felt that there were two aspects that could be strengthened in future iterations of the course. First, there were a number of missed opportunities to solidify the link between cell biology and sustainability as the class drew to a close. I find that the themes and information in a class often fall into place for students during the last few weeks of the course when they look back and synthesize information from the whole

Table 8.1 Planned steps to connect cell biology to pollinators

Frame the class in terms of pollinators and bee health
Establishing the emphasis of the class by interacting with live bees on the first day (with trips to see the apiary on the roof when weather permits)
Incorporate original research from the Snow lab into the readings
Start with cell biology experiments using bees before focusing on cells
Encourage students to think about the big picture when doing their final project

course. This is especially true when there is a cumulative assessment, which we have in this class in the form of both a final lab report and a lab final exam. Thus, I planned to spend additional time emphasizing the themes of sustainability covered throughout the course during the retrospective/synthesis period leading up to these assignments. In addition, I aimed to explicitly include these sustainability themes into the two final assignments to encourage student engagement through the end of the term. Second, any discussion of pollinators can be directly connected to concepts of environmental justice (Mohai et al., 2009). Making those associations in the class might represent an ideal opportunity to connect science and broader societal issues in the biology curriculum. Third, I did not have a method in place for gauging the success of the project in a more concrete and quantitative manner. In order to assess the success of using the Banks framework to bring sustainability in thought and action into the *Laboratory in Cell Biology* class, I hoped to solicit feedback in a more structured manner.

I have now taught the class for the fourth time. A survey administered to the most recent iteration (taught fall, 2021) class assessing the effectiveness of the five strategies named above was encouraging. When asked to rate the effectiveness of the strategies, students ($n = 17$, Mean ± SD) gave the following scores; framing (4.58 ± 1), interacting with bees (4.71 ± 0.99), incorporating original Snow lab research (4.41 ± 1), starting with experiments on bee tissues before cells (4.65 ± 0.79), and referencing the broader context in the final projects (4.47 ± 1.07) (where 5 = very effective, 4 = somewhat effective, 3 = neutral, 2 = somewhat ineffective, and 1 = very ineffective). Feedback in comment form was quite positive (e.g. "Thank you for a wonderful class."), but also provided some constructive suggestions for future iterations. The recommendations fell into two broad categories. First, a number of comments suggested spending more time discussing specific threats to pollinators (e.g. pesticides and infection) and bridging these insults to the UPR more concretely. A specific example was a request to emphasize Snow lab research more in recitation. Second, there were a number of calls to include even more bee species and more interactions with bees during the class the first day. For example, one student suggested using visits to the apiary and college green roofs to discuss bees, pollinators, and sustainability in a NYC-specific manner. These points represent positive and productive feedback that I hope to incorporate over the next few iterations of the class. One future goal commensurate with these student reactions is described in the next section.

8.3   Beyond Honey Bees

According to E. O. Wilson's Biophilia hypothesis, humans are inherently interested in learning about other life-forms (Wilson, 1984). Many scholars and activists alike have worried about the loss of this human attribute as we have increasingly less interaction with nature in the modern world (Pyle, 1993), leading to diminishing interest in conservation (Soga & Gaston, 2016). In a 2002 article, authors described the ability of schoolchildren to name approximately 50% more Pokémon characters than real species (Balmford et al., 2002). In agreement with this finding, most students I talk to think there are only two types of bees, "honey" and "bumble." Yet in fact there are ~20,000 other bee species on the planet (Michener, 2000). However, the vast majority of attention and resources are spent focused on the honey bee and to a lesser extent on other managed bee species (Geldmann & González-Varo, 2018). Of the myriad bee species, ~50 species are managed and only 12 are commonly used for pollination services (Potts et al., 2016). Agriculturally-important managed bees include select species of the following groups; bumble bees (genus *Bombus*), alkali bees (*Nomia melanderi*), alfalfa leaf-cutter bees (*Megachile rotundata*), and mason bees (genus *Osmia*) (Klein et al., 2018).

In addition to bees, many other animals are involved in pollination (Ollerton, 2017). A long-held belief has been that diversity in pollinators is of significant value, leading to higher levels of pollination even in the presence of managed pollinators (Kremen, 2018) and recent studies support this view (Garibaldi et al., 2013, 2016; Winfree et al., 2018). Reductions in all animal pollinators, including wild bees, have been documented in recent years (Potts et al., 2016). One study found a 50% reduction in the bee species visiting plants within a specific study region over the last century (Burkle et al., 2013). While difficult to quantify the effects, such pollinator losses could have significant impacts on human health (Smith et al., 2015). It is likely that these losses are largely due to human impacts (Winfree et al., 2011). This troubling trend of diversity loss as a direct result of our activities extends beyond pollinators to other invertebrates. For example, a 2017 study showed a 75% decrease in flying insect biomass over the last three-decade period in Germany (Hallmann et al., 2017). As a person interested in sustainability and conservation, I found these numbers disturbing, and I aspired to stretch my goals for the Laboratory in Cell Biology even further. Could I use the class to increase student's knowledge of other bee pollinators and invertebrates more broadly? Could I use the class to think about conservation beyond species directly relevant to the well-being of human society?

Charismatic megafauna is a term used to describe widely popular animal species, which are often used by conservation advocates to achieve environmental stewardship goals (Ducarme et al., 2013). There are pros and cons to using this model to involve the wider public in conservation. However, there is some evidence that people's natural interest in honey bees (reviewed in Snow & Rivera Maulucci, 2017) can be used to increase exposure to other wild species. Some argue that honey bee conservation on its own will not help wild bees and other pollinators (Geldmann & González-Varo, 2018) and some evidence even suggests that where introduced, managed bee colonies can have negative impacts on native bees (reviewed in Mallinger et al., 2017). In addition, it is clear that conservation of pollinators will have to be driven by forces beyond the crop pollination services these species provide (Kleijn et al., 2015). But if honey bee awareness could be the gateway to broader awareness of and interest in wild bees, other insect pollinators, and invertebrates more broadly, it could have significant beneficial consequences for the conservation of these species. For example, in March of 2017, the rusty-patched bumblebee became the first bumblebee, and only the eighth bee, to be protected under the Endangered Species Act. It is highly probable that increased awareness of the challenges facing honey bees and the importance of pollination may have contributed to this somewhat remarkable event.

To incorporate ideas of conservation beyond honey bees into this class, we are now revisiting the strategies from Table 8.1. First, when framing the class I now expand the topics to include wild bees and the challenges facing them by including additional reading about them (e.g., Klein et al., 2018). Second, when establishing the emphasis of the class we hope to include a second 'field' trip to the green roof located on the Diana Center here at Barnard. Wild bees are abundant in New York City (Matteson et al., 2008) and they are likely to be on the Barnard Campus on the first day of class in September. So, our first class might include a 'bee hunt' (Mueller & Pickering, 2010) in addition to an apiary tour. To make this exercise more meaningful, I plan to have students photo-document any pollinators they see for subsequent identification. Documentation of various organisms including pollinators has been part of multiple "citizen science" initiatives in recent years. One of the most successful has been the "bioblitz", which involves students and volunteer experts attempting to document and identify all the species in a given area over a specific time period (Ballard et al., 2017; also see O'Donnell & Brundage, this volume). While a true bioblitz is far beyond the scope of this class, documenting pollinating insects on our green roof over a 2-h period would work well. While data from citizen science initiatives have been shown to be somewhat incomplete (Kremen et al., 2011), the impact of such programs on understanding and appreciation of species diversity and conservation are apparent (Mitchell et al., 2017). Because we do not have experts in entomology available to help with species identification in this class, we can employ a widely used crowdsourcing site, iNaturalist (www.inaturalist.org, McKinley et al., 2017) for this task. iNaturalist is described as a "crowdsourced species identification system and an organism occurrence recording tool" and allows for additional time and map data to be recorded with observations. As proof of principle, I used the site to identify the picture of the bee in Fig. 8.6, taken on the Barnard campus in late July of 2017.

Fig. 8.6 Brown-belted bumblebee, Bombus griseocollis on Barnard's campus. (Photo credit: Rachel Snow)

The bee was identified by a number of experts as a brown-belted bumblebee, *Bombus griseocollis*, which is native to much of the United States except for the Southwest. It is my hope that even this limited exposure to wild bees and the unique challenges faced by them will shift the attitudes of the students in my class towards appreciation and conservation.

8.4   Full Circle from Teaching to Research

One of the many delights of being an educator is learning from your students or your own teaching. Thinking about how to incorporate sustainability into my Laboratory in Cell Biology class forced me to consider pollinators other than honey bees more than I had ever done in the past. Like with honey bee losses, native bee decreases are likely caused by a complex set of interacting stresses (Goulson et al., 2015). Yet, from a research perspective, especially at the cell and molecular level, we have, by orders of magnitude, more information about honey bees than we do about a given wild bee species (Trapp et al., 2017). One metric of the quantity of research on an organism or group of organisms is the number of publications about them. As an example, looking at specific bees, as of 2021 there are 4 scholarly publications in Web of Science about the bee species *Megachile centuncularis,* which is native to New York City. There are 795 publications on all bees of the genus *Megachile.* By contrast there are 4997 publications about the European honey bee *Apis mellifera* and 9521 publications on bees from the genus *Apis* for the last 5 years alone!

Another indicator of the degree to which an organism is being studied at the molecular and cellular level is whether its genome has been sequenced. Of the 20,000 bee species, we have genomes sequenced for only 15. I have in no way given up my interest in or commitment to honey bees. However, I hope to think about these other bee groups and be more inclusive when framing my own research questions and drawing conclusions. I have now expanded the organisms I study beyond honey bees to some wild bee species. As mentioned above, solitary bees of the genus *Megachile* have been studied (Pitts-Singer & Cane, 2011; James & Pitts-Singer, 2013) and tools have been developed for studying them (Fischman et al., 2017; Trapp et al., 2017). There is even some preliminary research looking at thermal stress and the immune response in these bees (Xu & James, 2012), which overlaps with the kind of cell stress pathways we typically explore in honey bees (McKinstry et al., 2017). We have now begun working with *M. rotundata* to better understand how what we learn in honey bees might apply to other bee species.

8.5   Conclusions and Reflections

Honey bees are critical pollinators for agriculture and are facing health problems of complex origin. From a research perspective, understanding them at the cellular level will be important in helping to improve their well-being and survival, with critical benefits to human nutrition. As a teacher, I am responsible for many courses focused on biology at the cell and molecular level. Biology is a continuum of different levels of organization, all of which are important and critically intertwined. Because of this, cell biology as a discipline can be framed in a variety of contexts. Although the default context is typically medicine, connecting the theory and practice of cell biology with the broader environmental issues of honey bee disease and the growing pollinator crisis is promising. Furthermore, I believe I can use the widespread appeal of honey bees to encourage students' interest in more bee species, other pollinators, and invertebrates in general with positive impacts on conservation and sustainability. Student feedback revealed that the sustainability aspects of the class were very well received, but also provided some constructive suggestions for changes that will be the basis of future improvement to the class.

References

Adames, T. R., Rondeau, N. C., Kabir, M. T., Johnston, B. A., Truong, H., & Snow, J. W. (2020). The IRE1 pathway regulates honey bee unfolded protein response gene expression. *Insect Biochemistry and Molecular Biology, 121*, 103368. https://doi.org/10.1016/j.ibmb.2020.103368

Auchincloss, L. C., Laursen, S. L., Branchaw, J. L., Eagan, K., Graham, M., Hanauer, D. I., et al. (2014). Assessment of course-based undergraduate research experiences: A meeting report. *CBE Life Sciences Education, 13*(1), 29–40. https://doi.org/10.1187/cbe.14-01-0004

Bach, D. M., Holzman, M. A., Wague, F., Miranda, J. L., Lopatkin, A. J., Mansfield, J. H., Snow, J. W. (2021). Thermal stress induces tissue damage and a broad shift in regenerative signaling pathways in the honey bee digestive tract. *Journal of Experimental Biology, 224*(18), jeb242262. https://doi.org/10.1242/jeb.242262

Ballard, H. L., Robinson, L. D., Young, A. N., Pauly, G. B., Higgins, L. M., Johnson, R. F., & Tweddle, J. C. (2017). Contributions to conservation outcomes by natural history museum-led citizen science: Examining evidence and next steps. *Biological Conservation, 208*(C), 87–97. https://doi.org/10.1016/j.biocon.2016.08.040

Balmford, A., Clegg, L., Coulson, T., & Taylor, J. (2002). Why conservationists should heed Pokemon. *Science, 295*(5564), 2367–2367. https://doi.org/10.1126/science.295.5564.2367b

Banks, J. A., & Banks, C. A. M. (2016). *Multicultural education: Issues and perspectives* (9th ed.). Wiley Global Education.

Burkle, L. A., Marlin, J. C., & Knight, T. M. (2013). Plant-pollinator interactions over 120 years: Loss of species, co-occurrence, and function. *Science, 339*(6127), 1611–1615. https://doi.org/10.1126/science.1232728

Cooper, K. M., Soneral, P. A. G., & Brownell, S. E. (2017). Define your goals before you design a CURE: A call to use backward design in planning course-based undergraduate research experiences. *Journal of Microbiology & Biology Education, 18*(2), 1–7. https://doi.org/10.1128/jmbe.v18i2.1287

Ducarme, F., Luque, G. M., & Courchamp, F. (2013). What are "charismatic species" for conservation biologists? *BioSciences Master Reviews.* https://doi.org/10.2307/24475553?refreqid=search-gateway:dbf404cb9df9dcf4f6058cbe21657151

Fischman, B. J., Pitts-Singer, T. L., & Robinson, G. E. (2017). Nutritional regulation of phenotypic plasticity in a solitary bee (Hymenoptera: Megachilidae). *Environmental Entomology, 46*(5), 1070–1079. https://doi.org/10.1093/ee/nvx119

Flores, M. E., McNamara-Bordewick, N. K., Lovinger, N. L., Snow, J. W. (2021). Halofuginone triggers a transcriptional program centered on ribosome biogenesis and function in honey bees. *Insect Biochemistry and Molecular Biology, 139*. 103667. https://doi.org/10.1016/j.ibmb.2021.103667

Garibaldi, L. A., Steffan-Dewenter, I., Winfree, R., Aizen, M. A., Bommarco, R., Cunningham, S. A., et al. (2013). Wild pollinators enhance fruit set of crops regardless of honey bee abundance. *Science, 339*(6127), 1608–1611. https://doi.org/10.1126/science.1230200

Garibaldi, L. A., Carvalheiro, L. G., Vaissière, B. E., Gemmill-Herren, B., Hipólito, J., Freitas, B. M., et al. (2016). Mutually beneficial pollinator diversity and crop yield outcomes in small and large farms. *Science, 351*(6271), 388–391. https://doi.org/10.1126/science.aac7287

Geldmann, J., & González-Varo, J. P. (2018). Conserving honey bees does not help wildlife. *Science, 359*(6374), 392–393. https://doi.org/10.1126/science.aar2269

Goulson, D., Nicholls, E., Botías, C., & Rotheray, E. L. (2015). Bee declines driven by combined stress from parasites, pesticides, and lack of flowers. *Science, 347*(6229), 1255957. https://doi.org/10.1126/science.1255957

Hallmann, C. A., Sorg, M., Jongejans, E., Siepel, H., Hofland, N., Schwan, H., et al. (2017). More than 75 percent decline over 27 years in total flying insect biomass in protected areas. *PLoS One, 12*(10), e0185809–e0185821. https://doi.org/10.1371/journal.pone.0185809

Hetz, C., Zhang, K., & Kaufman, R. J. (2020). Mechanisms, regulation and functions of the unfolded protein response. *Nature Reviews Molecular Cell Biology, 21*, 421–438. https://doi.org/10.1038/s41580-020-0250-z

James, R. R., & Pitts-Singer, T. L. (2013). Health status of alfalfa leafcutting bee larvae (hymenoptera: Megachilidae) in United States alfalfa seed fields. *Environmental Entomologist, 42*(6), 1166–1173. https://doi.org/10.1603/EN13041

Johnston, B. A., Hooks, K. B., McKinstry, M., & Snow, J. W. (2016). Divergent forms of endoplasmic reticulum stress trigger a robust unfolded protein response in honey bees. *Journal of Insect Physiology, 86*, 1–10. https://doi.org/10.1016/j.jinsphys.2015.12.004

Kleijn, D., Winfree, R., Bartomeus, I., Carvalheiro, L. G., Henry, M., Isaacs, R., et al. (2015). Delivery of crop pollination services is an insufficient argument for wild pollinator conservation. *Nature Communications, 6*(1), 550–559. https://doi.org/10.1038/ncomms8414

Klein, A.-M., Boreux, V., Fornoff, F., Mupepele, A.-C., & Pufal, G. (2018). Relevance of wild and managed bees for human well-being. *Current Opinion in Insect Science, 26*, 82–88. https://doi.org/10.1016/j.cois.2018.02.011

Kremen, C. (2018). The value of pollinator species diversity. *Science, 359*(6377), 741–742. https://doi.org/10.1126/science.aar7614

Kremen, C., Ullmann, K. S., & Thorp, R. W. (2011). Evaluating the quality of citizen-scientist data on pollinator communities. *Conservation Biology, 25*(3), 607–617. https://doi.org/10.2307/27976507

Lawson, L. (2016). Agriculture: Sowing the city. *Nature, 540*, 522–523. https://doi.org/10.1038/540522a

Lee, K. V., Partnership, F. T. B. I., Steinhauer, N., Rennich, K., Wilson, M. E., Tarpy, D. R., et al. (2015). A national survey of managed honey bee 2013–2014 annual colony losses in the USA. *Apidologie, 46*(3), 292–305. https://doi.org/10.1007/s13592-015-0356-z

Linn, M. C., Palmer, E., Baranger, A., Gerard, E., & Stone, E. (2015). Undergraduate research experiences: Impacts and opportunities. *Science, 347*(6222), 1261757–1261757. https://doi.org/10.1126/science.1261757

Mallinger, R. E., Gaines-Day, H. R., & Gratton, C. (2017). Do managed bees have negative effects on wild bees?: A systematic review of the literature. *PLoS One, 12*(12), e0189268–e0189232. https://doi.org/10.1371/journal.pone.0189268

Matteson, K. C., Ascher, J. S., & Langellotto, G. A. (2008). Bee richness and abundance in New York City urban gardens. *Annals of the Entomological Society of America, 101*(1), 140–150. https://doi.org/10.1603/0013-8746(2008)101[140,BRAAIN]2.0.CO;2

McKinley, D. C., Miller-Rushing, A. J., Ballard, H. L., Bonney, R., Brown, H., Cook-Patton, S. C., et al. (2017). Citizen science can improve conservation science, natural resource management, and environmental protection. *Biological Conservation, 208*(C), 15–28. https://doi.org/10.1016/j.biocon.2016.05.015

McKinstry, M., Chung, C., Truong, H., Johnston, B. A., & Snow, J. W. (2017). The heat shock response and humoral immune response are mutually antagonistic in honey bees. *Scientific Reports, 7*(1), 8850. https://doi.org/10.1038/s41598-017-09159-4

McNamara-Bordewick, N. K., McKinstry, M., & Snow, J. W. (2019). Robust transcriptional response to heat shock impacting diverse cellular processes despite lack of heat shock factor in microsporidia. *mSphere, 4*(3). https://doi.org/10.1128/mSphere.00219-19

Michener, C. D. (2000). *The bees of the world* (2nd ed.). Johns Hopkins University Press.

Mitchell, N., Triska, M., Liberatore, A., Ashcroft, L., Weatherill, R., & Longnecker, N. (2017). Benefits and challenges of incorporating citizen science into university education. *PLoS One, 12*(11), e0186285–e0186215. https://doi.org/10.1371/journal.pone.0186285

Mohai, P., Pellow, D., & Roberts, J. T. (2009). Environmental justice. *Annual Review of Environment and Resources, 34*, 405–430. https://doi.org/10.1146/annurev-environ-082508-094348

Mueller, M. P., & Pickering, J. (2010). Bee hunt! Ecojustice in practice for Earth's buzzing biodiversity. *Science Activities, 47*(4), 151–159. https://doi.org/10.1080/00368121003631637

O'Donnell, K., & Brundage, L. (this volume). It turned into a BioBlitz: Urban data collection for understanding and connection. In M. S. Rivera Maulucci, S. Pfirman, & H. S. Callahan (Eds.), *Transforming sustainability research and teaching: Discourses on justice, inclusion, and authenticity*. Springer.

Ollerton, J. (2017). Pollinator diversity: Distribution, ecological function, and conservation. *Annual Review of Ecology, Evolution, and Systematics, 48*(1), 353–376. https://doi.org/10.1146/annurev-ecolsys-110316-022919

Pitts-Singer, T. L., & Cane, J. H. (2011). The Alfalfa Leafcutting Bee, *Megachile rotundata*: The world's most intensively managed solitary bee. *Annual Review of Entomology, 56*(1), 221–237. https://doi.org/10.1146/annurev-ento-120709-144836

Potts, S. G., Imperatriz-Fonseca, V., Ngo, H. T., Aizen, M. A., Biesmeijer, J. C., Breeze, T. D., et al. (2016). Safeguarding pollinators and their values to human well-being. *Nature, 540*(7632), 220–229. https://doi.org/10.1038/nature20588

Powers, E. T., & Balch, W. E. (2013). Diversity in the origins of proteostasis networks—A driver for protein function in evolution. *Nature Reviews Molecular Cell Biology, 14*(4), 237–248. https://doi.org/10.1038/nrm3542

Pyle, R. M. (1993). *The thunder tree*. Oregon State University Press.

Rivera Maulucci, M. S. (this volume). Pathways to education for sustainability: Examples from science teacher education. In M. S. Rivera Maulucci, S. Pfirman, & H. S. Callahan (Eds.), *Education for sustainability: Discourses on justice, inclusion, and authenticity*. Springer.

Rodenbusch, S. E., Hernandez, P. R., Simmons, S. L., & Dolan, E. L. (2016). Early engagement in course-based research increases graduation rates and completion of science, engineering, and mathematics degrees. *CBE Life Sciences Education, 15*, ar20. https://doi.org/10.1187/cbe.16-03-0117

Rule, A. C. (2006). The components of authentic learning. *Journal of Authentic Learning, 3*(1), 1–10.

Samali, A., FitzGerald, U., Deegan, S., & Gupta, S. (2010). Methods for monitoring endoplasmic reticulum stress and the unfolded protein response. *International Journal of Cell Biology, 2010*(11), 1–11. https://doi.org/10.1074/jbc.M505784200

Schneider, I. (1972). Cell lines derived from late embryonic stages of Drosophila melanogaster. *Journal of Embryology and Experimental Morphology, 27*(2), 353–365.

Shih, S. R., Huntsman E. M., Flores, M. E., Snow J. W. (2020). Reproductive potential does not cause loss of heat shock response performance in honey bees. *Scientific Reports 10*(1), 19610 https://doi.org/10.1038/s41598-020-74456-4

Shih, S. R., Bach, D. M., Rondeau, N. C., Sam, J., Lovinger, N. L., Lopatkin, A. J., Snow, J. W. (2021). Honey bee sHSP are responsive to diverse proteostatic stresses and potentially promising biomarkers of honey bee stress. *Scientific Reports, 11*(1), 22087 https://doi.org/10.1038/s41598-021-01547-1

Smith, M. R., Singh, G. M., Mozaffarian, D., & Myers, S. S. (2015). Effects of decreases of animal pollinators on human nutrition and global health: A modelling analysis. *Lancet, 386*(10007), 1964–1972. https://doi.org/10.1016/S0140-6736(15)61085-6

Snow, J., & Rivera Maulucci, M. S. (2017). You can give a bee some water, but you can't make her drink: A socioscientific approach to honey bees in science education. In *Animals and science education* (4 ed., Vol. 2, pp. 15–28). Springer. https://doi.org/10.1007/978-3-319-56375-6_2

Soga, M., & Gaston, K. J. (2016). Extinction of experience: The loss of human-nature interactions. *Frontiers in Ecology and the Environment, 14*(2), 94–101. https://doi.org/10.1002/fee.1225

Taylor, R. C., Berendzen, K. M., & Dillin, A. (2014). Systemic stress signaling: Understanding the cell non-autonomous control of proteostasis. *Nature Reviews Molecular Cell Biology, 15*(3), 211–217. https://doi.org/10.1038/nrm3752

Trapp, J., McAfee, A., & Foster, L. J. (2017). Genomics, transcriptomics and proteomics: Enabling insights into social evolution and disease challenges for managed and wild bees. *Molecular Ecology, 26*(3), 718–739. https://doi.org/10.1111/mec.13986

Vabulas, R. M., Raychaudhuri, S., Hayer-Hartl, M., & Hartl, F. U. (2010). Protein folding in the cytoplasm and the heat shock response. *Cold Spring Harbor Perspectives in Biology, 2*(12), a004390–a004390. https://doi.org/10.1101/cshperspect.a004390

Walter, P., & Ron, D. (2011). The unfolded protein response: From stress pathway to homeostatic regulation. *Science, 334*(6059), 1081–1086. https://doi.org/10.1126/science.1209038

Wikipedia contributors. (2018, April 13). *Cell biology*. Retrieved 27 Apr 2018.

Wilson, E. O. (1984). *Biophilia*. Harvard University Press.

Winfree, R., Bartomeus, I., & Cariveau, D. P. (2011). Native pollinators in anthropogenic habitats. *Annual Review of Ecology, Evolution, and Systematics, 42*(1), 1–22. https://doi.org/10.1146/annurev-ecolsys-102710-145042

Winfree, R., Reilly, J. R., Bartomeus, I., Cariveau, D. P., Williams, N. M., & Gibbs, J. (2018). Species turnover promotes the importance of bee diversity for crop pollination at regional scales. *Science, 359*(6377), 791–793. https://doi.org/10.1126/science.aao2117

Xu, J., & James, R. R. (2012). Temperature stress affects the expression of immune response genes in the alfalfa leafcutting bee, Megachile rotundata. *Insect Molecular Biology, 21*(2), 269–280. https://doi.org/10.1111/j.1365-2583.2012.01133.x

Jonathan Snow is an Associate Professor of Biology at Barnard College. His research focuses on the cellular stress responses of the honey bee, a species that is crucial to agricultural and ecological systems and the infection of honey bees by the microsporidia species *Nosema ceranae*. He teaches courses in molecular and cellular biology.

Chapter 9
Finding the Most Important Places on Earth for Birds

Terryanne Maenza-Gmelch

9.1 Birds as Gauges of Changes in the Environment

Bird song is a signature of the arrival of spring and it can be used as a gauge of ecosystem health. In Rachel Carson's *Silent Spring* (1962), she writes, "And as the number of migrants grows and the first mists of green appear in the woodlands, thousands of people listen for the first dawn chorus of robins throbbing in the early morning light. But now all is changed, and not even the return of birds may be taken for granted." Looking and listening are ways we can observe the ecology around us. However, with the distractions of daily life, we may be more concerned with politics, writing book chapters, and deciding on what to cook for dinner, than listening for birds or thinking about birds as monitors of the environment. With these preoccupations in place, how can we convince ourselves, or others, to make time to enjoy birds and their leafy habitats which would naturally lead to concern for their well-being? In this chapter, the mutualistic relationship between birds and humans is explored. Perhaps we will learn to care about birds after we gain an understanding of the beneficial roles of birds in our lives, neighborhoods, and the world.

Observing species in nature is a good way to monitor issues of sustainability, including the impact of changes in climate, land use, and the way humans use resources. Birds, in particular, are commonly regarded as monitors of environmental change. You can make casual observations of birds, insects, plants, *etc.*, during hiking and gardening, among other things. The next level is to include where and when you made these observations, an important step in making the data valuable and worthy of contribution to scientific databases that scientists use such as eBird or iNaturalist apps. This is community science and bird watchers are active community scientists and therefore able to contribute to data-driven conservation activities.

T. Maenza-Gmelch (✉)
Department of Environmental Science, Barnard College, New York, NY, USA
e-mail: tmaenzag@barnard.edu

© The Author(s) 2023   147
M. S. Rivera Maulucci et al. (eds.), *Transforming Education for Sustainability*,
Environmental Discourses in Science Education 7,
https://doi.org/10.1007/978-3-031-13536-1_9

Birds are responsive, ubiquitous, easily surveyed, and have interesting ecological requirements. Noticing which birds are where and their relative abundances has great value and is the basis for important contributions to conservation policy by the general public as well as scientists. We know that birds like the Golden-winged Warbler are declining in part because of land-use issues such as habitat destruction and the natural succession of shrubby habitat to the forest (Confer et al., 1992). Carolina Wrens in the Northeast are expanding northward in range in response to climate change (Hitch & Leberg, 2007). Both nicely illustrate how birds can be monitors of change in the environment. In Rachel Carson's *Silent Spring* (1962), some of her case studies focused on declining bird populations in general and, specifically, the discoveries of dead birds in areas just after marshes and other habitats were sprayed with insecticides. This drew attention to and highlighted the use of birds as sentinels of environmental health and the interconnectedness of nature in general.

Historically, bird watchers and academics have been collecting bird presence and abundance data for over one hundred years, beginning with the first Audubon Christmas Bird Count in 1900. In spring, Breeding Bird Surveys have been conducted since 1966 across Canada and the United States to monitor bird populations in the region with the goal of applying the data to conservation. The Breeding Bird Surveys were started in response to the profligate use of pesticides and the concern that these chemicals posed threats to wildlife and humans. Trends revealed in bird monitoring data have historically shown declines of charismatic birds such as the Peregrine Falcon, the Osprey, and the Bald Eagle due to the pesticide DDT. These observations prompted actions, such as banning DDT and reintroducing the species. As a result, these species of birds have recovered.

In another example, a concerned group of women from Boston in the late 1800s brought attention to the decline of various wading birds due to the unfortunate use of the long and ornate feathers from species such as the Great Egret. The population of these large, white, wading birds was shrinking because they were collected and killed for their breeding plumes which were in high demand to decorate fashionable hats worn at the time. This episode of activism was the stimulus for the formation of the first chapter of the Audubon Society (Weidensaul, 2008, p. 157) and legislation to protect these species. Today, New York City Audubon monitors Great Egret populations in New York Harbor and the wetlands throughout the New York metro area where the birds are successfully breeding.

Currently, birders are entering their observational data directly into their phones while in the field using the ebird app (Ebird, 2021) which is a portal developed by Cornell University for collecting georeferenced bird data, a powerful tool that has great conservation value. Ebird allows visualizations of where birds are, their movements, and their abundances. One of my favorite products from eBird is the animated migration maps where entire northbound and southbound routes across the Western Hemisphere are shown for 118 avian species (Powell, 2016). The bird and habitat conservation that is informed by these data collection efforts aligns nicely with the Life on Land United Nations Sustainable Development Goal 15, which specifically states, "Protect, restore and promote sustainable use of terrestrial ecosystems, sustainably manage forests, combat desertification, and halt and reverse land degradation and halt biodiversity loss" (United Nations, 2021).

9.2 Our Relationship with Birds

In a mutualistic ecological relationship, the interacting species both benefit in that each provides a helpful service for the other. Some of the more familiar mutualism examples are mycorrhizae (the connection between plant root tips and fungal hyphae) where sugars made in photosynthesis, water, and minerals are shared between plants and fungi. And, of course, there are various pollination mutualisms. What could birds possibly do for us? What can we do for birds? There are some fundamental and beneficial conditions and commodities that are provided to humans that are due to the workings of the bird community. Drinking water quality, coffee bean quality, and seed dispersal are just a few of these services. Humans need to do more in return.

9.2.1 Bird Stories: Purification of Drinking Water

Let's say you are not moved to care that much about a bird's well-being. You can frame the issue in another way…what do the birds do for you? What do you get out of it? One answer is clean drinking water. Before reading on, can you assemble the puzzle pieces in your mind that support this truth? Here is a clue: think watersheds, as seen in Fig. 9.1.

In a forested watershed, trees – and the soil they are rooted in – purify drinking water. As rain falls, trees intercept precipitation which infiltrates the soil where tree roots and soil microbes remove impurities. When you turn on the tap in New York City, the high-quality water originated from and was purified in forested watersheds in upstate New York. What keeps the forests healthy? This is where birds fit in. Birds keep forests healthy through their consumption of forest insect pests. As a

Fig. 9.1 Students at Black Rock Forest exploring the area near Jim's Pond, a nice example of a forested watershed

matter of fact, in spring, migratory birds from the tropics head north through eastern North America following a green wave of tree leaf bud bursts. Along with those bud bursts, large quantities of leaf-eating caterpillars also hatch. Birds eat them as they move north. In this mutualism, trees provide bird habitat and the birds provide insect control. It has been shown that forest birds can reduce the defoliation of forest trees by 50% by eating these caterpillars (Marquis & Whelan, 1994). In their study, Marquis and Whelan wrapped trees with mesh that excluded birds but allowed insects free access. Unwrapped trees served as the control that birds and insects could get to. The wrapped trees experienced more biomass loss (leaves were eaten by the insects) compared to unwrapped trees, thus illustrating how birds keep herbivores (in this case – caterpillars) under control. Therefore, you can thank a bird for your clean drinking water, especially if you are in New York City.

9.2.2   Bird Stories: Coffee

To further the mission of convincing readers to care about birds and their habitat, let's talk about coffee. What can get more personal than that? There are two broad categories for coffee: sun-grown and shade-coffee. Sun-grown coffee is an ecological disaster. To grow it, tropical forests must be cut down and a monoculture of sun-grown coffee is planted. As is typical of a monoculture, disease can spread fast, so pesticides are applied. The result is tropical forest habitat loss with a reduction of associated biodiversity and a loss of forest ecosystem services. Plus, there is chemical contamination of the crop and watershed and dangerous exposure to chemicals for farmers. In comparison, shade-grown coffee is sustainable. The forest is not removed, coffee plants are grown in or near the understory. Since the structure of the forest is not damaged, the associated animals, for example, birds, do not move out. As a matter of fact, the birds keep the coffee plants healthy by eating coffee berry pests. This, in turn, reduces or eliminates the need for pesticides, decreases watershed contamination, and keeps farmers safer. Research has shown that birds reduce berry infestation and can do so up to 50% in agroforestry settings (Kellermann et al., 2008; Martínez-Salinas et al., 2016). This translates to an increase in coffee yield and farm income, a financial incentive for growers to choose shade-grown coffee. You can thank several species of warblers and vireos and other avian species for this service. Many of these birds are the same birds that migrate to North America in the spring and summer, for example, the American Redstart and the Ovenbird. In the winter they are in the tropics carrying out various ecosystem services like keeping coffee plants healthy. The bird-coffee berry story is another tangible example of ecological mutualism and why we should care about birds. Students can connect with these issues and get enthusiastic. Now that coffee woke everyone up, it is time to talk about migration.

9.2.3 Bird Stories: Migration of the Red-Eyed Vireo

Anyone that has spent time in eastern North American forests has heard the incessant song of the Red-eyed Vireo, consciously or not (check out Red-eyed vireo sounds, at https://www.allaboutbirds.org/, Cornell Lab of Ornithology). If you stop to look, you will see this lovely common forest songbird, with its black and white eyebrow and red eye, and hopefully wonder about its story. This species migrates to our eastern forests in spring and summer to pair up with a mate and raise young among lush trees and ample food. Many other species, like warblers and thrushes, *etc.*, do this too. We pretty much know exactly where in the tropics the Red-eyed Vireo spent its winter before coming here, which route it took to get here and where it stopped to rest. This has been made possible with geolocators (Stutchbury et al., 2009) which are lightweight devices that are attached to (and later retrieved from) the birds. These instruments record how light levels change along their journeys, over time, which translates to latitude and longitude information. For example, one particular Red-eyed Vireo left northwest South America on April 6, stopped over in Central America on April 29 for about a week, crossed the Gulf of Mexico in one night on May 8, rested in vegetation on the Gulf Coast from May 9 to 12 and finally arrived in Pennsylvania on May 18 (Callo et al., 2013). What was your immediate reaction to this information? I think it is amazing that these birds can do this and survive the trip and that we know the details. Many of the birds that cross the Gulf do not make it, especially if a rainstorm with winds from the north develops during their crossing. My reaction is – if they are coming all this way, it is our responsibility to protect the habitat here for them.

9.3 What Do Birds Need from Us?

Birds need us to protect their habitats. An important and ecologically supportive place for birds is one that has a structurally complex layering of vegetation: canopy, understory, shrubs, herbs, and ground layers – in other words, a diverse and inclusive environment. The location should have access to water (streams and ponds) and be large in area while being connected to other large patches of vegetation that have different characteristics due to differences in topography and/or land-use history. In short, healthy habitat for birds is an expanse of land that is a mosaic of habitats analogous to the squares in a quilt. The more diverse the habitats in the landscape the more "living" options for the birds and therefore the higher the bird diversity. When areas like this exist and the resident bird communities have been documented by bird clubs, conservationists, and academic groups, it is a perfect opportunity for humans to participate in their part of mutualism: land conservation. Humans have the responsibility to preserve this land from disturbances and degradation. Figure 9.2 shows some species found in northeastern North America.

Fig. 9.2 Eastern Screech Owl, Ovenbird, and Scarlet Tanager at Black Rock Forest

Let's get back to my earlier comment – if they (birds) are migrating all this way to spend a portion of their life cycles in North America, we had better protect the forests here for them, as well as forests in the tropics. Recall the long and potentially dangerous journey of the red-eyed vireo described above. Upon arrival, these birds need a habitat for resting and eating and, eventually, reproduction.

9.3.1 How to Identify Ecologically Important Places for Birds

Identifying ecologically important places for birds is what I do. Students and I have identified some of these important places and our work has successfully informed land conservation policy for the Audubon New York Important Bird Area (IBA) program (Audubon New York, n.d.). Through the IBA program, the most important places on Earth for birds are identified and conserved. IBAs have been identified throughout the United States. New York state has well over 100 IBAs alone. When nominating a region for IBA designation, there are three key components that must be investigated for the application. The first is whether there are significant populations of "listed" species. A listed species is a bird that in previous surveys has been shown to be declining in numbers and is therefore of conservation concern. Examples of some of the listed species in New York state forests are the Wood Thrush, Prairie Warbler, Cerulean Warbler, and Scarlet Tanager. The second criterion considered in the IBA application is the presence of all or most of the responsibility species assemblage. This is an assemblage of species that together typifies or represents a particular habitat. For example, an eastern deciduous forest is represented by various species of hawks, thrushes, vireos, warblers, and flycatchers, *etc.* The third requirement is large numbers of migratory birds.

Barnard students assisted me and colleagues from other institutions in the effort to collect data for the Black Rock Forest IBA application. Surveys revealed that the proposed area met all three criteria of the IBA program: significant populations of "listed" species, presence of all or most of the responsibility species, and large numbers of migratory birds. In particular, thresholds of key individual birds such as

Cerulean Warbler, Worm-eating Warbler, Wood Thrush, Blue-winged Warbler, and Prairie Warbler were well above the required levels. As a result of our research, Black Rock Forest was awarded IBA designation in June 2016 by Audubon New York (Maenza-Gmelch et al., 2016). Thus, to hold up the human end of our responsibility to birds and provide them with ecojustice, an inclusive set of circumstances is required: habitat diversity, organized research, authentic citizen input, organizational approval, as well as local management – described below.

9.4 Black Rock Forest

What is Black Rock Forest? Black Rock Forest is a 3914 acre, eastern deciduous forest field station that Barnard College, and other member institutions, have the opportunity to use for research and education endeavors. It is a gem of a field station because of its natural beauty and amazing staff. Black Rock Forest is about one hour north of Manhattan and is in the Hudson Highlands of New York State. The forest is part of a northeast-southwest trending corridor of preserves and parks that has tremendous wildlife and ecosystem service value. The region has high species and habitat diversity due to topographical sequences of ridgetops, slopes, ravines, lakes, and streams with more than 1000 feet of relief. Also contributing to this habitat diversity are chronological sequences featuring a mosaic of different aged forest patches within the forest matrix that is maintained through natural disturbance and thoughtful forest management. Figures 9.3 and 9.4 show such a diverse landscape. Vegetation structural complexity, a natural feature of a healthy deciduous forest, plays an important role in supporting a diverse animal population, especially if its integrity is maintained and enhanced through excellent timber and deer management plans, which is the case at Black Rock. A large contiguous area is also important and its almost 4000-acre size is comparable to more than four Central Parks.

Fig. 9.3 Example of the diverse landscape at Black Rock Forest as seen from the summit of Black Rock

Fig. 9.4 Examples of the plant diversity that arises from the topographical diversity in the forest. Shown here are the native shrubs: Pink Azalea, Choke Cherry, and Mountain Laurel

Fig. 9.5 Students measure herb diversity in an area of the forest that had been burned

9.4.1 Getting Students into the Field

My spring and summer research projects with students (Fig. 9.5) focus on the exploration of why Black Rock Forest supports a diverse bird community and investigation of possible threats. Questions considered have included: Where are the bird diversity hotspots in the forest? Is vegetation structural complexity important? Is plant diversity important? Is habitat heterogeneity important? What are the threats? Do roads and trails negatively impact bird communities? Is connectivity important? Can we avoid putting forest activities in the middle of biodiversity hotspots?

Our research plan started out by showing that unique habitats within the forest matrix support species that are not found in other parts of the forest, for example, the Chestnut-sided and Blue-winged Warblers in early successional patches, Acadian Flycatchers in hemlock ravines, and Cerulean Warblers in deciduous forest areas near streams. How did we do this? We developed a plan for authentic, direct observations. After dividing the forest into habitat types we did point-count bird surveys in each. In point-counts, you look and listen for birds for 5 or 10 min within a circle that has a 50-meter radius centered in the targeted habitat. You typically visit each survey site several times during the season. After surveys were completed, the data revealed that some birds were everywhere (generalists) and some birds were only found in certain habitats (specialists). For example, we found that Prairie Warblers are usually at higher elevation sites in the forest, such as the open tops of Mount Rascal and Black Rock Summit. Eastern Towhees like the shrubby areas typically found at Jim's Pond and by The Stone House. Other birds, such as the Red-eyed Vireo and the American Robin, seem to be everywhere.

In another study, we set out to determine how vegetation structural complexity, plant diversity, and habitat diversity (when the landscape has a mosaic of habitats – like the patches in a quilt) affect bird distribution. We began by surveying bird populations, again using point-counts, throughout the breeding season in six different habitats. Each habitat was visited several times spanning two breeding seasons for 5 or 10-min counts totaling 100 min at each habitat. The habitats were either structurally homogeneous or structurally heterogeneous. We conducted a vegetation survey with vertical and horizontal components for each habitat in order to calculate plant vertical density indices (as a measure of forest structural complexity) and plant species richness values (Fig. 9.6).

Our data suggest that structurally complex habitats at Black Rock Forest (BRF) support a higher diversity of birds than simpler ones. We also confirmed that landscape heterogeneity and habitat vertical complexity were positively correlated with bird species richness (Maenza-Gmelch & Gilly, 2015). The fifteen most abundant bird taxa during the breeding season at this location are American Robin, Red-eyed Vireo, American Crow, Veery, Cedar Waxwing, Common Yellowthroat, Gray Catbird, Chipping Sparrow, Scarlet Tanager, American Goldfinch, Great-crested Flycatcher, American Redstart, Baltimore Oriole, Yellow Warbler, and Eastern Towhee.

Another project was aimed at understanding the impacts of various road types and trails on bird abundance and diversity in order to inform land-use decisions in and near this forest for bird conservation purposes (Maenza-Gmelch & Wasmuth, 2017). A total of 692 detections were made of 57 bird species in this survey. Measures of bird species richness and diversity were not significantly different between paved roads, dirt roads, and trails. Since Black Rock Forest exists in a landscape matrix that is heavily forested in general, it is not surprising that the bird diversity and abundance near roads and trails may not differ significantly (Rodewald & Arcese, 2016). We plan to repeat this study with more sites and an expanded analysis that separates nesting and foraging guilds. Looking at nesting and foraging diversity, as opposed to just species diversity, in relation to habitat structure could reveal important relationships.

Fig. 9.6 Students use a vegetation complexity board to estimate vertical vegetation density

Conservation recommendations to preserve bird diversity (and biodiversity in general) at Black Rock Forest include: continue to maintain the mosaic of different aged forest patches within the forest matrix (burning at the Stone House, mowing at the reservoirs, *etc.*); continue maintaining/creating structural complexity during timber management and hemlock ravine restoration (planting white pine to replace hemlocks); continue to speak up about threats to the composition and configuration of landscapes around the forest; and continue with partnerships in land acquisition for the wildlife corridor (Anderson et al., 2012). The Wildlife Corridor Project, conducted by colleagues at the forest, aims to reveal how animals, such as bobcats and fishers, move around and use the land. Habitat fragmentation, such as land divided by highways, is a major threat (Tucker et al., 2018). The IBA designation, the Wildlife Corridor Project, and Conservation Easement status all serve as "clout" when proposing policy recommendations.

9.4.2   Data Collection Using the Soundscape

I am sitting in the forest writing this paragraph and I hear these sounds: ank-ank, peter-peter-peter, pee-ah-wee, and tea-kettle-ettle. Therefore, I know that a White-breasted Nuthatch, a Tufted Titmouse, an Eastern Wood Pewee, and a Carolina Wren are nearby. Since it is early August and these birds have been singing in the area since April, it is probable that they spent their breeding times in these woods.

Birders and academics have been birding by ear for a very long time. As a student, I began to learn bird songs by listening to <u>A Guide to Bird-song Identification</u>, a collection of bird sounds put together by Richard Walton and Robert Lawson. Along with their recordings, they offered a study tool which was an alphabetical list of phonetic phrases that were assigned to each bird song because they provided a memorable phrase that had the same cadence as the bird song. For a completely amusing moment, check out Lang Elliot's recording of the Eastern Towhee (The Music of Nature, 2010a) and see if you agree that this bird is saying "Drink your tea-hee-hee" and listen to his White-throated Sparrow recording and consider if you hear this bird say "Old Sam Peabody, Peabody." (The Music of Nature, 2010b).

Since students often join me to do bird projects at Black Rock Forest, I wanted to create a bird sound tutorial for the BRF website. With a small grant, I was able to hire and collaborate with an amazing audio archivist named Martha Fischer from the Macaulay Library at Cornell's Laboratory of Ornithology. Martha accompanied me to the forest one spring and brought along an impressive collection of sound recording equipment (Fig. 9.7). We set out before dawn to get in position to record general habitat soundscapes as well as specific bird songs in the Forest. We went to the bird hotspots that I knew of and to the unique habitats that harbored some avian specialists. On some occasions when students joined us, it was a challenge to keep perfectly quiet during recordings since everyone had the urge to swat at the mosquitoes that were landing on us. Keeping our feet completely still on the gravelly paths (which make a lot of noise when you move your feet) was also key. The result of this effort is the "Listen to the Forest" website (Listen to the Forest, 2021). This resource is used as a learning tool for students in my bird ecology classes and summer research students, as well as anyone else who wants to practice birding by ear in preparation for a walk-in the forest.

Fig. 9.7 The author is shown recording birds at Black Rock Forest

9.4.3   Land Protection Is Not the Only Issue We Need
to Address for Bird Conservation

Climate change, invasive species, and pollution are three additional threats that impact the well-being of all biodiversity, not just birds. Plants and animals are inter-dependent on each other due to their web-like links. Climate influences the timing of their natural life cycle events (phenology) like egg-laying, flowering time, and leaf fall. Thus climate change is predicted to impact the life cycles of plants and animals. Already there is a large and growing body of peer-reviewed research that reveals many of these impacts. One of the first studies, a rigorous meta-analysis by Terry Root and others (2003), showed that several groups of organisms (inverte-brates, plants, amphibians, and birds) are carrying out various natural life events several days earlier per decade in response to rising global temperatures. Each spe-cies responds to the changes in climate at a different pace resulting in the uncou-pling of interactions, such as those between pollinator and plant. For example, a plant might flower earlier and its pollinator may not arrive or hatch in time (Snow, this volume). This results in an ecological mismatch.

Over the last several years, much attention has been focused on the phenology data collected by Aldo Leopold and Henry David Thoreau. Not many people know that Thoreau kept careful field journals based on his natural history observations, many of which were precise recordings of when the plants near Walden Pond first flowered each spring. Ellwood et al. (2013) compared recent first flowering dates in Wisconsin and Massachusetts with Leopold's Wisconsin data from 1935 and Thoreau's Massachusetts flowering time data from 1852. Their analyses revealed that, at both locations, plants are flowering earlier now than in the past. This is important because of the potential to disrupt ecological interactions.

Plants are responding to climate change not only through changes in flowering time. Some plants cannot tolerate the new higher temperatures so they redistribute northward, if they can, via seed dispersal and successful establishment. Plants at their southern limit are stressed. Tree physiology research by Dr. Angelica Patterson at Black Rock Forest (2021) reveals that the various tree species, based on range distribution, experience different impacts on their photosynthetic effi-ciency (Patterson, 2021). Oaks, the most common tree in the Northeast, are less efficient at sequestering carbon which may weaken their ability to compete with migrant trees under a warming climate.

Plant migration results in new plant communities in which the interconnected insects and birds may or may not be able to tolerate the habitat changes. Audubon's Birds and Climate Change Report (Survival by Degrees, 2021) shows that "birds are on the move" as they keep up with shifting and shrinking habitat ranges. This report shows that 389 bird species are at risk of extinction. The scientists arrived at their conclusions by comparing United Nations estimates of climate change for 2050 and 2080 with Audubon's data from Christmas bird counts – based on citizen inputs – and spring breeding bird surveys. They created maps of current temperatures and associated bird geographical ranges. Then, they created maps with projected temperature changes and modeled how the birds would respond. The models indicate loss of ranges and shifting ranges which set the stage for new species interactions such as competition for limited resources or danger of predation. More recent research shows that bird abundance is declining, nearly three billion birds have been lost since 1970 and not just rare and threatened species, but also species that are common and widespread (Rosenberg et al., 2019). This is mainly due to habitat loss as well as climate change. In addition, insects, a very important part of the food web, are in decline. Habitat loss, invasive species, climate change, and chemical and light pollution are among the threats to insects (Wagner et al., 2021).

9.5 How Do I Engage and Prepare Students for Bird Projects?

In one of my classes at Barnard College called Bird, Plant, and Land-use Dynamics, I cover many topics in ecology and environmental science using bird and plant community ecology as the case studies. The students analyze content like ecosystem services, trophic mismatch, and invasion biology through lectures, readings, presentations, and field trips. They also "get trained" in bird identification by sight and by ear so that they can join my field projects later in the course to collect data for analysis and presentation. Some move on to a senior thesis in bird conservation. In this class, students take a journey themselves. The learning experience spans from January to May where they start by learning birds' names, they draw and color them, they memorize and are quizzed on bird songs and they hone their identification and surveying skills on winter days by observing the bird forms and behaviors in Riverside Park near campus. They read and present papers from the scientific literature. Then, the migratory birds arrive and the students are ready. Beginning at the end of March the class takes field trips around the region: Central Park, Jamaica Bay, The Great Swamp, Black Rock Forest, and Sterling Forest (Fig. 9.8). Here they meet the species they have been studying and observe their interactions at each habitat. At all sites, we are interested in observing and quantifying the relationship between bird diversity and habitat structural complexity. Students collect data on birds and vegetation, plot and analyze the findings and write reports. I typically bridge my research with the curriculum in my classes by having students assist with authentic data collection and/or use previous results as problem sets and case studies. These data are contributed to long-term databases and can also be used for projects by future students. Most years we present at one or more conferences and my collaborators are often students.

After discussing, anticipating the onset of, observing, and quantifying spring migration for several weeks during the semester, the final class assignment is posted: go to the top of the Empire State Building at night to count migrating birds. Specifically, go to the 86th-floor observation deck on an evening in late April or early May. Select an evening as late in the spring semester as possible, one close to the full moon, with a wind from the south or southwest. Be sure it is an evening with good

Fig. 9.8 Students on a class trip to Jamaica Bay National Wildlife Refuge in Queens, New York. Jamaica Bay is one of New York's most famous Important Bird Areas

visibility and pick a night when the building's lighting schedule calls for "white" lighting. Arrive just after sunset. Migration will start about one hour after sunset. Purpose: to document and count night-flying migratory songbirds as they head to their northern breeding grounds after spending the winter in the tropics. With all of these conditions required, it is a challenge to organize. However, on a visit one late April evening, all of the conditions were optimal. I positioned 9 students and myself 1050 feet above street level and waited. There was a noticeable breeze from the south and the sky was clear. We all faced south looking upward in the direction of lower Manhattan along the Hudson River, even the Statue of Liberty could be spotted. At 8:45 pm the first birds flew over, with 200 in the first 15 min. Then, the pace picked up. Heading north and passing just over the peak of the building at 1472 feet, another 261 birds were counted in a 15-min period. Students estimated that 800–1000 north-bound migratory birds flew over the building per hour that night when conditions were right. The birds cannot be identified under these conditions, just size class can be determined, for example, songbird, heron, raptor, *etc*. During a trip to Sterling Forest (40 miles north of the Empire State Building) 2 days later, the class found numerous newly arrived migrant birds like Wood Thrush, Prairie Warbler, Scarlet Tanager, and Great-crested Flycatcher. These species could possibly have been among the flocks that were flying over the Empire State Building.

As the semester closes, if a student was not already a bird conservation sympa-thizer, they are now. In course evaluations, students report increased awareness and tuning in to a more inclusive appreciation of the environment around them through bird songs. One student wrote, "The course has taught me to appreciate the birds and trees around me. Sometimes I pay careful attention when I walk outside or when I am sitting near my window and I am surprised to find myself listening and spotting many of the birds we learned about." Another wrote, "I have found a new appreciation of birds. I've found myself going more often to Riverside Park and trying to test my knowledge and notice the wildlife around me in ways I didn't before. These are skills I want to continue to develop." A third said, "I had no experience with and very little prior interest in (birds) and it turned into one of my major passions." The students finish the course knowing the answers to these questions: What could birds possibly do for us? What can – and should – we do for birds? And I believe they will move on to their own various careers and somehow share what they learned in this class. Mission accomplished.

References

Anderson, M. G., Clark, M., & Sheldon, A. O. (2012). Resilient sites for terrestrial conservation in the Northeast and Mid-Atlantic region. *The Nature Conservancy, Eastern Conservation Science, 289*.

Audubon New York. (n.d.). *Important bird areas: Identifying and protecting the habitat birds need to thrive and survive is the key to conservation*. https://ny.audubon.org/conservation/important-bird-areas. Retrieved 25 Dec 2021.

Black Rock Forest. (2021, September 3). *Listen to the forest*. http://blackrockforest.org/listen-forest. Retrieved 5 Dec 2021.

California Society of Sciences and National Geographic. (n.d.). *A community for naturalists: iNaturalist*. https://www.inaturalist.org/. Retrieved 5 Dec 2021.

Callo, P. A., Morton, E. S., & Stutchbury, B. J. (2013). Prolonged spring migration in the Red-eyed Vireo (*Vireo olivaceus*). *Auk, 130*(2), 240–246.

Carson, R. (1962). *Silent spring*. Houghton Mifflin.

Confer, J. L., Hartman, P., & Roth, A. (1992). *Golden-winged warbler: Vermivora chrysoptera*. American Ornithologists' Union.

Cornell Lab of Ornithology: All About Birds. *Red-eyed vireo sounds, all about birds*. https://www.allaboutbirds.org/guide/Red-eyed_Vireo/sounds. Retrieved 5 Dec 2021.

eBird. (2021). *An online database of bird distribution and abundance* [web application]. eBird, Cornell Lab of Ornithology, Ithaca, New York. Available http://www.ebird.org. Accessed 5 Dec 2021.

Ellwood, E. R., Temple, S. A., Primack, R. B., Bradley, N. L., & Davis, C. C. (2013). Record-breaking early flowering in the eastern United States. *PLoS One, 8*(1), e53788. https://doi.org/10.1371/journal.pone.0053788

Hitch, A. T., & Leberg, P. L. (2007). Breeding distributions of North American bird species moving north as a result of climate change. *Conservation Biology, 21*(2), 534–539. https://doi.org/10.1111/j.1523-1739.2006.00609.x

Kellermann, J. L., Johnson, M. D., Stercho, A. M., & Hackett, S. C. (2008). Ecological and economic services provided by birds on Jamaican Blue Mountain coffee farms. *Conservation Biology, 22*(5), 1177–1185. https://doi.org/10.1111/j.1523-1739.2008.00968.x

Maenza-Gmelch, T., & Gilly, S. (2015). Bird diversity in relation to vegetation composition and structure at Black Rock Forest, Cornwall, New York. *Abstracts of the 100th Ecological Society of America Annual Meeting*, August 10–14, 2015, Baltimore, Maryland.

Maenza-Gmelch, T., & Wasmuth, M. (2017). Impact of road types and trails on bird abundance and diversity at Black Rock Forest, Hudson Highlands, New York. *Abstracts of the 102th Ecological Society of America Annual Meeting*, August 7–11, 2017, Portland, Oregon.

Maenza-Gmelch, T., Schuster, W. S. F., & Kenyon, C. (2016). Hudson Highlands west important bird area: Harriman and Sterling to Black Rock and Storm King, New York. *Abstracts of the North American Ornithological Conference*, August 16–20, 2016, Washington, DC.

Marquis, R. J., & Whelan, C. J. (1994). Insectivorous birds increase growth of white oak through consumption of leaf-chewing insects. *Ecology, 75*(7), 2007–2014. https://doi.org/10.2307/1941605

Martínez-Salinas, A., DeClerck, F., Vierling, K., Legal, L., Vílchez-Mendoza, S., & Avelino, J. (2016). Bird functional diversity supports pest control services in a Costa Rican coffee farm. *Agriculture Ecosystems & Environment, 235*(2016), 277–288. https://doi.org/10.1016/j.agee.2016.10.029

Patterson, A. E. (2021). *Seeing the forest for the trees: The physiological responses of temperate trees in a warmer world.* Doctoral dissertation, Columbia University, ProQuest Dissertations Publishing. https://doi.org/10.7916/d8-c1cx-8e45

Powell, H. (2016, January 20). *Mesmerizing migration map: Which species is which?* All About Birds.   https://www.allaboutbirds.org/mesmerizing-migration-map-which-species-is-which. Retrieved 5 Dec 2021.

Rodewald, A. D., & Arcese, P. (2016). Direct and indirect interactions between landscape structure and invasive or overabundant species. *Current Landscape Ecology Reports, 1*(1), 30–39. https://doi.org/10.1007/S40823-016-0004-Y

Root, T. L., Price, J. T., Hall, K. R., & Schneider, S. H. (2003). Fingerprints of global warming on wild animals and plants. *Nature, 421*(6918), 57. https://doi.org/10.1038/nature01333

Rosenberg, K. V., Dokter, A. M., Blancher, P. J., Sauer, J. R., Smith, A. C., Smith, P. A., Stanton, J. C., Panjabi, A., Helft, L., Parr, M., & Marra, P. (2019). Decline of the North American avifauna. *Science, 366*(6461), 120–124. https://doi.org/10.1126/science.aaw1313

Snow, J. (this volume). What does cell biology have to do with saving pollinators? In M. S. Rivera Maulucci, S. Pfirman, & H. S. Callahan (Eds.), *Transforming sustainability research and teaching: Discourses on justice, inclusion, and authenticity.* Springer.

Stutchbury, B. J. M., Tarof, S. A., Done, T., Gow, E., Kramer, P. M., Tautin, J., Fox, J. W., & Afanasyev, V. (2009). Tracking long-distance songbird migration by using geolocators. *Science, 323*(5916), 896–896. https://doi.org/10.1126/science.1166664

Survival by degrees: 389 bird species on the brink. Audubon. (2021, December 5). https://www.audubon.org/climate/survivalbydegrees

The Music of Nature. (2010a, July 19). *Eastern towhee.* YouTube. https://www.youtube.com/watch?v=mWVa08fpnXg. Retrieved 5 Dec 2021.

The Music of Nature. (2010b, February 10). *White-throated sparrow: Whistler of the north.* YouTube. https://www.youtube.com/watch?v=sL_YJC1SjHE. Retrieved 5 Dec 2021.

Tucker, M. A., Böhning-Gaese, K., Fagan, W. F., Fryxell, J. M., Van Moorter, B., Alberts, S. C., Ali, A. H., Allen, A. M., Attias, N., Avgar, T., Bartlam-Brooks, H., Bayabaatar, B., Bellant, J. L., Bertassoni, A., Beyer, D., Bidner, L., van Beest, F. M., Blake, S., Blaum, N., et al. (2018). Moving in the Anthropocene: Global reductions in terrestrial mammalian movements. *Science, 359*(6374), 466–469. https://doi.org/10.1126/science.aam9712

United Nations. (2021). *Goal 15 | Department of Economic and Social Affairs.* United Nations. https://sdgs.un.org/goals/goal15. Retrieved 5 Dec 2021.

Wagner, D. L., Grames, E. M., Forister, M. L., Berenbaum, M. R., & Stopak, D. (2021). Insect decline in the Anthropocene: Death by a thousand cuts. *Proceedings of the National Academies of Sciences, 118*(2). https://doi.org/10.1073/pnas.2023989118

Weidensaul, S. (2008). *Of a feather: A brief history of American birding.* Houghton Mifflin Harcourt.

Terryanne Maenza-Gmelch, Ph.D., is a Senior Lecturer in Environmental Science at Barnard College. She is a palynologist, forest ecologist and ornithologist. Research and teaching interests include paleoecology of the Hudson Highlands and the effects of climate and land-use change on bird populations. In 2021, Terryanne was a recipient of the Barnard College Teaching Excellence Award. Terryanne loves to be at Black Rock Forest.

Chapter 10
Going Up: Incorporating the Local Ecology of New York City Green Roof Infrastructure into Biology Laboratory Courses

Matthew E. Rhodes, Krista L. McGuire, Katherine L. Shek, and Tejashree S. Gopal

10.1   Introduction

We will be sampling soil from green roofs from around NYC. Yes, we will be going up on roofs in January. Come to class on the 19th prepared to spend a considerable amount [of time] outside in cold, windy conditions. Dress warm!!!! This is mandatory except if you have a medical or other excused absence. (Correspondence between Dr. Matthew Rhodes and the Barnard Lab in Molecular Biology Class, January 5th, 2017)

With that email, I was committed. For the Spring 2017 semester, the Barnard molecular biology lab course would conduct a project lab investigating the microbial communities underlying New York City green roofs. The work was a continuation of another Barnard project lab conducted by Dr. Krista McGuire a few years earlier surveying the microbial communities inhabiting different types of green infrastructure around New York City. By necessity, soil samples had to be acquired the first week of class, in late January, in New York City, atop a roof, along the Hudson River. *What if it snowed? What if it was 10 °F? What if there were 60 MPH winds?* I said a quick prayer, began checking the weather incessantly, and barring actual physical danger to the students, committed the class to a field day.

M. E. Rhodes (✉)
Department of Biology, College of Charleston, Charleston, SC, USA
e-mail: rhodesme@cofc.edu

K. L. McGuire
Biology and Institute of Ecology and Evolution, University of Oregon, Eugene, OR, USA

K. L. Shek
Department of Biology, Barnard College and University of Oregon, Eugene, OR, USA

T. S. Gopal
Barnard College, New York, NY, USA

© The Author(s) 2023
M. S. Rivera Maulucci et al. (eds.), *Transforming Education for Sustainability*,
Environmental Discourses in Science Education 7,
https://doi.org/10.1007/978-3-031-13536-1_10

As it turned out, January 19, 2017, was an unseasonably warm, 50 °F and sunny day. Thank luck. Thank global warming, climate change, and the urban heat island effect. Fifteen Barnard and Columbia students sampled soils from the roofs of the US Postal Service Morgan Processing and Distribution Center and the Jacob Javits Convention Center. Despite the relatively mild conditions, by 4 PM, students were huddling for warmth. Had it been ten degrees colder, there would have been a mutiny; twenty and I would have been teaching a course to approximately two students. Dr. McGuire, in her wisdom, had scheduled her project lab for the fall semester, thereby necessitating field excursions in early September as opposed to mid-January.

Why then was I standing atop a roof in January with fifteen students, carrying three down jackets in my backpack? Why did I gamble on a winter field excursion? What goal did I hope this would achieve? For both of our project labs, we intended to expose the students to real-world ecological challenges threatening the sustainability of modern cities, efforts at both mitigation and sustainable design practices, and novel and cutting-edge ecological research. We hoped that this exposure to sustainable, responsible, and equitable urban development would motivate the students to lifetimes of climate justice awareness or even advocacy and inspire them to explore sustainability-related research endeavors. Inherent within our chosen projects were first-hand exposure for the students to climate exacerbated inequities, such as combined sewage overflow, and active attempts at remediation via green infrastructure. In turn, we intended for our students to gain a deeper understanding of these ecological challenges in the city in which they lived and achieve a more personal connection to their laboratory research projects.

This gamble and these hopes seemed to pay dividends as Tejashree (Teja) S. Gopal, a student in the Spring 2017 Molecular Biology Laboratory, later reflected:

> I must admit, when Dr. Rhodes first announced that we were taking a field trip to gather soil samples for our class, I was a little apprehensive. As a neuroscience major, my knowledge of plants and soil biodiversity was limited to what I had learned freshman year in Introduction to Biology. However, I was intrigued by the research nature of the course and was eager to see how we were going to develop the project. We met the first day of class in our classroom but soon were whisked away from campus to collect soil samples from two green roofs in New York City.

The successful field excursion provided a memorable introduction to the course, even for students with vastly divergent backgrounds. The uncertainty and novelty of the research project nature of the course promoted student buy-in that persisted throughout the semester.

10.2   Background

10.2.1   The Problem with Cities

For better or worse, our planet's urbanization seems destined to continue unabated in the coming decades (United Nations DESA/Population Division, 2017). Cities certainly provide many advantages to a global population numbering over seven billion, such as the energy efficiency of mass transit systems and their concentration of human population densities, leaving more land for other species and purposes. However, current urban practices are also at the root of many of the most egregious socioeconomic and environmental catastrophes facing the planet. Many Barnard students, residing and studying near Harlem, New York, are all too aware of these issues that they regularly encounter, despite living in one of the preeminent cities in a highly developed country.

Every time a Barnard student encounters people who are homeless mere blocks from campus, they face the abject poverty and income inequality faced by many New York City residents. Every time it rains heavily and streets and subway stations flood with storm water contaminated by raw sewage, they face the inadequacy of the New York City sewer system. Every time an uncomfortable 95 °F day amplifies to a sweltering 105 °F by seas of dark asphalt, they encounter the worsening realities of global climate change and the urban heat island effect. These are but a few problems facing New York City and virtually all other major metropolitan areas (Wu, 2014). The poorer the city, or even the poorer the neighborhood, the more pronounced these issues become.

10.2.2   Green Infrastructure to the Rescue

Green infrastructure installations, or areas designed to contain vegetation, have been implemented across urban environments to alleviate some of the ecological challenges caused by human development. Cities are now investing in large-scale green infrastructure projects such as urban parks, roadside plantings known as bioswales, and green roofs.

In the quintessential concrete jungle, impermeable urban substrates like conventional roofs, roads and sidewalks, and parking facilities absorb little precipitation. Instead, nearly the entirety of rainfall rapidly funnels to urban sewer systems where it mixes with raw sewage. Thus, during periods of heavy rain, especially in low and poorly drained areas, the sewer system can rapidly become overwhelmed (Fig. 10.1). In these "combined sewer overflow events," a combination of storm water and raw sewage is dispelled into surrounding streets and waterways.

The soils and growing media comprising green infrastructure have far greater water holding capacity than traditional, impermeable urban surfaces such as asphalt and concrete (Carter & Jackson, 2007). During heavy rainfall, more water is retained in the green infrastructure soils and subsequently returned to the atmosphere via plant transpiration, reducing runoff and the likelihood of costly and dangerous combined sewer overflow episodes (Carter & Jackson, 2007). For this reason, green infrastructure figured prominently in New York City Mayor Michael Bloomberg's forward-thinking sustainability plan for the future of New York, PlaNYC (*Mayor's Office of Recovery & Resiliency*, 2017). More specifically, PlaNYC presented an effort to prepare New York City for future population increases and climate change, focusing on improved housing, energy efficiency, water infrastructure, and air

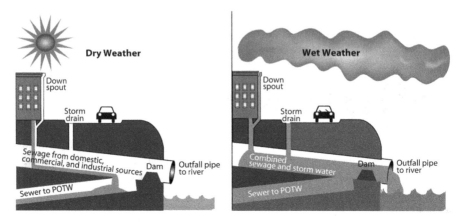

Fig. 10.1 Combined sewage system functioning under dry conditions (left panel) and wet conditions (right panel). The addition of storm drainage combines with raw sewage and overwhelms existing infrastructure causing raw sewage to bypass the Publicly Owned Treatment Works (POTW) and discharge directly into waterways. (Reproduced from the United States Environmental Protection Agency, Report to Congress: Impacts and Control of CSOs and SSOs 2004 (US EPA, 2004))

quality, amongst other aims. In addition to combatting sewage overflow, green roofs can help contribute to virtually all of these PlaNYC objectives.

Another key benefit of green infrastructure is reducing the urban heat island effect. Much of a city's surface, such as asphalt roads and shingles, are dark radiation-absorbing surfaces. These surfaces capture far more solar radiation than rural and suburban areas. On a typical summer evening, cities like New York City can be upwards of ten degrees warmer than surrounding suburbs (Solecki et al., 2005; US EPA, 2017). Green roofs absorb significantly less heat than traditional building materials. This feature combines with the natural evapotranspiration of plants, whereby plants utilize solar energy to raise and evaporate water, greatly reducing the heat transmitted to the underlying building. Estimates suggest that green roofs placed upon buildings, such as the one atop the Jacob Javits Convention Center have the potential to save New York City upwards of a net 16 million dollars annually (Akbari & Konopacki, 2005) in reduced cooling costs as well as reducing carbon emissions required to generate the electricity used for cooling.

10.2.3 Green Infrastructure Around New York City

Various green infrastructure installations are spread throughout the New York City area, each with a particular ecology and environmental challenges (McGuire et al., 2013). Roadside bioswales offer narrow, fragmented, and often disturbed sanctuaries of various trees and flowering plants. In contrast, urban parks such as Central Park or Riverside Park offer some of the few contiguous vegetated areas in New York City. Green roofs such as the one atop the Javits Center must contend with relatively high winds as well as significant direct sunlight, exposing the plants and soils to temperature spikes and high levels of solar radiation (Takebayashi & Moriyama, 2007).

Like most green roofs in the US, the Morgan Center and Javits Center green roofs incorporate plants from the genus *Sedum*. Though not native to the New York City environment, succulent *Sedum* spp. have proven highly resilient to the seasonal temperature fluctuations, daily temperature fluctuations, irregular rainfall, and high winds present atop northeastern United States rooftops (Monterusso et al., 2005). A select few New York City green roofs utilize native species, and they offer similar levels of water retention and temperature reduction compared to *Sedum* spp. green roofs (Lundholm et al., 2010). However, an additional benefit of native species green roofs is that they can offer valuable repositories of plant and microbial biodiversity and provide forage and habitats for wildlife, including insects and birds.

10.2.4 Don't Forget the Microbes

Naturally, as both of the professors teaching these courses were trained as environmental microbiologists, both the Microbiology Laboratory course and the Molecular Biology Laboratory course placed a significant emphasis on microorganisms, an often undervalued component of every ecosystem. Microorganisms comprise more biomass on this planet than all animal life combined (Whitman et al., 1998) and figure prominently in virtually every planetary chemical cycle (Falkowski et al., 2008). The sustainability and continuing effectiveness of the performance of green infrastructure ties intimately to the quality of the underlying soils and growing media (McGuire et al., 2013), which affects and is affected by the health of the plant and microbial ecosystems.

Soil and growing media quality are a complex interaction of biotic and abiotic processes. One pivotal component of soil and growing media is the soil microbiome. Microbes, including fungi and bacteria, play various roles in belowground ecology, including decomposition, nutrient cycling, and water cycling. In the case of green roofs, for example, low-quality growing media will detrimentally affect plant health, thereby reducing the water carrying and temperature reducing capacity of the green roof, two of the critical purposes of green roofs. Therefore, a better understanding of the factors that shape green infrastructure soil microbial communities is imperative for optimizing the long-term effectiveness of green infrastructure.

10.3 Course Descriptions

10.3.1 Microbiology Project Laboratory (Fall 2013)

In 2013 a project laboratory was designed by Dr. McGuire for students enrolled in Microbiology Laboratory (BC3321) to evaluate microbial communities in green infrastructure across New York City. A total of 15 students enrolled in the course.

The purpose of the project lab was to provide students with training in traditional and cutting-edge microbiological techniques in conjunction with direct field experience sampling soils and growing media from parks, green roofs, and tree pits. The course syllabus included 3 weeks of field sampling and additional sampling by the teaching assistant (TA; Seren Gedallovich, Barnard class of 2013) and student volunteers. The students and the TA sampled 15 green roofs, six parks, and 16 tree pits. The park samples were taken from forest sites, suburban parks, and urban parks to traverse an urban to rural gradient. Using these samples, students were trained in sterile sieving techniques, culturing, microbial biomass assays, mycorrhizal fungal quantifications, DNA extractions, amplifications of microbial DNA (PCR), and data analysis.

The ultimate goal of the class project was to produce a manuscript-style publication assessing the biogeography of microbial communities in green infrastructure across NYC and across the urban to rural gradient. Each student chose a subset of the work within this larger project framework to address their own research questions for their final project. For example, one student was interested in whether or not microbial biomass shifted across the urban to rural gradient, and another student was interested in quantifying human pathogens across different green infrastructure types. A collaborative effort of Columbia and Barnard students chose to explore the impact of nitrogen deposition from dog urine on the soil microbiome of roadside green infrastructure. This in-class project evolved into a senior thesis and then a co-first author publication for these students (Lee et al., 2019). The course's final session was a conference-style day of presentations in which each student made a poster highlighting their research projects, data analyses, and results.

Since students did not complete the processing of all the samples necessary for the final publication, a postdoc from the McGuire lab, an undergraduate researcher from the Microbiology project lab who enrolled in research for credit (Adam Abin), and an additional undergraduate thesis student, finished the remaining analyses (Gill et al., 2020). In the process, they added additional samples to the overarching project, which included road medians and bioswales.

10.3.2   Molecular Biology Project Laboratory (Spring 2017)

Building on Dr. McGuire's 2013 Microbiology Laboratory course results, together we decided that for the Spring 2017 Molecular Biology Laboratory course, the class would focus exclusively on New York City green roofs. In particular, the class would seek to compare the microbial communities underlying *Sedum* spp. Versus native green roofs and test their response to typical levels of heat stress found atop a New York City roof in the summer. Thus, on January 19th, the class sampled soils from the Morgan Center and Javits Center *Sedum* spp. roofs.

The course TA, Katherine (Kaye) Shek (Barnard class of 2015), sampled the Jackie Robinson, Hansborough, and Chelsea recreation center roofs the same week. The recreation center roofs were all comprised of native rather than *Sedum* species. Together, the class and Kaye collected almost 200 soil samples. The soil samples were then processed according to the protocols of the Earth Microbiome Project that standardize the characterization of environmental microbial communities (Caporaso et al., 2011, 2012).

We began the semester with a fully developed syllabus and plan that rapidly needed adjustment due to equipment malfunctions or experimental failures. Built-in flexibility was a must. Looking back upon the course organization, Teja described her experiences:

> As the class progressed, Dr. Rhodes introduced us to the molecular biology techniques that we would be using to look at the diversity of the soil samples we had collected. The syllabus Dr. Rhodes distributed the first day of class had a clear plan of the semester, detailing which experiments we would be conducting each week. However, as with any research project, things did not go according to plan. When one experiment failed, Dr. Rhodes was required to come up with a new plan for the following week's class, and thus began the semester's journey, shaped by our work. I found the flexibility of the project refreshing, given that many other classes in college are meticulously planned out. Something about the exploratory nature of this project fostered such a wonderful class dynamic, where we all, including Dr. Rhodes and our TA, Kaye, worked together to collectively reach our goal.

Thus, overcoming obstacles that often occur during scientific investigations added elements of stress and authenticity to the semester.

For a final project, the students utilized the QIIME (Quantitative Insights into Microbial Ecology) bioinformatics platform to analyze the DNA sequences (Caporaso et al., 2010). Each group of four students worked with only 1/200th of the data set to facilitate data analysis and reduce the time and computer resources required. With modern DNA sequencing technology, this reduced data set still comprised hundreds of thousands of DNA sequences. During the last day of class, as the QIIME analysis revealed results that confirmed and contradicted our original hypotheses, numerous students expressed feelings of accomplishment and satisfaction. On a side note, it is one of our favorite teaching endeavors to introduce students to the powers of computing beyond Microsoft Word and the graphical user interface. To the uninitiated, even limited computing expertise comes across as "magic." Eliminating the mental roadblocks that many students place between themselves and their computer is inevitably a rewarding experience.

Another distinct advantage of the project lab format, as noted below by student and TA, Kaye Shek, is that the entire semester was dedicated to furthering a single objective. This single objective enabled students to place their efforts in a broader context and to observe firsthand how the techniques that they were mastering could be utilized to answer questions and obtain both informative and potentially actionable results.

10.4   Course Results

10.4.1   Microbiology Course Results

Results from the 2013 project lab can be briefly summarized by three main conclusions: (1) both fungal and bacterial communities vary predictably across different types of green infrastructure (green roofs, parks, bioswales, etc.) and across geographic location, (2) microbial communities are more abundant and diverse in non-urban soils, and (3) increased N deposition from sources such as canine urine are detrimental to microbial biomass and diversity (Lee et al., 2019). Think about the important functions of green infrastructure the next time you see someone's dog urinating on them.

Despite the harshness of the urban environment, the comparatively small size of green roofs, and the inherent spatial isolation of urban green infrastructure, considerable microbial diversity was uncovered in all types of infrastructure. Green roofs, in particular, had high levels of bacterial diversity and, in some cases, had greater numbers of taxa than ground-level infrastructure. Interestingly, microbial communities from engineered bioswales had a greater diversity of both fungal and bacterial communities compared to other types of ground-level infrastructure, possibly due to increased resource inputs in the form of storm water. In bioswales, storm water is intentionally directed into the engineered installation, which brings a plethora of resources to feed microbes. By contrast, parks, tree pits, and green roofs primarily receive rainwater and less street-side runoff (and no runoff in the case of green roofs).

10.4.2   Molecular Biology Course Results

The results of the QIIME analyses proved to be both informative and unexpected. The class first looked at the overall species diversity (how many species of microbes were present) in the different green roofs. Our initial hypothesis was that the two *Sedum* spp. green roofs would display lower levels of diversity when compared to the native species green roofs due to the homogeneous nature of the overlying plant life. We also hypothesized that heat stress, experimentally imparted by heating soil samples prior to microbial analysis, would reduce the microbial diversity of the respective roofs.

Our analysis did show the native species green roofs to have more species diversity than the Javits Center *Sedum* spp. green roof shown in Fig. 10.2. However, the Morgan Center *Sedum* spp. green roof shown in Fig. 10.3 appeared to be more akin to the native green roofs than to the Javits Center green roof. The cause for this is unclear. However, students noted an increased level of plant species that were growing spontaneously on the roof in addition to deliberately planted vegetation atop the Morgan Center green roof. They also noted a shallower soil depth in the Javits

Fig. 10.2 The Jacob Javits Convention Center *Sedum* sp. green roof

Fig. 10.3 United States Postal Service Morgan Center Processing Plant *Sedum* sp. green roof showing considerable contamination of non-Sedum grasses

Center green roof compared to the other four roofs. Both observations suggest a reduction in available microbial habitats in the Javits Center roof in comparison to the other four roofs.

The overall level of microbial diversity does not necessarily indicate how alike two different communities are. Two communities may have similar levels of microbial diversity but be comprised of entirely different assemblages with highly distinct ecological functions. We hypothesized that the Javits Center and Morgan Center roofs would appear similar and that the three native roofs would map together. Once again, unexpectedly, the Javits Center green roof appeared distinct from the other four roofs, with the Morgan Center roof demonstrating similar species composition to the native green roofs. A fuller investigation of 32 New York City green roofs did show distinct differences between those with *Sedum* spp. versus native consortia (Hoch et al., 2019).

10.5   Reflections on an Ecological Project Lab

10.5.1   From the Professors

There is always the element of the unknown and the unknowable with any novel laboratory experiment. Will the experiment work? Will the results be meaningful? If not, was it due to a faulty hypothesis, a faulty setup, or human error? It is a common source of anxiety for any researcher from undergraduate through the tenured professor. When career and course outcomes are thrown into the mix, they only compound the anxiety. Contingency plans can be made, but they are just that, contingency.

Teaching a project lab for the first time was a humbling and a learning experience. From the outset, there were difficult questions to be addressed. Where was the budget going to come from? The Javits Center graciously donated the funds to support the class project and additional green roof research. What was the overarching goal? A class authored Nature paper would be nice but would anything that enhanced the learning environment be enough to justify the increased cost and stress?

These, in turn, raised additional questions. Principally, how ambitious could we make the project design? For the molecular biology course, we opted for five rooftops, three native and two *Sedum* spp., which seemed like a reasonable amount of work for the students (and the professor along with the course TA) to accomplish. Those also were the roofs that we were able to gain access to. Although five roofs would not be sufficient to publish a peer-reviewed research publication, we wanted the results to be informative for larger forthcoming green roof projects. As stated above, the course design would also enhance the students' learning experience. Ultimately, both hopes became a reality.

With that perspective, we would consider the Barnard Molecular Biology Green Roof Project Lab an unmitigated success. We obtained both unexpected and informative results. Many students openly expressed appreciation for venturing beyond the traditional laboratory experience of predetermined experiments and outcomes. Despite the chill, the students overwhelmingly appreciated the field excursion. It was a fantastic tone-setting experience for the semester and provided the class with a close-up view of the latest attempts to combat urban ecological problems. Students also expressed satisfaction with the knowledge that they were conducting research that had never been done.

Nevertheless, perhaps we were still overly ambitious. In retrospect, a project lab would be a more guaranteed success if it was not an independent research project in and of itself but rather an offshoot of ongoing work already well underway in the lab. Other students, researchers, and technicians dedicated to the project could help eliminate many of the risks and stressors associated with running a project lab. Preliminary data, well-established in-house protocols, and having additional hands on deck could facilitate the project and aid with contingency plans. As assistant and visiting professors respectively at the time, we had significant hurdles to overcome in attempting to conduct a successful project lab. For an adjunct, the barriers would be nearly insurmountable. It is yet another example of the disservice done to students by the academic world's increasing reliance on a revolving door of adjunct faculty.

Not surprisingly, a few of the class experiments failed outright, such as our attempt at a gel extraction. Given the time constraints of the course, these experiments could not be repeated. Others worked sufficiently to demonstrate the concept but insufficiently to obtain meaningful results, such as our attempt at constructing a library of DNA clone sequences to correspond to a representative population of arbuscular mycorrhizal fungi sampled across all roofs. A silver lining is that some students, Teja included, actually expressed appreciation for being exposed to the frustrations associated with novel scientific research. But enough of the experiments worked sufficiently well to provide meaningful results to analyze later. Any successful outcome for the course would have been impossible without the tireless efforts of the course TA, Kaye Shek, whose perspective both as an undergraduate student involved in green roof research and as a TA for the course is provided below. Kaye worked tirelessly with the students and behind the scenes to ensure that samples were processed and ready for class.

Since conducting their respective project labs in the Fall 2013 and Spring 2017 semesters Dr. McGuire and Dr. Rhodes have moved on to different institutions. Dr. McGuire is now an Associate Professor at the University of Oregon. Having the resources of a high throughput lab at a Research 1 university has facilitated Dr. McGuire in incorporating elements of a project lab in guiding undergraduate research projects in her lab, often mentored by graduate students. She continues to engage undergraduates in fieldwork, including green infrastructure research in Portland and Eugene, OR, and has also taken several students to Malaysia, Puerto Rico, and Panama. Dr. Rhodes is now an Assistant Professor at the College of Charleston. Dr. Rhodes teaches multiple microbiology laboratory sections per semester with more of a medical emphasis at a very large (10,000 undergraduates) liberal arts college. These responsibilities have precluded the incorporation of project lab elements into the microbiology laboratory curriculum. However, faculty at the College of Charleston who teach the molecular biology laboratories often structure the lab as a project lab by attempting to isolate and characterize novel virus strains. The anticipated reintroduction of an upper-level environmental microbiology course to the College of Charleston's curriculum could once again offer the opportunity to design a project lab course emphasizing the potential impacts of environmental microbiology research to global sustainability and equity.

10.5.2   From a Student

Tejashree Gopal, Barnard class of 2018, was enrolled in the Molecular Biology Laboratory in Spring 2017.

Looking back on the class, I think I most appreciate how I grew as a scientific writer. Being assigned to write about something that I had so little prior knowledge of was a truly eye-opening experience, not only to the topic of biodiversity itself but also to the fundamental process of writing a research paper. In many of my other classes, I have been asked to write research papers but have often been able to rely on my knowledge of the subject. When writing this research paper, I was, for the first time, challenged to learn everything that I could about the urban heat island effect, climate change, heat shock proteins, soil biodiversity, and many other related topics that I would never have had the opportunity to learn about were it not for this class.

I must admit that I struggled a lot along the way. The first writing assignment we had – to write an introduction for the first set of experiments we conducted – was one of the most challenging tasks I have confronted in my time at Barnard. I floundered, unsure of where I should even start. But as the class went on, I learned more effective techniques to find sources, extract information from dense publications, and write in a concise manner. Soon, I was writing papers of which I was tremendously proud.

As I started to write my thesis for my senior year at Barnard, I was incredibly grateful for the skills that I learned in this class. I felt extremely privileged to have had the experience of carrying out a research project from the initial stage of collecting samples to the final product of a published research paper, illustrating the research project from start to finish.

10.5.3   From Both Sides of the Table

Katherine (Kaye) Shek, Barnard class of 2015, was an undergraduate research student in the McGuire lab, and Teaching Assistant (TA) for the Molecular Biology Laboratory in Spring 2017.

As an undergrad, I took as many laboratory courses as I could. I wanted to understand my interests and strengths better, and what that meant for my career trajectory (or even to help me understand what to do after graduating). As most young scientists do, I thought about my future goals in the context of global issues: find a cure for cancer, stop global warming, feed the hungry – make a difference. It's this drive to make a difference that I believe boosts many young students into STEM fields and is important to consider when planning and designing undergraduate courses. This reason, among many, is why I consider the project-based laboratory courses in Molecular Biology and Microbiology at Barnard invaluable experiences to not only my personal growth as a scientist but also to my undergraduate peers and the students I worked with as a teaching assistant.

The immersive nature of a project lab course gives students an opportunity to make a real contribution to original research. This value was tangible in the Molecular Lab classroom: students were more confident to ask questions throughout every lab technique we covered to better place each activity and its purpose into the big picture. There was excitement and pride amongst classmates as we obtained results, and genuine curiosity and interest when certain procedures failed. As a TA, it was rewarding to have a classroom full of engaged students and it felt realistic to teach the methods and techniques through troubleshooting failures and celebrating successes. As a student in the course, I felt like I was making a useful contribution not only to the actual research project, but also to my own personal skills and experiences to build upon outside of class. (Frankly, as a student, the project lab felt both like a learning experience and a "resume builder," as I suddenly had many additional molecular biology methods that I could include on my CV. This was especially comforting as a soon-to-be college graduate with no plan. However, it was the learning experience that has paid off with more meaning in my life and nascent career.) Simply put, when compared to the more classical upper-level Biology lab course design, the project-based courses challenge students in a more fulfilling and realistic way.

In general, the 'classical' lab course syllabi include individual mini-experiments each week to focus on a particular organism or technique, not to be revisited until the final exam or an independent project at the end of the term. As a student in these courses, I enjoyed the opportunity to flex my creative muscles and design my own hypotheses to test in an independent project; however, as a TA in one of these lab courses, an issue with this particular design became apparent. By the time the final 2 weeks of the term rolled around and students were starting to design their independent experiments, nearly every student wanted to ask research questions that would take a full dissertation (and a hefty grant) to work towards an answer. Once their research questions narrowed down to a feasible degree, applying the methods they learned throughout the term outside of the context of the mini-experiments was a challenge to most students. I found myself running around to each student, helping them envision the methodological basis behind various techniques we had covered in depth earlier in the term. In contrast, the project laboratory allows students to answer broader research questions similar in scope and creativity to those they originally asked, spread out over a whole term, and learn new techniques along the way.

Obviously, choosing a general course design for effective teaching in Biology varies on a case-by-case basis: the ability to incorporate original research into a course depends entirely on the subject matter. In molecular biology, cell biology, or microbiology for instance, for the aforementioned reasons, I believe a project lab is an excellent means by which a young scientist can immerse themselves in learning new material that is otherwise challenging to grasp and retain, while simultaneously practicing the scientific method in a realistic context. Saying goodbye to the students walking out of the classroom on the last day with more than just a letter grade on their transcripts – specifically, with the ability to think and write like a scientist - was proof enough to me that the project lab environment leads to more meaningful and educational success.

10.6   Would We Do This Again? (Concluding Remarks)

Whether communicating basic scientific research to funding agencies, students, or the general public, one recurring challenge is placing the research into a broader and socially relevant context. The more esoteric and removed from a direct impact the research is, the more difficult this challenge can become. One advantage of designing both labs as project labs was that all laboratory exercises, from sample collection, to laboratory manipulations, through data analysis were all focused on a single theme. In this manner, students were given a more direct and comprehensive window into the broader scientific process and its potential impacts on society. In both the Microbiology Laboratory and Molecular Biology Laboratory courses, students observed how green infrastructure is attempting to solve global problems and societal inequities. As a whole, the final papers for both classes demonstrated a thorough understanding of the methodologies employed, the results obtained, and the background context motivating the project. For Teja in particular, "by the final writing assignment, [she] was impressed with [her] own ability to research completely new topics, gather information from a variety of sources, and reflect on how it related to the data we had collected in class." Furthermore, individuals from both courses could see the work to fruition and provide suggested steps toward improving the diversity, sustainability, and microbial health of green infrastructure.

There were successes, and there were failures. Progress was incremental. Let's face it; we were not going to solve the inequities, injustices, and dangers associated with global climate change in a semester. However, by sampling in the field, by processing those samples, by collecting data, by analyzing the data, and by writing up the results, students could experience first-hand the scientific process from start to finish. The site visits and field sampling provided a unique and memorable experience to illustrate the scientific questions and societal challenges we attempted to address. The data analysis and write-up provided an opportunity for students to measure progress and forced them to place their research in a broader context once again.

For us, the overwhelming sentiments of pride and accomplishment in completing an authentic research project, combined with a tinge of frustration at the end of the semester, signaled a successful project lab experience. The most significant impact for our students was that they have now participated in the non-linear march of scientific progress and its desired beneficial impact on society. To some who want to contribute to social justice causes, the uncertainty of basic scientific research may push them toward or solidify their desire for a career in medicine or other similar fields. Nevertheless, exposure to project-driven sustainability-related research during their college experience may contribute to a greater appreciation for basic research and help shape the finer details of their careers. But to others, experiences with authentic research may transform vague inspiration into the specific abilities and confidence required to support and motivate them to pursue a Ph.D. and a career in the basic sciences. We see both results in the students who have contributed to this chapter. The experiences as a TA for the 2017 molecular biology project lab solidified Kaye Shek's aspirations toward a Ph.D. and a career in environmental microbiology research. She is now completing her 5th year as a Ph.D. student with Dr. McGuire at the University of Oregon, investigating how variation in soil management practices alters vineyard microbiome structure and function. Meanwhile, for Tejashree Gopal, her experiences as a student in the 2017 molecular biology project lab contributing to basic research from project conception through publication have stuck with her into medical school.

The question then is not "would we do this again?" But instead, "when will we do this again?" Connecting molecular biology and microbiology laboratory techniques that are generally conducted in a laboratory with larger ecological and environmental principles, helps to focus the class on a broader purpose. Projects that give students direct exposure to sustainability concerns and allow them to directly observe the intended impacts of research enhance the authenticity of the experience.

Unfortunately, not every research project is an appropriate medium for a project lab. Not every course or institution is the appropriate venue. Nor will every topic have a strong undercurrent of sustainability or social justice. Often the questions addressed in the pursuit of scientific knowledge are on the esoteric side or have limited immediate applicability. Our chosen course topics on green infrastructure incorporated sufficient aspects of eco-justice, relevance, authenticity, and impact to be inclusive and spark the interest of the vast majority of our students. There are certainly numerous areas of scientific research that could serve this purpose to a similar extent, if not even more so. In that vein, when we have subsequently been called upon or will be called upon to teach biology laboratory courses in the future, we do so with the project lab structure in the back of our minds.

References

Akbari, H., & Konopacki, S. (2005). Calculating energy-saving potentials of heat-island reduction strategies. *Energy Policy, 33*(6), 721–756. https://doi.org/10.1016/j.enpol.2003.10.001

Caporaso, J. G., Kuczynski, J., Stombaugh, J., Bittinger, K., Bushman, F. D., Costello, E. K., Fierer, N., Peña, A. G., Goodrich, J. K., Gordon, J. I., Huttley, G. A., Kelley, S. T., Knights, D., Koenig, J. E., Ley, R. E., Lozupone, C. A., McDonald, D., Muegge, B. D., Pirrung, M., et al. (2010). QIIME allows analysis of high-throughput community sequencing data. *Nature Methods, 7*(5), 335–336. https://doi.org/10.1038/nmeth.f.303

Caporaso, J. G., Lauber, C. L., Walters, W. A., Berg-Lyons, D., Lozupone, C. A., Turnbaugh, P. J., Fierer, N., & Knight, R. (2011). Global patterns of 16S rRNA diversity at a depth of millions of sequences per sample. *Proceedings of the National Academy of Sciences, 108*(Suppl 1), 4516–4522. https://doi.org/10.1073/pnas.1000080107

Caporaso, J. G., Lauber, C. L., Walters, W. A., Berg-Lyons, D., Huntley, J., Fierer, N., Owens, S. M., Betley, J., Fraser, L., Bauer, M., Gormley, N., Gilbert, J. A., Smith, G., & Knight, R. (2012). Ultra-high-throughput microbial community analysis on the Illumina HiSeq and MiSeq platforms. *The ISME Journal, 6*(8), 1621–1624. https://doi.org/10.1038/ismej.2012.8

Carter, T., & Jackson, C. R. (2007). Vegetated roofs for stormwater management at multiple spatial scales. *Landscape and Urban Planning, 80*(1), 84–94. https://doi.org/10.1016/j.landurbplan.2006.06.005

Falkowski, P. G., Fenchel, T., & Delong, E. F. (2008). The microbial engines that drive Earth's biogeochemical cycles. *Science, 320*(5879), 1034–1039. https://doi.org/10.1126/SCIENCE.1153213

Gill, A. S., Purnell, K., Palmer, M. I., Stein, J., & McGuire, K. L. (2020). Microbial composition and functional diversity differ across urban green infrastructure types. *Frontiers in Microbiology, 11*, 912. https://doi.org/10.3389/FMICB.2020.00912

Hoch, J. M. K., Rhodes, M. E., Shek, K. L., Dinwiddie, D., Hiebert, T. C., Gill, A. S., Salazar Estrada, A. E., Griffin, K. L., Palmer, M. I., & McGuire, K. L. (2019). Soil microbial assemblages are linked to plant community composition and contribute to ecosystem services on urban green roofs. *Frontiers in Ecology and Evolution, 7*, 198. https://doi.org/10.3389/fevo.2019.00198

Lee, J. M., Tan, J., Gill, A. S., & McGuire, K. L. (2019). Evaluating the effects of canine urine on urban soil microbial communities. *Urban Ecosystem, 22*(4), 721–732. https://doi.org/10.1007/S11252-019-00842-0

Lundholm, J., MacIvor, J. S., MacDougall, Z., Ranalli, M., Mitsch, W., Cronk, J., Wu, X., Nairn, R., Stottmeister, U., Wiener, A., Kuschk, P., Kappelmeyer, U., Kastner, M., Darlington, A., Dixon, M., Dietz, M., Clausen, J., Brenneisen, S., Balvanera, P., et al. (2010). Plant species and functional group combinations affect green roof ecosystem functions. *PLoS One, 5*(3), e9677. https://doi.org/10.1371/journal.pone.0009677

Mayor's Office of Recovery & Resiliency. (2017). http://www1.nyc.gov/site/orr/index.page

McGuire, K. L., Payne, S. G., Palmer, M. I., Gillikin, C. M., Keefe, D., Kim, S. J., Gedallovich, S. M., Discenza, J., Rangamannar, R., Koshner, J. A., Massmann, A. L., Orazi, G., Essene, A., Leff, J. W., Fierer, N., Carter, T., Jackson, C., Del Barrio, E., Mentens, J., et al. (2013). Digging the New York City skyline: Soil fungal communities in green roofs and city parks. *PLoS One, 8*(3), e58020. https://doi.org/10.1371/journal.pone.0058020

Monterusso, M. A., Rowe, D. B., & Rugh, C. L. (2005). Establishment and persistence of *Sedum* spp. and native taxa for green roof applications. *HortScience, 40*(2), 391–396. https://doi.org/10.21273/HORTSCI.40.2.391

Solecki, W. D., Rosenzweig, C., Parshall, L., Pope, G., Clark, M., Cox, J., & Wiencke, M. (2005). Mitigation of the heat island effect in urban New Jersey. *Environmental Hazards, 6*(1), 39–49. https://doi.org/10.1016/j.hazards.2004.12.002

Takebayashi, H., & Moriyama, M. (2007). Surface heat budget on green roof and high reflection roof for mitigation of urban heat island. *Building and Environment, 42*(8), 2971–2979. https://doi.org/10.1016/j.buildenv.2006.06.017

United Nations DESA/Population Division. (2017). *World population prospects – Population Division – United Nations.* https://esa.un.org/unpd/wpp

US EPA. (2004). *Report to Congress: Impacts and control of CSOs and SSOs.* https://www3.epa.gov/npdes/pubs/csossoRTC2004_chapter02.pdf

US EPA. (2017). *Heat island effect.* https://www.epa.gov/heat-islands

Whitman, W. B., Coleman, D. C., & Wiebe, W. J. (1998). Prokaryotes: The unseen majority. *Proceedings of the National Academy of Sciences, 95*(12), 6578–6583. https://doi.org/10.1073/pnas.95.12.6578

Wu, J. (2014). Urban ecology and sustainability: The state-of-the-science and future directions. *Landscape and Urban Planning, 125*, 209–221. https://doi.org/10.1016/j.landurbplan.2014.01.018

Matthew E. Rhodes is an Assistant Professor of Microbial Genetics at the College of Charleston. In his capacity as an educator he enjoys engaging with his students and the general public in a creative and informative manner. He often gets on the proverbial soapbox about climate change and childhood vaccination.

Krista L. McGuire is an Associate Professor of Biology at the University of Oregon. Her research examines community assembly processes of plant-microbe associations across gradients of human disturbance. She is particularly interested in leveraging basic microbial ecology research for the sustainable management of farms, cities, and forests while accounting for environmental justice.

Katherine L. Shek received her BA in Biology (minor Chemistry) from Barnard in 2015, and is currently a Ph.D. candidate in the Institute of Ecology and Evolution, University of Oregon. Her research aims to understand plant-microbial feedbacks under compounded disturbances related to human activity and climate change, using vineyards in Oregon as a model system.

Tejashree S. Gopal graduated from Barnard College with a degree in Neuroscience and continued on to conduct a clinical project concerning the diagnosis of acute kidney injury in Pondicherry, India under a Fulbright scholarship. She is now a medical student at the Vagelos College of Physicians and Surgeons at Columbia University.

Chapter 11
It Turned into a Bioblitz: Urban Data Collection for Building Scientific Literacy and Environmental Connection

Kelly L. O'Donnell and Lisa A. Brundage

11.1 The Macaulay Honors College

There is a question we like to pose at the start of our science pedagogy workshops, and it's one you can answer while you are reading this. Think for a minute about your favorite moment in learning science. Where was it? How old were you? Was it in school? Were you with others or alone? What was the root of your joy in that moment?

When we ask this question, we find – overwhelmingly – that the experiences many participants note happened early in life, sometimes in a classroom but just as often, outside of one. Even though many of the workshop participants are scientists, they seldom select a moment from high school or undergraduate courses. The annual Macaulay Honors College BioBlitz, an essential component of our honors curriculum, is an attempt to bring that spark, that frisson of excitement, to nearly 500 students a year by immersing them in a massive data collection event and research experience that connects them to the ecology of New York City in a way that few students and citizens ever experience. Part of Macaulay and CUNY's mission is to use the city as a classroom and laboratory, including both its institutions and the wide variety of habitats across the five boroughs, from bits of old-growth forests, to wetlands, to grasslands and more.

The City University of New York is the largest public urban university in the United States, serving about a quarter million matriculated undergraduates (and close to another quarter million in graduate, professional, and continuing education programs) (CUNY Office of Institutional Research, 2019; CUNY, 2021). It has 25 campuses across New York City's five boroughs. The Honors College was founded in 2001 (and became Macaulay Honors College in 2006) as a highly selective

K. L. O'Donnell (✉) · L. A. Brundage
Macaulay Honors College, City University of New York, New York, NY, USA
e-mail: kelly.odonnell@mhc.cuny.edu

© The Author(s) 2023
M. S. Rivera Maulucci et al. (eds.), *Transforming Education for Sustainability*,
Environmental Discourses in Science Education 7,
https://doi.org/10.1007/978-3-031-13536-1_11

program to serve exceptional undergraduates at CUNY. We currently admit approximately 520 students annually, all of whom co-matriculate at Macaulay and one of our eight consortial CUNY campuses (Baruch College, Brooklyn College, The City College of New York, Hunter College, John Jay College, Lehman College, Queens College, and the College of Staten Island). Macaulay serves a diverse student body whose students are primarily drawn from the New York City metro area, including many graduates of NYC's public high schools. Many years, over half are bilingual and a large number of students are immigrants, children of immigrants, and/or the first in their family to attend college. Despite this, Macaulay is still not representative of the overall student body at CUNY, and consistently seeks ways to address this discrepancy. We recently began accepting our first transfer students, an opportunity available exclusively to associate degree students already within the CUNY system. Macaulay students who are New York State residents receive a tuition scholarship, and all students receive a laptop, dedicated advisors, access to funding for unpaid internships or study abroad, and, of course, the opportunity to participate in our seminars and learning events. In 2021, we also launched the Justice and Equity Honors Network with Barrett, The Honors College at Arizona State University. The program engages students in online discussion sessions pertaining to current events and justice, including climate and environmental justice. Because of its consortial model, Macaulay faces some distinctive structural challenges that also speak to one of the core tenets of our mission: to provide the benefits of a small, liberal arts education combined with the resources and opportunities of a large research university.

Because of our distributed model, we need to create opportunities for Macaulay students from all eight campuses to form bonds with one another and develop their identities as Macaulay students. To that end, all of our students take the same sequence of core honors seminars in their first 2 years of study, starting with the Arts in New York City, and then continuing to the People of New York City, Science Forward, and Shaping the Future of New York City. All of our seminars (capped at 20 – a rarity at CUNY) are focused on critical thinking in the context of lived experience, and the vast resources and challenges that New York City offers. Taught mostly by faculty on the home campuses, these core courses form a common experience for our students and allow them to be trained in critical thinking in a variety of disciplines.

In addition to the curriculum, Macaulay develops "common event" programs for each of the seminars, giving students across our eight campuses a chance to come together for memorable experiential learning activities that augment both their academic and social development at Macaulay. As students begin their time at Macaulay in the Arts in New York City, they participate in Night at the Museum, during which they have dedicated access to the Brooklyn Museum for one evening. Prior to attending, they also participate in workshops to help them develop skills in doing close readings of, and having conversations about, works of art. Even though the majority of our students reside in New York City, many are not routine museumgoers, and Night at the Museum gives them skills and confidence to engage with museums and art in a personally meaningful way. After Night at the Museum, students typically craft reflections on their time there, using photographs, film snippets, and sometimes audio recordings of conversations that they had at the museum to create projects about their experience. They also continue to visit art institutions and performances throughout their seminar.[1] Surveys of our senior students show that many of them, from numerous majors, consider Night at the Museum a positive and beneficial experience.

After our success with launching Night at the Museum, we wanted to create an event for Science Forward that would give students a personal, authentic connection to science and make a lasting impression, as well as be useful in the seminar classroom. To do this, we mount our most ambitious event: our annual BioBlitz held at the start of the fall semester. Our BioBlitz is a 24-h biodiversity survey of a park or green space of New York City, introducing ecological fieldwork to almost 500 students, and requiring the work of dozens of scientists and naturalists to lead and supervise the data collection. To date, we have organized BioBlitzes for Macaulay students in Central Park (2013, Manhattan), New York Botanical Garden (2014, Bronx, where our event took place primarily in the Thain Family Forest, the largest old-growth forest in New York City), Freshkills Park (2015, Staten Island and built atop Fresh Kills Landfill, once the largest in the world), Brooklyn Bridge Park (2016, Brooklyn, constructed over rehabilitated industrial piers), Alley Pond Park (2017, Queens), Inwood Hill Park (2018, Manhattan, site of the only old-growth forest on the island, the Shorakapok Preserve), and Green-Wood Cemetery and the Gowanus Canal (2019, Brooklyn, where the Gowanus has been designated a Superfund site). Our 2020 and 2021 BioBlitzes were held remotely and students collected iNaturalist observations from wherever they were located to contribute to a geographically-distributed, whole-class dataset. iNaturalist is a free, openly available database and social network where people can upload geotagged and time-stamped photos of living things and have an engaged community of users help them with identification (iNaturalist, 2021a).

At the BioBlitz, students participate in authentic data collection, and in turn, use those data during the rest of the fall semester to complete research projects, culminating in a cross-campus poster session at the end of the semester. The BioBlitz is a resource-intensive, but beloved, tradition at Macaulay and is the product of our commitment to student-centered pedagogical innovation, and our hard work to bring together a logistically complex event. Macaulay understands that it occupies a position of privilege within CUNY, but we believe that our model demonstrates best practices and potential outcomes for *all* students in a public university, when properly supported and resourced, not just honors students. We maintain that a BioBlitz is the type of experience that can be brought to a variety of students and educational settings, regardless of the student population, and such an event is also adaptable to different budgetary and pedagogical concerns. In this chapter, we will describe how we developed our Science Forward skills curriculum and how the BioBlitz is an integral part of that framework. We will detail how this massive data

---

[1] Due to the COVID-19 pandemic, Night at the Museum could not be held in 2020 and 2021.

collection event becomes an authentic research experience in the classroom and how you can conduct your own BioBlitzes. Finally, we will describe some responses to the BioBlitz from our students.

11.2   The Birth of Science Forward and the Macaulay BioBlitz

Science Forward is the second iteration of Macaulay's third sequential honors seminar, which students always take in the fall of their sophomore year. The first version was "Science and Technology in New York City" and while all of our courses have approved and shared learning outcomes, the science course became unwieldy to coordinate as the Honors College program grew from just a few sections in its first years to over 20 by 2012. Because of our distributed nature, the science course suffered from curricular drift; faculty across the campuses made many different and wonderful courses, but they were often in very narrow fields of science or sometimes were not even science courses at all. We want our students to become scientifically literate, to be able to encounter science, any science, and be able to understand why the people involved are doing what they are doing. Students were not always getting this experience, so in the beginning of 2013, then-Dean Mary Pearl (herself a primatologist and conservation biologist) set out to redesign and refresh the curriculum with two main goals: (1) creating a more engaging experience for a diverse student body, and (2) focusing on scientific skills in the context of multiple fields of life and physical sciences. We want our students to think about how the process of science and our current scientific knowledge can help to address urgent local and global challenges facing our society. We also want our students to address justice by exploring the intersection of science, ethics, and equity issues in addition to harms groups have faced on the way to scientific discovery. This new curriculum framework would be called Science Forward and we would create a companion open educational resource (OER) to share it with the broader community.

We wanted to revamp the curriculum so that students on all of our campuses could have similar experiences and platforms for conversation, without putting overly prescriptive mandates on our faculty. We also needed the course to productively serve both STEM and non-STEM majors together in the same classroom. We used these goals to reflect on the type of course we wanted our students to take, and what skills and knowledge we wanted them to gain that they don't typically get elsewhere in their college careers. The course is meant to explore science in the tradition of liberal arts, with critical inquiry at its heart. To realize this ambitious goal, Macaulay needed to draw upon the expertise of faculty from across the university but also needed a realistic implementation plan. We needed someone to take this initial proposal and develop it into a full-fledged curriculum to be taught for the first time in only a few short months. Dr. O'Donnell, an evolutionary ecologist with experience in pedagogical innovation and active learning in science education, was hired for this task in June of 2013. The university's central administration committed support to the project under its CUNY Advance[2] initiative, which was a program dedicated to supporting innovative pilot projects at the university. During its existence, Dr. Brundage – who holds a Ph.D. in English and gained expertise in digital pedagogy and experiential learning events as an Instructional Technology Fellow at Macaulay during her doctoral training – served as CUNY Advance's director and provided project management and pedagogical consultation for its portfolio. The partnership that began then has continued, as Dr. O'Donnell has moved into a full-time role as Director of Science Forward, and Dr. Brundage transitioned first to Director of Teaching, Learning, and Technology, and then Director of Academic Affairs, at Macaulay.

Both of us have landed at Macaulay in non-traditional academic roles that are at once administrative and directly student-serving. Both of us also received robust training in pedagogical methods throughout our doctoral and postdoctoral work, and draw on years of classroom experience that underlies and inspires how we approach our work as non-faculty academics and teachers. While we both teach courses regularly, our day-to-day jobs have responsibilities that are typically outside of the scope of faculty lines. The types of work we do are certainly possible for faculty through summer and release time, but sustaining ongoing activities related to developing and scaling a course such as Science Forward and having an annual event like the BioBlitz would be challenging for a tenure-track faculty team and impossible for contingent faculty.

As Dr. Brundage (with Gregory and Sherwood) has argued elsewhere (2018), in the rise of alternative academic or "alt-ac,"[3] roles, "makers" and "disruptors" are often highly valued, but universities also need to cultivate connectors: those who do the carework required to bring together disparate participants, stakeholders, departments, and other branches necessary to do and sustain complex, innovative projects such as creating the companion Science Forward OER video series and unique events like the BioBlitz. While BioBlitz itself invites reflections on environmental sustainability, our work also highlights sustainability as a labor issue. Here, we assert that the BioBlitz draws upon our background as project managers *and* pedagogues, and is enriched by our academic training and our cross-disciplinary approach to problem-solving and intellectual exploration. We are also firm in our belief that the technical execution of the BioBlitz cannot be separated from the emotional labor – the carework built into project management – that is required to provide outstanding and inclusive experiences for our students and team leaders. Because we, as a partnership, do the soup-to-nuts planning and implementation (everything from site selection to bug spray) we are able to constantly build, adapt, and iterate from year to year during planning, and minute to minute during the

---

[2] CUNY Advance was funded from 2013 to 2016 and was not affiliated with NSF ADVANCE.

[3] "Alt-ac" is a term that has been in circulation since approximately 2010, and refers to careers for PhDs that are off the tenure track, and may be within or outside of academia, but draw upon expertise gained in PhD training (Manzo & Renner, 2015).

BioBlitz itself. The multiple types of services we provide are facilitated by being in alt-ac connector roles at Macaulay.

Securing both of us to work on the Science Forward curriculum project added diversity of career types to the planning committee tasked with building the Science Forward framework. Because we knew we were going to be building an interdisciplinary science curriculum, it was important for the committee to include faculty from both the life and physical sciences. There were faculty from chemistry, wildlife biology, ecology, astronomy, and climate science included. In addition to having multiple perspectives from within science, our committee also included representatives from the humanities and social sciences. All were interested in improving undergraduate science education in the context of Macaulay's liberal arts curriculum and this diversity of career-type and academic background aided our efforts.

The committee was tasked with stepping out of their discipline-specific silos and agreeing on a framework that highlights the common skills that scientists employ when they do their work. The scientists were able to enumerate the skills they use in their fields and the topics to be used in the course as the context for the skills. A key aspect to the functioning of this committee was the inclusion of perspectives from outside of the sciences. Our humanities representatives (including Dr. Brundage) had experience teaching survey courses that are similar to what we were trying to build for science and they helped to focus our efforts on making the common narrative thread (science skills) clear throughout the curriculum.

Required undergraduate science courses are often burdened by the sheer volume of content faculty must cover with students and are almost always tied to specific disciplines. Many instructors acknowledge that teaching scientific skills is important but feel that they need to prioritize covering the course content over teaching about the process of science (Coil et al., 2010). However, evidence suggests that students learn better with a more student-centered, active approach that focuses on core concepts and competencies (Petersen et al., 2020). In a typical, lecture-based science course, there is scant time to explain and reflect on how a simple 101-level experiment relates to the larger research work in climate change or drug development, for example. At the same time, the amount of scientific information that we (generally) encounter and need to evaluate daily is staggering, touching on intimate aspects of life like the food we eat, water we drink, air we breathe, medications we take, and climates we experience. In an age of fake news and science denial, creating a scientifically literate population is a critical need. Our Science Forward curriculum does just that, and the BioBlitz and experiential learning are cornerstones of how we connect the dots from science as an academic pursuit to science as a human endeavor that has meaning and context in lived experiences.

After several committee meetings and many spirited debates, we defined the backbone of our Science Forward framework, the Science Senses, which includes those scientific skills that we wanted our students to learn and practice. Having Science Sense means using these scientific skills to distinguish science from non-science and from pseudoscience; to recognize how people collect and process facts into knowledge; to question and evaluate information that is presented as scientific. Our overarching learning goal was to produce students who are informed

Science Sense	Skills include
NUMBER SENSE	Being able to apply basic mathematical reasoning Having a sense of scale Making order of magnitude estimates
DATA SENSE	Making measurements Measuring uncertainty Recognizing bias Using proxies Managing data sets Doing statistical analysis Using mathematical models Finding relationships and trends Visualizing data Interpreting graphs
KNOWLEDGE SENSE	Asking a scientific question Using proper experimental design Communicating results to scientists & the public Understanding how science makes progress Thinking critically Being reasonably skeptical Making evidence-based arguments Applying scientific knowledge Understanding the intersection of science & ethics Distinguishing science from pseudoscience Acknowledging equity issues in science

Fig. 11.1 A sample of the general types of skills that fall into each of the Science Sense categories (Number Sense, Data Sense, Knowledge Sense) in the Science Forward curriculum framework.

consumers, evaluators, and/or practitioners of science. The committee categorized the Science Sense skills into Number Sense, Data Sense, and Knowledge Sense (see Fig. 11.1 for a sample of some of the Science Sense skills). Number Sense skills focus on basic numeracy, unit conversion, scales, etc. Data Sense skills include what a scientist does with those numbers: collecting data, doing analyses, visualizing results, etc. Knowledge Sense skills are those that deal with the creation of knowledge from scientific evidence and also include skills related to the philosophy of science: what does it mean to ask a scientific question or to be reasonably skeptical, for example. Science Forward requires an interdisciplinary context so that students understand that these skills are common across all fields of science. A scientist needs to know why it's important to establish a hypothesis before looking at the results of an experiment whether that experiment involves invasive plants or photons. Likewise, we consider providing all students with the foundational knowledge to navigate the scientific information they will encounter in day-to-day life to be a pressing social justice matter, whether they will become scientists or not. We support the implementation of our novel curriculum with an openly available and accessible[4] video library and web resource we created for it, the Science Forward OER (O'Donnell et al., 2014a).

Much research has suggested that explicitly teaching science skills helps undergraduates succeed in their future coursework, but despite faculty recognition of this, few programs exist with this purpose (Coil et al., 2010 and references therein). Training in these skills is clearly important for students who plan on becoming scientists and based on comments we have heard from our students, they do appreciate this. Dr. O'Donnell has had students tell her that they like that they can get into the reasoning behind experiments and research programs and that they do not have the opportunity to do this in their introductory science classes that are more focused on memorization. Additionally, these skills are important for all students, regardless of major, because all students are going to encounter scientific information or information presented as scientific in their daily lives. Students should be able to critically evaluate claims that they see in their social media feeds, for example.

With the interdisciplinary skills framework set, the second major innovation of Dr. Pearl's original proposal was the inclusion of a massive data collection event to start the course: a BioBlitz. Because of time and space constraints across the different campuses, there is no capacity for this course to have a formal lab component, but we felt it was very important to give our students hands-on experience with science. What better way to have students hone their scientific skills than by having them actually engage in authentic data collection and interaction with actual scientists? Having the BioBlitz included as an essential experiential learning component of Science Forward helps us achieve three goals. First, it creates a student-generated data set that can be used in class to practice the Science Sense skills. Second, it allows our students to interact with naturalists/scientists in the field who can model for them observational skills and how to think of the city as a laboratory. Finally, the BioBlitz builds inclusive connections in our student community; all students from our eight campuses are shuffled together on teams at the event.

11.3   How the BioBlitz Happens

The BioBlitz is a part of our authentic research experience in Science Forward, but what is a BioBlitz? In general, a BioBlitz is an intense species survey that takes place in a specific location in a defined period of time, typically 24 h However, there is no real hard and fast definition of BioBlitz. The term "BioBlitz" was first used by the US National Parks Service in 1996 when they surveyed Kenilworth Park in Washington, DC; no one person or organization owns the term "BioBlitz" itself

---

[4] The Science Forward videos come with closed captioning, written transcripts, and a version of the video with audio descriptions of the visual images.

(Droege, 1996). The National Parks Service continues to hold annual BioBlitzes and, for their centennial celebration in 2016, they held many BioBlitzes across the country (National Parks Service, 2019). BioBlitzes are often meant to be celebrations of nature and, usually, these are public events where "citizen scientists" can aid in cataloging the biodiversity of their local parks. We use "citizen science" here to indicate a global citizen of our whole society or a member of the public, not as a citizen of a particular nation. There are ongoing discussions about figuring out a new term for this type of amateur participation in the scientific process due to the exclusionary nature of the word "citizen" (see discussions in Eitzel et al., 2017; Cooper et al., 2021).

Our BioBlitz is slightly different from a typical, citizen science event in that participation is restricted to our own students, due to space limitations on our survey teams. The Macaulay BioBlitz is for all Macaulay sophomores (approximately 500 students annually) and is the kickoff event for their required Science Forward seminar. The Macaulay BioBlitz is typically held in early September, and students use the data they collected during BioBlitz and other relevant data sets to complete a research project and present a poster at our STEAM Festival in December. STEAM stands for Science, Technology, Engineering, Arts, and Technology. At Macaulay, the STEAM Festival is attended by first-year students in our Arts in New York City seminar and sophomores in Science Forward. Students present their final projects in poster sessions or gallery-style displays and can engage in conversations that bridge arts and sciences and link our curricula across years and students across campuses.

The BioBlitz is set within a series of courses and events that connect Macaulay students to New York City and ask them to deepen their understanding of how the city works, their role in creating a more resilient and just city. The curriculum and events invite them to make full use of the many resources, spaces, and institutions that New York has. In addition to the aforementioned Arts in New York City, which all students take in the fall semester of their first year, students take the People of New York City (a social sciences course) in the spring of their first year, followed by Science Forward and then Planning the Future of New York City (a public policy and urban planning course) in the fall and spring of their sophomore year, respectively. Most of our students have spent their entire lives in New York City, and these four core seminars are meant to introduce basic methods in humanities, social sciences, life and physical sciences, and public policy. We see this series of courses as an invitation to our students to understand that they are entitled to the spaces and resources of New York, but also carry the responsibility to be stewards of and ensure a more sustainable, equitable future for NYC. We believe this is best accomplished through a mixture of classroom work and hands-on direct work in these spaces.

In order for the BioBlitz to achieve such broad and ambitious goals, we have a structured logistical organization that ensures student-to-student and student-to-scientist interaction. We have our students sign up for 3-h shifts to be at the BioBlitz. This does not include travel time to and from the park, but in some years, we provide charter buses to take our students from a central location to the site. During their shift, students are assigned to teams led by taxonomic experts and they are tasked with finding and documenting as many species as possible within that shift as they can. BioBlitzes can have generalist or specialist teams. Our teams are taxon-specific; in 2017, for example, we had teams searching for ants, bats, bees, beetles, birds, fish, flies, herps, lepidopterans, lichens, microbes, odonates, plants, pond invertebrates, spiders, and even one abiotic team monitoring water quality. All data are collected on paper and/or via the iNaturalist app and are made public for all to share. Science Forward faculty are also invited to participate at the BioBlitz.

Preparing for the Macaulay BioBlitz takes months of planning and coordinating with the host location. We usually begin by identifying a willing host organization and initiating legal agreements for individual and shared responsibilities. Since our students are from all over New York City, we have made a point to include all boroughs in our BioBlitzes: Central Park in Manhattan in 2013, New York Botanical Garden in the Bronx in 2014, Freshkills Park on Staten Island in 2015, Brooklyn Bridge Park in Brooklyn in 2016, and Alley Pond Park in Queens in 2017. We then started to cycle back through boroughs with Inwood Hill Park (Manhattan) in 2018 and Green-Wood Cemetery and the Gowanus Canal (Brooklyn) in 2019. We were in the midst of planning a 2020 BioBlitz at Randall's Island (technically part of Manhattan, but situated between the shores of Manhattan, Queens, and the Bronx) when the COVID-19 pandemic hit. In 2020 and 2021, we shifted to fully remote BioBlitzes, inviting students to submit observations via iNaturalist from wherever they were located during the month of September. In the remote setting, students are required to attend one of several webinars we set up to include panels of taxon leaders who describe how to make iNaturalist observations of their particular group of interest. Students get a chance to talk with the taxon leaders in this setting.

After a location is determined, we begin to call on our network of scientists, naturalists, and local nature lovers with taxonomic expertise to lead the teams at our BioBlitz. From our experience, we have noticed that people who choose to lead teams are generally people who, in addition to being experts in a particular type of organism, love to share their knowledge with our students and demonstrate how they go about making scientific observations. We have many returning taxon leaders every year, including some that have been with us for nearly all of our BioBlitzes! We build the day's schedule based on the taxon leaders' availability, park restrictions, sunrise/sunset, and tides. We also survey students for any ADA or other accessibility needs. We are committed to creating solutions that will allow the full participation of all students. The result is multiple concurrent taxon-specific teams distributed throughout the area we are surveying (see Fig. 11.2 for an example). We open student registration for the BioBlitz before classes start and students can choose to register for any open shift. This way we can control how many students are at the BioBlitz at any given time and we get a mix from all of our campuses to build a sense of community among our distributed students.

In addition to our taxon leaders, the BioBlitz requires Macaulay staff to be onsite coordinating the day. We (Drs. O'Donnell and Brundage) are present for the entire duration of the BioBlitz, plus set up and clean up. We have a cache of equipment that our leaders and students need to facilitate their work, with everything from

Fig. 11.2 The first few hours of the 2016 Macaulay Honors College Brooklyn Bridge Park BioBlitz as seen in the Sched app showing staff shifts, student check-in, and concurrent taxon survey teams

microscopes, to heavy-duty boots, to UV protection goggles to use at our insect light trap, to aerators for fish boxes. We also have pop-up tents; signage; tables and chairs; T-shirts for our students, leaders, and staff; first aid kits; snacks; water; sunscreen; and bug spray. Equipment is inventoried and replenished every summer, and usually moved to the survey site for storage sometime in the week before the event, with pick up the week after. During the BioBlitz, Macaulay staff work in shifts checking in students, providing assistance to taxon leaders, and providing general logistical support. We sometimes attract the attention of the local media and field questions and coordinate with them on the kinds of reporting they would like to do. We also ensure that we do a complete cleanup of the areas we use. On the Science Forward OER, faculty are provided with lesson and project ideas for incorporating personal reflections and scientific analysis of the BioBlitz experience and data into their classes (O'Donnell et al., 2014b).

11.4 The Macaulay BioBlitz as an Authentic Research Experience

Experiential learning is one of Macaulay's key curricular themes. The BioBlitz represents the key experience in Science Forward and it was proposed to integrate an authentic research experience into the core requirements of our students. We want our students to learn how science works by engaging in an investigation based on the BioBlitz data, but led by their own lines of inquiry. Although Science Forward is an interdisciplinary course, we followed recommendations of reports such as Vision and Change (AAAS, 2011), which states that whether or not students who participate in a research experience go on to careers in scientific research, the experience itself improves their ability to understand how scientists work. The experience also prepares them to better evaluate scientific claims that they may encounter in their daily lives.

The positive outcomes of integrating a research experience into a science course are supported by a growing body of research. For example, Gasper and Gardner (2013) showed increased critical thinking ability and understanding of the nature of science after an authentic research experience was integrated into their introductory biology course. But what does it mean to be "an authentic research experience?" In a large national survey of biology faculty conducted by Spell et al. (2014), it was shown that faculty define authentic research experience in many different ways, however, common themes do emerge, such as incorporating experimental design, data collection, data analysis, and presentation. When developing Science Forward, we wanted to touch on all of these themes as they directly relate to the skills training we were hoping to instill in our students.

The Macaulay BioBlitz serves as the data collection opportunity for the authentic research experience in Science Forward. We want our students to be thinking about making meaningful observations and to interact with enthusiastic experts. We also recommend that students think about what kinds of empirical questions they could ask about the creatures and environment they are observing. Both students and faculty can access some BioBlitz data immediately as observations are uploaded to iNaturalist. About 2–3 weeks after the event, the full data set, including all of the data collected on paper data sheets, are made available to the faculty first to allow students time to come up with their questions before seeing the full data set.

The rest of the research experience takes place in the classroom during the semester. The Science Forward framework requires that the BioBlitz experience be used during the course. We strongly encourage the ~30 faculty who teach the course to use the BioBlitz experience as the basis for an original student research project, which in most cases, will make up the largest portion of a student's grade. Given the length of the semester and the placement of the BioBlitz in the schedule, these projects usually are not highly complex studies, but they do give our students the opportunity to take ownership of a part of the BioBlitz data and think deeply about what those data mean.

Science Forward faculty are provided with a recommended project scaffold to use so that students can take time to explore different aspects of the scientific process (available at the Science Forward OER (O'Donnell et al. 2014a)). Students were out in the urban habitat and should then think of testable questions they could answer using the data gathered at the BioBlitz. They may also incorporate other open data sources or even revisit the BioBlitz location to collect more data. The students then decide on how they want to analyze the data: what comparisons will they make and what statistical analyses will they perform. In the project scaffolding, we recommend that students write this part up as a brief project proposal. We have had students compare bat activity in different areas of a forest using chi-square analysis, calculate diversity indices for bird communities, and compare diversity across different years of the BioBlitz, for example. We have also had students collect more data to investigate lichen diversity across an assumed pollution gradient, human impact on bird abundance, and even survey data on their classmates to gauge response to the event itself.

The final step in this research experience is communicating results at our semester-closing common event for Science Forward: the Macaulay STEAM Festival. At this event, all (approx. 500) of our Science Forward students share the results of their semester-long projects as research posters. In class, our students determine how to clearly communicate what they have found out in the visual format of a poster. The activity models preparation for a professional meeting and calls on specific skills related to communicating scientific information to a scientific audience at the STEAM Festival. Our students get the opportunity to tell their classmates about what they have done and to interact with our science faculty and graduate students who are interested in their findings.

By having our students collect data, ask their own questions, design their analyses, and communicate their results, we are hitting major themes of an authentic research experience (Spell et al., 2014). Our students are participating in the actual process of science and employing many of the Science Sense skills that we want them to be honing in the Science Forward course.

11.5 "So, You Want to Do a Bioblitz…"

Our annual BioBlitz will continue into the future and there are many other BioBlitzes occurring around the world at any given time. At Macaulay, Dr. O'Donnell also organizes the New York City arm of the City Nature Challenge, which is an annual global BioBlitz for urban biodiversity run by the California Academy of Sciences and the Natural History Museum of Los Angeles County (Young & Higgins, 2021). If the reader is interested in starting their own BioBlitz, we can offer a few tips. The first thing to do is determine what the outcome of the BioBlitz is going to be: will it be a research experience integrated into class? Will it be for public engagement with nature? Determining your outcomes beforehand allows you to structure the event in a way that best achieves these goals.

The scope of your BioBlitz depends on your purpose and participants. Our BioBlitz is tightly scheduled because we are guaranteed about 450–500 student participants and simply do not have the capacity to plan for and distribute teams if the students showed up whenever they wanted. Students sign up for a specific session and know their time commitment upfront. We also choose to have taxon-specific teams to facilitate a more diverse species list. Students are assigned to teams once they arrive, which minimizes uneven placement and eliminates requests to swap teams. Depending on our site's proximity to public transportation, we sometimes provide shuttle bus service to the event, so we need to figure out how large each wave of students will be, and how many buses we need to accommodate them. This may be less of a challenge if you are in an area where most students use personal vehicles as their primary mode of transportation, and your sites have ample parking.

Many BioBlitzes are less structured than ours; they may have a smaller number of participants so it might make sense for them to have more generalist teams so that there are fewer gaps in taxonomic coverage. With the broader use of iNaturalist, we are also seeing more and more open and unstructured BioBlitzes like the City Nature Challenge. BioBlitzes like these require a robust local community of iNaturalist users to help with identifications synchronously or asynchronously. You can even participate in a "Personal BioBlitz" (Pollock et al., 2015) which takes place over many weeks and seeks to increase biodiversity awareness through the documentation of a personal life list of species in a friendly competition with other users.

Whatever your scope, start planning early. Yes, it is true that we planned the 2013 Central Park BioBlitz in about 3 months, but this is not ideal for finding experts to lead teams or for the organizers' mental health. Ironing out the legal agreements alone can take 6 or more months for some locations, even if the hosts are all on board already. It is vital to have formal agreements about the event, delineating specific logistical and monetary responsibilities of each party, including access to buildings and spaces at off-hours, safety information and waivers, and who has key decision-making powers, such as event cancellation due to inclement weather. Additionally, you must also consider governmental and institutional permits and timelines for the survey activities you are conducting, such as state/city research permits and Institutional Animal Care and Use Committee (IACUC) authorization. With the 2015 Freshkills Park BioBlitz, for example, we were holding a CUNY event in a city park with its own conservancy on land jointly held by the NYC Parks Department and the NYC Department of Sanitation and students were checking in at a ferry terminal operated by the Department of Transportation. We had to have agreements and/or waivers in place for all of those involved institutions. Then for the surveying activities, we had a NYS DEC License to Collect and Possess – Scientific, an NYC Parks Research Permit, and IACUC approval from City College.

At our BioBlitzes, we strive to provide easy and positive experiences for our students and our taxon group leaders, which means a meticulous organization of both logistics and objects. We get T-shirts printed for all of our participants, with students in one color and scientists, staff, and volunteers in another color. It is incredibly helpful to have all participants visually identifiable and to be able to tell at a glance who is a student and who is working. If you cannot afford to do this, we recommend that you try to have T-shirts for your leaders and staff so that students (and you!) can easily locate them. We also assign students a team number, rather than a team name. Faced with a choice like "flies or fish" students generally show some vertebrate bias and choose fish. We learned the hard way that students can "forget" which teams they have been assigned if they believe their assigned taxa are unfamiliar, uninteresting, or frightening. But what we have also witnessed is that an enthusiastic leader can absolutely sway attitudes about the organisms in team survey work! We do, however, ask for allergy information and won't place a student allergic to bees on the bee team, for example. We also pre-assign any students with disabilities to the team we have mutually agreed will be best for them. We work with our taxon leaders to determine the ideal group size, assign students in accordance with their wishes, and pre-determine where in the park their work will take place. Taking this step is especially important for certain groups like the birding teams, which need to be small to remain quiet.

If you are going to incorporate a BioBlitz as a recurring event, consider your budget and space for equipment. We have acquired over 35 large plastic boxes (upwards of 90 cubic feet) of equipment over the past BioBlitzes. We are at capacity for storage in our building. Borrow equipment when you can. We also carefully inventory the equipment with a checklist and keep the bins organized and clearly labeled by taxon. Having the equipment (and other materials for registration etc.) carefully organized relieves a lot of stress onsite, when time is short and the registration area is crowded. Do what you can to help teams self-start! Also plan to thoroughly familiarize yourself with the area you will survey: time how long it takes to walk from the check-in area to the survey areas, note any obstacles, and anticipate challenges such as moving, guarding, and setting up equipment. If leaders will be working with two consecutive groups of students, will they be able to come back to registration to meet the new group, or will they need to stay with equipment in the field? We find it useful to print a poster-sized map and physically move objects around it representing teams and times to get a feel for how the day will stitch together. While we keep a detailed, private spreadsheet of all the logistical information, we have also used sched.com (an event organization tool, see Fig. 11.2) and other graphics to make an easily readable and shareable timeline of the day. Whatever form your BioBlitz takes, make sure to build some enjoyment in for yourself, too! Enjoy your surroundings, log some observations, and appreciate the experience you are bringing to your students.

11.6   The Value of the Macaulay BioBlitz

Developing curricula that instill scientific literacy, including procedural knowledge, evidence evaluation, and that take into account science's role in everyday life is critical across industry, social action, and the personal realm (Duggan & Gott, 2002). With our new Science Forward framework, we have created a course for our students, both STEM and non-STEM majors alike, that focuses on these skills and the role of science in the world. Our annual BioBlitz is integrated into this framework as a highly engaging research experience for our students who get a chance to participate in the actual process of science. In class, students explore how to ask an empirical question of these data, how to manage ecological data, what statistical analyses are appropriate, and how to communicate clearly to both a public and a scientific audience. Our BioBlitz is a worthwhile logistical challenge that fosters these opportunities.

Besides skills training and engagement with the scientific process, there are other important outcomes from the BioBlitz experience. Unlike many research experiences, such as those that take the form of apprenticeships in labs, Science Forward projects are usually a singular exploration rather than an iterative one. However, our data do live on in their open format. Anyone can visit our BioBlitz website[5] to access the data or download BioBlitz observations from iNaturalist. Using the iNaturalist platform also means that those BioBlitz observations that become what iNaturalist calls "Research Grade"[6] can be seen and used by a wider audience through the Global Biodiversity Information Facility (GBIF), an open repository of ecological datasets (GBIF.org, 2021). Research Grade iNaturalist observations get uploaded weekly to GBIF where they can be accessed by researchers looking for species occurrence data (Ueda, 2021). Our students are contributing to a worldwide dataset that any scientist can use to ask ecological questions about our changing planet. For example, iNaturalist data were used to document butterfly decline in the western US (Forister et al., 2021). iNaturalist data can even be used to address issues regarding sustainable urban planning (see Cambria et al., 2021 for a European example).

In the process of collecting these data for our BioBlitz, students are potentially also engaging in biodiscovery, documenting new species or interactions in a particular area (Hitchcock et al., 2021). For example, at our first BioBlitz in 2013, several students on the plant team led by Daniel Atha of the New York Botanical Garden, documented species that had never been reported at Central Park before, including *Plantago rugelli* observed on iNaturalist by Erisa Gjinaj (2013), *Persicaria extremiorientalis* observed by David Kleyman (2013), and *Cyperus iria* observed by Jonathan Chevinsky (2013) (Atha, personal communication). If one considers the impact on the iNaturalist database itself, our BioBlitzes often have students making the only records of some species in the area and our BioBlitz observations can make up a large percentage of a host park's total observations (see Hitchcock et al., 2021). Our BioBlitzes have also brought new, knowledgeable users to iNaturalist when our taxon leaders sign up as they prepare to lead teams for us. For example, Daniel Atha started the New York City EcoFlora project, an effort of NYBG to engage New Yorkers with the plant species around the city as they follow monthly themed quests to make iNaturalist observations (Boom et al., n.d.). Susan Hewitt, mollusk

---

[5] http://macaulay.cuny.edu/eportfolios/bioblitz/

[6] Research Grade observations are those where 2/3 or more of the iNaturalist users who contribute identifications agree on the species (iNaturalist, 2021b).

taxon leader at the 2014 BioBlitz, is the ninth most prolific observer on all of iNaturalist (as of this writing) with over 93,000 verifiable observations (she's twelfth for all observations) (iNaturalist, 2021c). Our BioBlitz has also been the inspiration for further surveys of the Central Park flora (Atha et al., 2020) and lichens at Freshkills Park (Allen & Howe, 2016). At a large scale, our students can feel more connected to a global community of naturalists through the contributions they make via iNaturalist and the interactions they have with the users of that database.

Another important outcome is students seeing and interacting with scientists in one of their natural habitats. Granted, our sample is not representative of all scientists, but students are getting the opportunity to have meaningful conversations about the work that these scientists do in an intimate group outside of the lecture hall. We hope that this shatters the misconception of all scientists being isolated people in sterile labs wearing white coats and safety goggles.

The BioBlitz experience also serves to instill a sense of place in our students at local and global levels. As a common event, the BioBlitz brings together our eight campuses and fosters a sense of community within the entire class. As an event that is focused on the green spaces of New York City, the BioBlitz connects our students to the urban habitat. They have even had the opportunity to explore areas not yet open to the public: the Hallett Nature Sanctuary in Central Park was not open yet when we had the 2013 BioBlitz, and Freshkills Park is still not entirely open to the public. Opportunities like these make the event special for our students and introduce them to ongoing stewardship efforts that the city supports. Students can feel proud that they are able to contribute to the protection of these natural areas that are a part of their urban homes. In a few cases, our students and naturalists are establishing the baseline species list in areas that had never been surveyed at this level before. Data like these may be particularly useful as more research effort is directed at questions about the importance of urban biodiversity in global conservation efforts. There remain many questions to be answered about not only the ecological processes happening across networks of urban green spaces (Lepczyk et al., 2017), but also how to effectively use that information in the context of the social, economic, and other human considerations that factor into effective urban green space management (Aronson et al., 2017).

Students can also make connections between their BioBlitz experience and other aspects of our Science Forward and honors curricula. All in-person Macaulay BioBlitzes have been held on the unceded territory of the Lenape people and we want students to acknowledge that fact and pay proper respect to the original inhabitants of the land our campuses sit upon. We incorporated our first formal land acknowledgment (drafted by members of Macaulay's Teaching and Learning Collaboratory) into our 2019 BioBlitz, and Dr. O'Donnell has created our current version (O'Donnell, 2021). We want students to think about how the landscape has changed with the impacts of colonization and we make sure to point them to resources like The Lenape Center (2021), *Mannahatta* (Sanderson, 2009), and The Welikia Project (WCS, 2008) so they can find out more. We also want students to think about how land use has continued to change even beyond colonization and which groups have been harmed by that process. BioBlitz host sites that are in the process of transformation from heavily human-damaged spaces to parkland are good examples of our attempts to recover from this destruction (Freshkills Park and the Gowanus Canal, for example). We also encourage faculty to use the BioBlitz experience as a way to broach themes of environmental justice in their classes.

Participating in BioBlitz is usually unlike any other science course experience students have had and gets many positive responses from our students. Because the course serves sophomores, even our STEM majors have generally not yet participated in any research activities. In September 2016, we informally surveyed 286 students when they registered to attend the BioBlitz and 82.5% were not currently involved in any research in a science lab, computer lab, or in the field, and 32.8% considered themselves unlikely or highly unlikely to engage in scientific research outside of their coursework. In addition to knowing that the event boosts students' research experiences, the qualitative feedback we get from students is also valuable to us. We have seen students take tentative, wader-encased steps into the East River, and then be amazed at what a seine net hauls in. We have witnessed students literally yelling with excitement after catching an insect in a net, and watching with amazement as a scientist delicately handles a bat and shows them how to assess its health. We hear from our team leader scientists that the BioBlitz gives them a unique way to interact with and relate to students, sharing the work that they love with students. The BioBlitz has lasting effects on students, too. In senior surveys (administered over 2 years after the graduating students' BioBlitz experience), we get comments such as, "BioBlitz was very interesting because I got to learn about the natural environment within the city, instead of thinking of wildlife as only something you can find out in the wilderness." Another graduating student articulated the benefit of interpersonal skills that experiential learning imparts: "Going to unique Macaulay opportunities such as … bioblitz, museums, etc., has opened my mind to a different way of thinking. If I didn't go to these events, I would be forced to only do coursework and activities relevant to my major. Because I've explored so many different fields like art, science, history, and more, I have more skills and knowledge which makes conversing with different people so much better." Events like the BioBlitz help our students gain academic skills, but also make them informed citizens with deeper connections to the places in which they live.

We also know the BioBlitz has the potential to alter career tracks for our students, like Allegra DePasquale. She told us, "BioBlitz was one of my favorite events, and changed the course of my academic career. Surveying mammals with Dr. Pearl showed me how fulfilling fieldwork can be, and inspired me to create my own major in zoology and conservation through CUNY BA [a degree program for individualized study]. I'm now applying to Ph.D. programs in primatology. I really wouldn't be where I am if I hadn't discovered my love [of] animal biology and ecology at the BioBlitz" (private correspondence, quoted with permission). Allegra's story shows how taking students out of the classroom and providing opportunities for them to interact with scientists in a casual, personal way like they do at BioBlitz can make scientists and the professions of science personally relatable. STEM and non-STEM majors alike deserve the chance to have authentic experiences with the scientific process and to experience moments of discovery and genuine joy.

At BioBlitz, students get to know faculty and scientists in ways that are not attainable in a typical course setting, which can open personal and professional pathways. But highlighting the human face of science is a useful goal in itself, especially in our age of science skepticism. In a study of public perceptions of scientists and trust in scientific information, Fiske and Dupree (2014) explained that trustworthiness is based on competence *and* warmth. Fiske and Dupree found that while most Americans view scientists as competent, fewer view them as warm, which affects public perceptions: "Overall, communicator credibility needs to address both expertise and trustworthiness. Scientists have earned audiences' respect, but not necessarily their trust." Fiske and Dupree conclude, "Rather than persuading, we and our audiences are better served by discussing, teaching, and sharing information, to convey trustworthy intentions." Our experience with BioBlitz is that it serves these same goals, connecting scientific inquiry to the fabric of the city and students' lives. Increased trust in science as a process and scientists as people is an important aspect of developing scientific literacy – a way of knowing that we insist they have a right to develop, no matter what their backgrounds, majors, career goals, or future may be.

Acknowledgments We would like to acknowledge all who have had a hand in supporting both Science Forward and the Macaulay BioBlitz including Mary Pearl, Joseph Ugoretz, Janet Fu, the Science Forward Planning Committee of 2013, and the Macaulay staff who work at the BioBlitz every year. Of course, we could not have a BioBlitz without the help of the many knowledgeable taxon leaders that take our student teams out to explore. And thanks to the Macaulay students who collect the data and make iNaturalist observations. Science Forward has been funded by CUNY Advance, with thanks to the Central CUNY Administration, Ann Kirschner, and George Otte; and by the CUNY OER Initiative.

References

AAAS. (2011). *Vision and change in undergraduate biology education: A call to action.* American Association for the Advancement of Science.

Allen, J. L., & Howe, N. M. (2016). Landfill lichens: A checklist for Freshkills Park, Staten Island, New York. *Opusc Philolichenum, 15,* 82–91.

Aronson, M. F., Lepczyk, C. A., Evans, K. L., Goddard, M. A., Lerman, S. B., MacIvor, J. S., Nilon, C. H., & Vargo, T. (2017). Biodiversity in the city: Key challenges for urban green space management. *Frontiers in Ecology and the Environment, 15*(4), 189–196. https://doi.org/10.1002/fee.1480

Atha, D., Alvarez, R. V., Chaya, K., Catusco, J., & Whitaker, E. (2020). The spontaneous vascular plant flora of New York's Central Park. *Journal of the Torrey Botanical Society, 147*(1), 94–116. https://doi.org/10.3159/TORREY-D-19-00024

Boom, B., et al. (n.d.). *New York City EcoFlora.* New York Botanical Garden. Retrieved November 8, 2021, from https://www.nybg.org/science-project/new-york-city-ecoflora/

Brundage, L. A., Gregory, E., & Sherwood, K. (2018). Working nine to five: What a way to make an academic living? In E. Losh & J. Wernimont (Eds.), *Bodies of information: Feminist debates in digital humanities* (pp. 305–319). University of Minnesota Press.

Cambria, V. E., Campagnaro, T., Trentanovi, G., Testolin, R., Attorre, F., & Sitzia, T. (2021). Citizen science data to measure human use of green areas and forests in European cities. *Forests, 12*(6), 779. https://doi.org/10.3390/f12060779

Chevinsky, J. [jcevinsky]. (2013). *Cyperus iria.* iNaturalist. Retrieved December 16, 2021, from https://www.inaturalist.org/observations/378874

Coil, D., Wenderoth, M. P., Cunningham, M., & Dirks, C. (2010). Teaching the process of science: Faculty perceptions and an effective methodology. *CBE Life Sciences Education, 9*(4), 524–535. https://doi.org/10.1187/cbe.10-01-0005

Cooper, C. B., Hawn, C. L., Larson, L. R., Parrish, J. K., Bowser, G., Cavalier, D., Dunn, R. R., Haklay, M. (Muki), Gupta, K. K., Jelks, N. O., Johnson, V. A., Katti, M., Leggett, Z., Wilson, O. R., & Wilson, S. (2021). Inclusion in citizen science: The conundrum of rebranding. *Science, 372*(6549), 1386–1388. https://doi.org/10.1126/science.abi6487

CUNY. (2021). *Continuing education.* City University of New York. Retrieved November 8, 2021, from https://www.cuny.edu/admissions/apply-to-cuny/continuing-education/

CUNY Office of Institutional Research. (2019). *Student Data Book.* City University of New York. Retrieved November 8, 2021, from https://www.cuny.edu/about/administration/offices/oira/institutional/data/

Droege, S. (1996). *BioBlitz: A tool for biodiversity exploration, education and investigation.* Patuxent Wildlife Research Center. Retrieved November 12, 2021, from https://web.archive.org/web/20161222181745/https://www.pwrc.usgs.gov/blitz.html

Duggan, S., & Gott, R. (2002). What sort of science education do we really need? *International Journal of Science Education, 24*(7), 661–679. https://doi.org/10.1080/09500690110110133

Eitzel, M. V., Cappadonna, J. L., Santos-Lang, C., Duerr, R. E., Virapongse, A., West, S. E., Kyba, C. C. M., Bowser, A., Cooper, C. B., Sforzi, A., Metcalfe, A. N., Harris, E. S., Thiel, M., Haklay, M., Ponciano, L., Roche, J., Ceccaroni, L., Shilling, F. M., Dörler, D., et al. (2017). Citizen science terminology matters: Exploring key terms. *Citizen Science: Theory and Practice, 2*(1), 1. https://doi.org/10.5334/cstp.96

Fiske, S. T., & Dupree, C. (2014). Gaining trust as well as respect in communicating to motivated audiences about science topics. *Proceedings of the National Academy of Sciences, 111*(Supplement 4), 13593–13597. https://doi.org/10.1073/pnas.1317505111

Forister, M. L., Halsch, C. A., Nice, C. C., Fordyce, J. A., Dilts, T. E., Oliver, J. C., Prudic, K. L., Shapiro, A. M., Wilson, J. K., & Glassberg, J. (2021). Fewer butterflies seen by community scientists across the warming and drying landscapes of the American West. *Science, 371*(6533), 1042–1045. https://doi.org/10.1126/science.abe5585

Gasper, B. J., & Gardner, S. M. (2013). Engaging students in authentic microbiology research in an introductory biology laboratory course is correlated with gains in student understanding of the nature of authentic research and critical thinking. *Journal of Microbiology and Biology Education, 14*(1), 25. https://doi.org/10.1128/jmbe.v14i1.460

GBIF.org. (2021). *What is GBIF?* GBIF. Retrieved November 8, 2021, from https://www.gbif.org/what-is-gbif

Gjinaj, E. [erisagjinaj]. (2013). *Plantago rugelli.* iNaturalist. Retrieved December 16, 2021, from https://www.inaturalist.org/observations/379176

Hitchcock, C., Sullivan, J. J., & O'Donnell, K. L. (2021). Cultivating bioliteracy, biodiscovery, data literacy, and ecological monitoring in undergraduate courses with iNaturalist. *Citizen Science: Theory and Practice, 6*(1), 26. https://doi.org/10.5334/cstp.439

iNaturalist. (2021a). *About.* iNaturalist. Retrieved November 8, 2021, from https://www.inaturalist.org/pages/about

iNaturalist. (2021b). *Help.* iNaturalist. Retrieved November 8, 2021, from https://www.inaturalist.org/pages/help

iNaturalist. (2021c). *Observations.* iNaturalist. Retrieved November 9, 2021, from https://www.inaturalist.org/observations?place_id=any&subview=map&view=observers

Kleyman, D. [kleymandavid]. (2013). *Persicaria extremiorientalis.* iNaturalist. Retrieved December 16, 2021, from https://www.inaturalist.org/observations/378792

Lepczyk, C. A., Aronson, M. F. J., Evans, K. L., Goddard, M. A., Lerman, S. B., & MacIvor, J. S. (2017). Biodiversity in the city: Fundamental questions for understanding the ecology of urban green spaces for biodiversity conservation. *Bioscience, 67*(9), 799–807. https://doi.org/10.1093/biosci/bix079

Manzo, R., & Renner, B. R. (2015). *Altac (Alternative Academic) careers: What is Alt-ac?* UNC Health Sciences Library. Retrieved November 12, 2021, from https://guides.lib.unc.edu/altac

National Parks Service. (2019). *National Parks BioBlitz.* National Parks Service. Retrieved November 8, 2021, from https://www.nps.gov/subjects/biodiversity/national-parks-bioblitz.htm

O'Donnell, K. L. (2021). *Land acknowledgement.* BioBlitz. Retrieved December 17, 2021, from https://eportfolios.macaulay.cuny.edu/bioblitz/land-acknowledgement/

O'Donnell, K. L., Brundage, L. A., & Ugoretz, J. (2014a). Science Forward open educational resource. Retrieved November 8, 2021, from https://eportfolios.macaulay.cuny.edu/science-forward/

O'Donnell, K. L., Brundage, L. A., Ugoretz, J. (2014b). *BioBlitz tools.* Science Forward open educational resource. Retrieved November 8, 2021, from https://eportfolios.macaulay.cuny.edu/science-forward/bioblitz-tools/

Petersen, C. I., Baepler, P., Beitz, A., Ching, P., Gorman, K. S., Neudauer, C. L., Rozaitis, W., Walker, J. D., & Wingert, D. (2020). The tyranny of content: "Content coverage" as a barrier to evidence-based teaching approaches and ways to overcome it. *CBE Life Sciences Education, 19*(2), ar17. https://doi.org/10.1187/cbe.19-04-0079

Pollock, N. B., Howe, N., Irizarry, I., Lorusso, N., Kruger, A., Himmler, K., & Struwe, L. (2015). Personal bioblitz: A new way to encourage biodiversity discovery and knowledge in K – 99 education and outreach. *Bioscience, 65*(12), 1154–1164. https://doi.org/10.1093/biosci/biv140

Sanderson, E. W. (2009). *Mannahatta: A natural history of New York City.* Abrams.

Spell, R. M., Guinan, J. A., Miller, K. R., & Beck, C. W. (2014). Redefining authentic research experiences in introductory biology laboratories and barriers to their implementation. *CBE Life Sciences Education, 13*(1), 102–110. https://doi.org/10.1187/cbe.13-08-0169

The Lenape Center. (2021). *The Lenape Center.* Retrieved November 9, 2021, from https://thelenapecenter.com/

Ueda, K. (2021). *iNaturalist research-grade observations.* GBIF.org. Retrieved September 27, 2021, from https://doi.org/10.15468/ab3s5x

Wildlife Conservation Society (WCS). (2008). *The Welikia project.* Retrieved November 9, 2021, from https://welikia.org/

Young, A., & Higgins, L. (2021). *City nature challenge.* Retrieved December 17, 2021, from https://citynaturechallenge.org/

Kelly L. O'Donnell is an evolutionary ecologist and science educator and the Director of Science Forward at Macaulay Honors College, CUNY. She is interested in the flora of NYC and is passionate about getting students to notice the ecology around them. She has experience creating educational videos to promote scientific thinking, including those videos that make up the Science Forward Open Educational Resource.

Lisa A. Brundage is Director of Academic Affairs at Macaulay Honors College, CUNY. In graduate school, she studied twentieth century English literature, while developing a specialty in instructional technology. Her background in interdisciplinary pedagogical work has helped her partner with colleagues to tell the stories of science.

Chapter 12
Syndemic: Using Game-Based Learning to Engage Students in the Human Microbiome

Emma Ruskin and Tal Danino

An enormous revolution is occurring in our field due to the changes in technology: shortly, there will be no more culture plates but bacterial signatures based on mass spectrometry and minimal manual work because of automation. There will be great need to use the electronic tools to communicate new findings, to teach what the new technology does, and how to transition from the classical to the "new" microbiology. We need to keep our minds, eyes, and ears open, as only the future will tell what we will need to learn and teach new generations and what electronic tools we will use. Jeannette Guarner and Silvia M. Niño, Microbiology Learning and Education Online, 2016

12.1 Introduction

The average adult has over 10 trillion microorganisms located in and on their body (Sender et al., 2016). This staggering total is more than the number of human cells. Thus, the commensal archaea, bacteria, fungi, protozoa, and viruses, which collectively form the human microbiome, outnumber the human cell population. Although it was largely believed that these microbes only inhabit and affect our gut, mucosal surfaces, and skin, recent research has led us to understand that our microbiomes play far larger roles in our health and well-being than previously thought (Khanna & Tosh, 2014). Disruptions of the formation and function of our microbial communities can have critical consequences for host health, including gastrointestinal disorders, allergies, obesity, and stress. The composition of these microbes not only influences the balance between health and disease, but also are markers of our genetic makeup, geography, and the environments we live in (Guarner & Niño,

E. Ruskin
David Geffen School of Medicine, University of California Los Angeles, Los Angeles, USA

T. Danino (✉)
Department of Biomedical Engineering, Columbia University, New York, NY, USA

© The Author(s) 2023
M. S. Rivera Maulucci et al. (eds.), *Transforming Education for Sustainability*,
Environmental Discourses in Science Education 7,
https://doi.org/10.1007/978-3-031-13536-1_12

2016). In fact, research revealing the importance of the human microbiome has not only dramatically impacted the scientific community's understanding of human health, but also environmental health and sustainability of planetary life as well. Many studies have shown the microbiomes of animals and plant species on which human civilization depend are critical for maintaining sustainable systems. However, much of this research on microbiomes has yet to become accessible to the general public.

The lack of communication between scientists and the public can be seen as a social justice issue in light of recent studies showing that we are failing to produce scientifically literate citizens. Since 2000, the Program for International Student Assessment (PISA) has measured the academic achievement of 15-year-old students in dozens of countries around the world every 3 years. In their most recent survey, conducted in 2015, PISA assessed science literacy in the educational systems of 73 countries. Although the United States spends more money on each student's education than most countries, students' scores on measures of science literacy are below average (Kastberg et al., 2016). Smaller studies tracking science achievement, specifically in the United States, have found that students' attitudes towards science, in addition to their scientific literacy, have significantly declined over the past few decades (Kelly et al., 2013). Osborne et al. (2003) recognized the decline and undertook a comprehensive review to better understand students' attitudes towards science. The study concluded that the decreasing number of students choosing to pursue science should be remedied by a dramatic pedagogical shift to teaching science in a more accessible rather than alienating way. For science learning outcomes to increase, students must be actively engaged in the material and see how their classroom learning applies to real-world situations. Syndemic, a video game inspired by the current research on the microbiome and its potential to engage students, attempts to fill this need. In this chapter, we will use the process of field testing Syndemic in the US and Nicaragua as a lens to explore:

1. How Digital Game-Based Learning can be an innovative tool to engage and motivate students in science
2. Lessons learned from our mission to make Syndemic inclusive and accessible to diverse populations in global contexts
3. Ways undergraduates, as novice researchers, can have authentic real-world fieldwork experiences

12.2 Microbial Ecology and Human Health: Catching the Curriculum Up to the Research (Tal Danino)

As we study the coevolution of the microbiome alongside humans, we see the progression of specific microbial communities developing in different environmental "niches" throughout the body. As these microorganisms both impact and adapt to their environments, they reach a state of dynamic equilibrium, which is influenced constantly and profoundly by modern society's innovations in food, medicine, and agriculture. It is now widely accepted that the majority of these microbes are not detrimental to human health, but rather exist in a largely mutualistic relationship with us, in which both parties benefit from the symbiotic interaction. The formation and function of microbial communities changes as human lifestyle evolves and alters the microbial environment. Moreover, given the natural co-evolution of humans and their microbes, there is generally a beneficial and protective effect. Researchers are currently investigating how many commonly used products could affect a microbial balance both within the body and in the environment.

Understanding our relationships to the microscopic world is critical to guiding our behaviors towards the environment and ourselves. Unlike animal and plant diversity visible at the macroscale, the microbial world is more difficult to perceive with the naked eye. Yet there are profound issues caused by local extinctions or excessive proliferation of microbes inside humans or in their environments, analogous to the loss of animals such as keystone predators or the proliferation of pests, deforestation, or problematic invasion of non-native plants. For example, bees are cornerstones for terrestrial ecosystem stability and critical components in agricultural productivity (see also Snow, this volume). Honey bee population declines have been partially attributed to increased parasite loads which are prevented by prophylactic use of antibiotics. This practice of using antibiotics such as tylosin results in persistent effects on both the size and composition of the honeybee microbiome, which increases susceptibility to opportunistic pathogenic infections, leads to antibiotic-resistance, and decreases survivorship (Raymann et al., 2017). Understanding these complex macro- and micro- relationships is critical to developing sustainable solutions for the health of bee pollinators and across many other industries.

At the Synthetic Biological Systems Laboratory at Columbia University, our research focuses on understanding and designing the microbes within the human body. As principal investigator, I first became interested in bacteria as some of the simplest organisms that could be genetically engineered. There was excitement in the newly established field termed, Synthetic Biology, that focused on engineering new functions and behaviors in microorganisms. In one of my research projects, I was able to engineer some bacteria that controlled the coordination of a bacterial community to produce a fluorescent protein in synchrony, all by manipulation or 'programming' of their DNA. We realized that because the many bacteria in and on our bodies are intertwined with human health, we could design and engineer them in the hopes of diagnosing and treating disease.

One of our lab's major goals is to specifically understand microbes in relation to cancer, and engineer bacteria to detect and treat this dreaded disease. From the scientific perspective, the complex interplay of different fields here is often difficult to communicate to many audiences. Our work includes elements of microbiology, human anatomy, cancer biology, engineering, and biotechnology. While doing demonstrations for kids at science fairs and museums, I found that visual representations of our research were the most engaging. In particular, the microscopic images and 'movies' we made garnered the most attention, not just for kids but also for a general audience and scientists. I began to focus my efforts on making our work as visually exciting as possible, teaming up with artists such as Vik Muniz and Anicka Yi. I later did two artist residencies to focus on making visual art. It was this first inclusion of art with science that I found attractive to young scientists and I have been continuing to develop engagement and learning programs based on this ever since.

Beyond communicating science to the general public, developing "bioartworks" has become a thrust of our lab. I share the vision of our research by creating art from materials in the laboratory and reflect more broadly on the themes in biotechnology that are shaping society (taldaninoart.com). Beyond engaging with people at a fundamentally more profound level, the hope is that these works will allow for inquiry and self-reflection, connection with the human experience, and help shape dialogue and future policy for how science integrates with society. For example, many projects utilize soil microbes, which form collective, fractal-like patterns on petri dishes (one instance is a series called *Microuniverse,* shown in Fig. 12.1).

These images reflect on the beauty of the microscopic world and the ecosystems of soil health. We depend on soil for the vast majority of our foods, use it as an energy source, and it is fundamental to our existence. Soil can be considered in some ways more valuable than oil. While critical to our society, soil has become unappreciated in the increasingly city-dwelling world that is disconnected from the land beneath it. Many farming practices such as deforestation and overgrazing are some of the causes of degraded soil. Additionally, the use of agrochemicals such as pesticides that help increase yields in the short term inevitably disrupt the balance

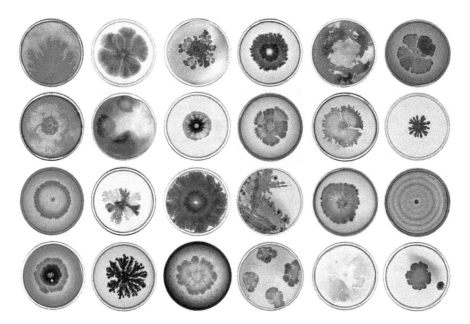

Fig. 12.1 Microuniverse. Various bacteria species grown on petri dishes. (Courtesy of Tal Danino)

and quantities of microorganisms in the soil and stimulate the growth of harmful versus beneficial bacteria. Degraded soil now covers 15–30% of the Earth's land surface, making them less useful for food and fiber production and carbon dioxide sequestration, which significantly impacts climate change. A broader appreciation of the need for soil erosion management and the development of effective legal frameworks are increasingly needed. By directly exploring the microbial species in the soil and other ecosystems, our projects bring attention to the connection between our collective health and that of the planet.

Similarly, in our cancer-art projects, we utilize cancer cells to print images with the theme of the "medium is the message." In *Colonies*, we created images of mandalas out of HeLa cells (with artist Vik Muniz), and in *Cancer Medium,* screen-printed cancer cells to explore cancer history, advocacy, and disparities. Using immortalized cells of Henrietta Lacks' (HeLa) cells to create an image of a mandala, a symbol of life, creates a conversation about mortality and bioethics and allows basic researchers to connect with the patient experience on a deeper level. In *Cancer Medium*, we charted the landscape of communication surrounding cancer research and advocacy, where producing large poster-sized prints with cancer cells as the medium provides a model of integrated communication on policy and disparity. Here the artworks create conversations about researchers' responsibilities regarding appropriate biospecimen consent, patient privacy, and supporting concrete actions to improve research participants' rights, as was brought to light in the case of HeLa cells. Furthermore, they highlight disparities in cancer research that are known to exist across geographic regions and ethnicities. For example, certain ethnic minority groups experience poorer cancer outcomes, which can be due to factors such as healthcare access barriers, stigmatization, under-representation in clinical trials, cancer literacy, and cultural beliefs, all of which can delay help-seeking and affect timely diagnosis. Here the universal language of art can transcend cultural and economic boundaries to raise awareness and provide accessible benefits for patients.

I first learned about Syndemic when I was searching for games that related to bacteria in the human body. What immediately struck me about the video game was how attractive the visuals were and that it was fun to play. I contacted the game's creator, Alex May, to see if he would be interested in collaborating to make the game more related to our research and to have scientific, educational value. We worked together to create levels in the game where bacteria would swim through the body to find and kill cancer cells, an aspect that was very similar to our laboratory's research. In this way, Syndemic is an excellent introduction to our laboratory's work and Synthetic Biology in a hands-on and interactive format. The game creates an authentic learning experience for students by accurately mirroring the work of our laboratory. While it specifically focuses on microbes swimming through the human body, the principles are fairly universal and could be applied to teach about microbiomes in other animal and plant species, as well as about microbiomes that are important in planetary health. Furthermore, the game teaches synthetic biology and biotechnological principles in the way that bacteria are engineered to produce new behaviors when they pick up DNA elements. Thus, it can lead teachers and audiences to think about the use of microbes in applications for the degradation of plastics, improvement of material yields, and other biofabrication and bioeconomy principles. More broadly, the game embodies our teaching philosophy that science education is about discovery and experience rather than textbook memorization. This philosophy is not only relevant to our work inside the lab, but it is also what led me to use the beta testing of Syndemic as an educational opportunity to engage undergraduates out in the field.

12.3   The Testing Phase: Bringing Syndemic to Students (Emma Ruskin)

Syndemic is named after a term coined in the 1990's that refers to ways that diseases co-occur and interact depending on their social, cultural, economic, and geographic contexts. The video game takes this concept to the microscopic level by exploring the synergistic interactions between microbes and their environments. Using the concept of syndemics as a framework to explore the relationship between humans and the microbiome is particularly timely given the on-going COVID-19 pandemic. In fact, some argue that COVID-19 should be more accurately described as a syndemic, rather than a pandemic. Instead of viewing the spread of COVID-19 from solely an infectious disease lens, its characterization as "syndemic" forces us to acknowledge the interplay of social and economic disparities that dictate its impact (Horton, 2020). The COVID-19 syndemic also showcases the need for a scientifically literate population that understands abstract concepts such as prevention and hygiene. By making the relationship between humans and microbial communities visually accessible, Syndemic is a critical tool to improve our current collective scientific and public health engagement.

Syndemic takes place in the future where a genetically-engineered microbe has been designed to aid the immune system in killing bacterial infections, metastasizing cancer cells, chronic allergies, and other health-related antigens. The characterization of this microbe as a useful aid helping the body promotes the idea that not all microorganisms are harmful to human health and that many are actually helpful. Moreover, this format encourages players to practice building the skills of imagination and intentional creation of a desired future. Although the game is not set in the present reality, the graphics are scientifically realistic and accurate. Cell biologist and Syndemic creator, Alex May, set the game within a stylized version of the real microbial/immunological world that is hidden from the naked eye in order to allow the learning to come from gameplay itself. This approach is in direct response to May's growing frustration that, "Many educational games simply ask questions and expect answers as if they were written exams with good graphics" (personal interview, 2017).

In our mission to meet the needs most current and relevant to the educational community, Syndemic directly builds on the following learning objectives outlined in the New York City Department of Education's Science Scope and Sequence for grades 6–12:

- Complex and microscopic structures and systems can be visualized, modeled, and used to describe how their function depends on the shapes, composition, and relationships among its parts; therefore, complex natural and designed structures/systems can be analyzed to determine how they function (7th grade, unit 4).
- Living things are composed of cells. Cells provide structure and carry on major functions to sustain life. Cells are usually microscopic in size. Cells take in nutrients, which they use to provide energy for the work that cells do and to make the materials that a cell or an organism needs (6th grade, unit 3; 7th grade, unit 3).

Some of the game's visuals were closely based on images from the Synthetic Biological Systems Laboratory at Columbia University in order to make the world of Syndemic as scientifically realistic as possible (see Fig. 12.2).

In each game-play session, players choose to enter either a resistance mode or a transmission mode. In resistance mode, the objective is to maximize high scores by harvesting nucleic acids from the environment and defeating infectious pathogens. The player can also upgrade their character by converting nucleic acids into new genes, which aids in their survival, energy, speed, and weaponry. In transmission mode, players can enter a more story-focused atmosphere, which involves curing various human hosts of their diseases. The player jumps from person to person, fighting bacterial infections, destroying cancer cells, and preventing tissue damage caused by allergies. In this mode, players can collect weapons from the environment, including new experimental antibiotics and plasmids. In order to help players understand the relationship of the game to real-world applications, there are various "reality checks." These include factual statements such as:

- Viruses are actually much smaller than bacteria and can range in size from 1/10 to 1/100 of a bacterium.
- Bacteriophages are viruses that target bacteria specifically. They are highly specialized to attack individual strains.

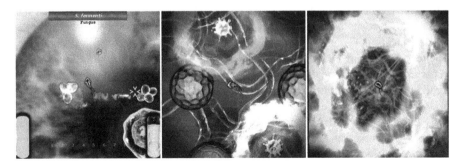

Fig. 12.2 Screenshots from Syndemic. (Courtesy of Alex May)

- The presence of pathogens in the bloodstream is a serious and potentially fatal condition called Septicemia. If untreated, it can lead to septic shock and organ failure.

Syndemic uses a variety of game design techniques to maximize player engagement. Instead of relying wholly on statistics and numbers (such as weapon power) to determine the player's "success," the player's reflexes and scenario anticipation skills are rewarded. Unlike the typical action shooter game, where pressing the fire button kills the target, Syndemic pushes its user to adapt to the host environment by requiring them to use specific tools against each antigen.

12.4   Testing Syndemic Across New York City

In the fall of 2016, a demo of Syndemic was ready for testing. As an undergraduate Biology major at Barnard, I had the opportunity to implement the beta testing phase of Syndemic. I started by contacting schools that pride themselves on their use of technology since I thought that computer availability for each student to play the game individually would be my biggest hurdle to testing Syndemic. However, I soon found that the more "technologically advanced" schools were also the most hesitant to allow any sort of research into the classroom, even though I had obtained approval from the Barnard College Institutional Review Board. I was surprised to find that getting administrators to allow research in their schools would be a minor hurdle compared to getting teachers on board. I showed Syndemic to numerous educators and consistently heard negative responses. In trying to understand their reaction, I met with a team of middle school science teachers at a college prep K-12 school on the Upper East Side. After watching a walk-through of the game, most of the teachers understood why kids would enjoy playing it, but did not see the game's educational value. After the meeting, one teacher wrote in an email, "I don't see how we could use it [Syndemic] – a lot of shooting – I really don't see how they will learn about the immune system from that." I understood their concern. Educators did not develop syndemic, and it was not a didactically instructive game. For example, names of bacteria would appear when the user scrolled over them, but the game did not make names or definitions an integral part of the experience. Instead, the game immersed its users in a world that represented scientific accuracy, whereas most "educational" games of this type often employ cartoonish unrealistic imagery. The responses from teachers made me question the game's value. Does Syndemic's departure from traditional educational tactics lessen its pedagogical value? Does Syndemic's use of shooting enemies engage users in the game or make it less appropriate for classroom use? Past research helped provide some answers. A growing body of academic work supports the value of games like Syndemic in educational contexts. This research has produced a pedagogical movement called Digital Game-Based Learning (DGBL).

12.4.1 Digital Game-Based Learning in STEM Education

DGBL directly responds to mounting concern that our educational system is failing to provide adequate STEM education to the greatest numbers and diversity of young people (Snow & Dibner, 2016). DGBL defined as, "a type of gameplay with defined learning outcomes" has been identified as a successful educational tool that can help improve student attitudes towards science (Plass et al., 2015). Researchers have evaluated variables in game-based learning such as gender (Papastergiou, 2009), level of collaboration (Chen et al., 2015), and contextual integration (Miller et al., 2011) while measuring knowledge gained through pre-and post-game tests. These studies have shown that DGBL can significantly enhance overall learning and knowledge acquisition under testing conditions (i.e., Chen et al., 2015). Despite this breadth of work, a meta-analysis of DGBL studies concluded that a gap exists between the possible benefits of game-based learning and the outcomes being assessed. Li and Tsai (2013) found that only a few studies measured student engagement and motivation, even though what makes digital games most appealing is their ability to create an enjoyable and engaging user experience. Since Syndemic is not a didactic game, I focused on assessing student experience and changes in learning motivation. This research attempts to fill the gap identified by Li and Tsai (2013) between the potential benefits of using digital games and the outcomes being measured.

12.4.2 Designing and Implementing the Study

After my lack of success with classroom teachers, I tried another approach: after-school programs. If teachers didn't feel that Syndemic belonged in the classroom, perhaps it would be easier to integrate the game into after-school hours. This approach proved to be a far more effective approach. Between December and April of 2016, I conducted workshops with a total of 94 students at four afterschool programs affiliated with New York City Department of Education public schools. The groups included a STEM and arts (STEAM) program, a Kids Who Code program, and a general afterschool program in Jamaica, Queens. Of those 94 students, 63 attended both workshops and fully completed both the pre-and post-surveys. Data from these 63 students were analyzed during our study of Syndemic.

The overall goal of the study, to assess students before and after their gameplay, was accomplished in two sessions at each afterschool location. Each student filled out an anonymous pre-survey with questions regarding demographic information and computer experience, including their frequency of computer use, liking of computer games, and computer gaming experience. Questions were adapted from the Science Motivation Questionnaire (SMQ-II) developed by Shawn M Glynn (2011) to assess a baseline for the student's level of motivation to learn science. Students were prompted with the statement: "In order to better understand what you think and how you feel about your science courses, please respond to each of the following statements from the perspective of 'When I am in a science course...'" The questionnaire used a Likert scale to measure: intrinsic motivation, self-efficacy, self-determination, and career motivation.

After completion of the pre-survey, students played Syndemic individually for 30 min. A week later, students played Syndemic for the same amount of time and took a post-survey directly after their second session. The post-survey contained the same measures of science motivation but excluded the background information on demographics and computer usage. In addition, the pos-survey measured the students' game playing experience. In past studies, Csikszentmihalyi's (1975) flow theory has been adapted to assess gameplay experience quantitatively. Flow is a psychological state that is intrinsically motivating, enjoyable, and challenging. Bressler and Bodzin (2013) use a qualitative analysis to show that the common features of flow experiences are: a feeling of discovery, a desire for improved performance, and flashes of intensity. The flow state can be assessed quantitatively by a short questionnaire developed by Csikszentmihalyi (Lester et al., 2014). This questionnaire was adapted for the post-survey used in the study of Syndemic. Theoretical underpinnings, such as flow theory, have increased the legitimacy of DGBL research and allowed studies to distinguish between games that are unable to engage users versus games that simply fail to deliver educational content. Therefore, in studying Syndemic, we attempted to add to the existing literature by using previously tested methods and questionnaires. Figure 12.3 shows an overview of the study design and protocols.

12.4.3 Results

The majority of the students surveyed were male (59%) and many of them were Hispanic (49%). Although there were significant demographic differences between the students in the four afterschool program groups surveyed, the Science Motivation Questionnaire results indicated no statistically significant differences among groups, motivating us to present combined data (Fig. 12.4). Students responding on the post-survey more frequently agreed with all statements, including questions like "Learning science is interesting" to evaluate intrinsic motivation, "I believe I can master science knowledge and skills" to evaluate self-efficacy, or "I study hard to learn science" to evaluate self-determination. For survey items in all three of these categories, post-surveys documented statistically significant increases ($p \leq 0.005$), but items assessing career motivation were not significantly increased in post-workshop surveys ($p = 0.105$). Despite these varying significance levels, all four measures of motivation showed increases from pre- to post-survey of a similar magnitude. The mean differences (post-score minus pre-score) across all categories of motivation were between 0.15 and 0.26,suggesting 15–26% of all students responded with answers at least one integer higher on the Likert scale (0–4) after playing the Syndemic game (see Fig. 12.4).

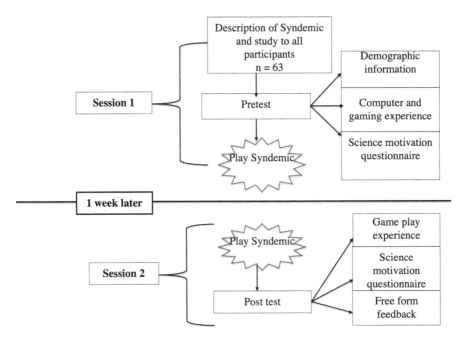

Fig. 12.3 Study design for syndemic evaluation

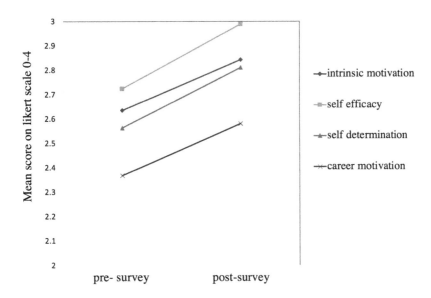

Fig. 12.4 Pre-and post-survey responses for all groups on the Science Motivation Questionnaire

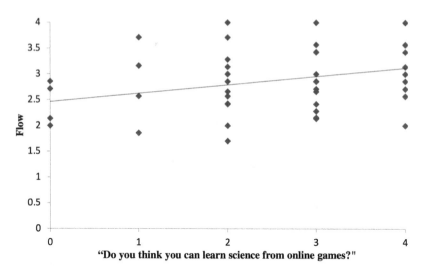

Fig. 12.5 Correlation between flow and student belief that online games can teach science ($R^2 = 0.303$, p = 0.0360)

When looking at student engagement, investigated using an additional seven items on the post-survey, 67.5% of students reported "strongly agree" or "agree" across the full set of questions, with even larger proportions reporting agreement about two questions: "When I was using the game, I wanted to do as well as possible" (80.4% selected "strongly agree" or "agree") and "I was able to make the game do what I wanted it to" (70.1% of students selected "strongly agree" or "agree"). Variation in flow scores was not explained by gender, race, gaming experience, computer usage, or any of the variables measured on the Science Motivation Questionnaire. However, as shown in Fig. 12.5, there was a significant positive correlation between the way students answered the question, "Do you think you can learn science from online games?" and how they scored on measures of flow experienced while playing Syndemic ($R^2 = 0.303$, p = 0.0360).

12.4.4 Analysis

Significant increases in survey scores from pre- to post-assessment indicate that Syndemic can increase students' intrinsic motivation, self-efficacy, and self-determination to learn science. However, career motivation did not significantly improve from pre- to post-survey. Career motivation was assessed using statements such as "my career will involve science" and "learning science will help me get a good job." This category aims to determine the level of relevance students feel science has to their career paths (Glynn et al., 2011). It is not surprising that this study, with its limited timeframe, could not create such a fundamental shift and significantly impact student responses to questions in this category. Responses to career-related questions may be inherently higher in variability among middle-school students, or even in an individual student's response from day to day. Further investigation, in which students have increased exposure to Syndemic over a more sustained period of time, could help to reveal such variability while also discerning whether Syndemic can potentially impact students' career motivations. If Syndemic still has no impact on student career motivation after increased gameplay, additions to the game itself would likely be needed. Some of these changes are already in progress. Alex May is currently completing an updated version of Syndemic to address the issue of career motivation. Possible additions include pop-ups explaining the jobs of the scientists who discovered each type of bacteria and more fundamental adjustments such as features allowing students to design their own microbes and experiments.

Although the majority (67.5%) of students experienced maximal amounts of flow, the flow scores reported by students in this study are over 10% lower than those found using a forensic science game called SSI (Bressler & Bodzin, 2013). The lower rates could be due to many factors, including differing subject matter and implementation settings; however, it is worth noting that, unlike Syndemic, SSI utilizes a collaborative game design. SSI allows students to play and problem-solve in small groups, while a single student can control the microbe while playing Syndemic. Students clearly noticed the limitations of this format, as requests for a multiplayer version were the most frequent suggestion written in the free-response portion of the post-survey. The request to create a more collaborative gaming environment, like SSI, reflects students' desire to maximize their own flow experience. Both Meluso et al. (2012) and Chen et al. (2015) explicitly tested the role of collaboration during gaming. The studies employed similar experimental procedures in which the control group played a game independently, and the experimental group played the same game in small groups. Both studies did not find a significant difference in learning outcomes or motivation between the independent and collaborative groups. However, Chen et al. (2015) acknowledge the need for further research on games specifically designed for cooperative play. Our goal is to make a multiplayer version of Syndemic, allowing for the results of this study to be directly compared to a collaborative game design. Testing a multiplayer version of the game would add to the existing literature as past studies have only tested the same game under different scenarios.

We tested both the Science Motivation Questionnaire and flow assessment results for correlations to the demographic information gathered from the pre-survey. There was a strong precedent that gender would predict the variance in science motivation or flow experience because Syndemic is a science game. Females have typically been less responsive than male peers to many STEM education initiatives (Tsai et al., 2012). However, gender did not significantly correlate with either flow or science motivation. These results are supported by numerous studies in DGBL, which have not found differences in how genders engage with games that create a narrative-driven virtual environment (i.e., Papastergiou, 2009). Therefore, DGBL provides a uniquely inclusive approach to engaging boys and girls. However, not all games have been able to create equal engagement. Inal and Cagiltay (2007) found that a game's narrative is critical to allowing females to achieve flow, while the rules and features of a game are more significant to males' flow experience. Therefore, Syndemic's ability to create equal engagement in this study is likely a direct reflection of its narrative-driven design.

The only significant correlation identified was between flow state and students' belief that online games have the ability to teach science. The more a student believed that games possess educational value, the more engaged the student was while playing Syndemic. Although Syndemic had a demonstrable impact on their motivation to learn science, students most often responded that games are only "sometimes" useful for teaching science. This hesitancy to acknowledge the potential of games as learning aids mirrors data reported by Anderson and Barnett (2013) on the ability of an online game to teach middle school students abstract concepts in physics. Anderson and Barnett (2013) found that although the game significantly improved learning outcomes, students did not consider playing the game to be a learning experience. This disconnect between the game's potential benefits and its perceived value is not limited to students. I saw it firsthand in teachers' responses when we asked them to consider using Syndemic with their students. Shaffer et al. (2005) comment specifically on this trend and explain, "there is a lot being learned in these games. But for some educators, it is hard to see the educational potential of the games because these virtual worlds aren't about memorizing words or definitions or facts. But video games are about a whole lot more."

Our society has deeply ingrained perceptions of video games as a "waste of time," notions that are undoubtedly true of some games but that must shift to address the proven educational value of others. Shaffer et al. (2005) argue that our education system must catch up with technology and stop perpetuating the harmful perspective claiming that games are only "entertainment." These pervasive societal beliefs mean that it is not enough to present middle school students with games and expect them to be accepted, without bias, as new learning tools. Instead, schools, parents, and teachers must integrate DGBL effectively by helping students recognize the power of games as a new facet of their educational experience. Over a decade ago, Shaffer et al. published their piece and predicted that the next challenges to DGBL would be helping learning environments acknowledge the power of games and integrate them effectively. The results of testing Syndemic illustrate that the challenges Shaffer et al. identified a decade ago are still relevant to the current climate and future success of DGBL.

12.5 Expanded Testing: Syndemic Goes to Nicaragua (Emma Ruskin)

12.5.1 Why Nicaragua?

I remember the cries of disgust that continued each time a new group was given the mouthwash. I tried to explain that microscopic bacteria live in our mouths and that mouthwash is a way to prevent the harmful ones from causing problems. This idea was met almost universally with blank faces. On the count of three, the children tipped the little cups into their mouths, ridden with skepticism, and gurgled. After encouraging some vigorous swishing, we instructed the children to spit out the foreign liquid. Sometimes, the volunteers also handed out toothbrushes. These were met with fewer complaints, but the children's sentiments towards them were made plain by the litter of toothbrushes left behind in the dirt after dismissal.

After my sophomore year at Barnard College, I volunteered with a small NGO in Nicaragua for the summer. I focused on executing the tasks given to me as part of their effort to improve hygiene among students in some of the poorest neighborhoods outside of Granada. The volunteers treated any complaints like typical children's shenanigans and tried to hand out as much mouthwash and as many toothbrushes as possible. I did not reflect on this aspect of my experience in Nicaragua for months afterward until I found myself testing Syndemic in afterschool programs across New York City. As I began to work with kids using Syndemic, I started to think more and more about the children's repulsion towards the mouthwash and toothbrushes donated by La Esperanza. I realized that the children's "shenanigans" in response to the toothbrushes and mouthwash were not a nuisance to be ignored but a key for understanding why a game like Syndemic might be an invaluable teaching tool. The explanation I had attempted to give when the children asked me why they needed to use this gross-tasting mouthwash was accurate but wholly inadequate. Without having any conception of the microscopic world, talking about the role of bacteria in their bodies and environment was meaningless and confusing to the children, just like handing out cups of mouthwash and toothbrushes. How can children be expected to participate in these critical acts of hygiene if they have no idea why these actions are important? How can children comprehend the need for these behaviors when the microorganisms responsible remain unseen, allowing them no conception of the microbiome?

As I began to wrap up my work with Syndemic in New York, I reached out to La Esperanza to see if they would be interested in my coming back again to test Syndemic and expand on the data gathered from after school programs in New York. The organization was excited to participate in the initiative, mainly because they had recently opened their first learning center, equipped with donated computers. I received funding from the Barnard Biology Department, which allowed me to include research on Syndemic in a larger biology education project I was conducting through Davis Projects for Peace. The game's creator, Alex May, set to work devising a way to input translated text into the game, and I printed translated copies of the pre- and post-surveys I had used during the workshops in New York. The computers at the center were offline, which allowed me to download Syndemic but not to use SurveyMonkey to collect data, as I had done in New York. Besides these technological limitations, I did not foresee too many other hurdles.

12.5.2   Syndemic Workshops at the Learning Center

Although the organization supported the idea of my work at the learning center, the reality of the project was met with skepticism similar to when I was handing out mouthwash 2 years before. I was expecting the local staff to be excited when I showed them Syndemic and the other materials I had brought to use with the kids. Instead, I met with hesitation and confusion. Unlike many of the classroom teachers I encountered in New York, who were hesitant because they did not see Syndemic's educational value, these educators were wary because I was treading in unfamiliar territory. I soon learned that children and local staff at the learning center had never been taught science as an academic subject. To get them on board, I needed to introduce the concept of microbiology and then place Syndemic within this framework so that they could understand its relevance. As a novice researcher, I was learning on the ground that making this learning tool accessible to a new geographic setting would require thoughtful changes in study design to meet the needs of this cultural context.

As much as I wanted to gather data on Syndemic that would allow me to draw fruitful comparisons to the data from after-school programs in New York, I was ill-equipped to make the major modifications needed to translate the study to this environment. One hurdle was that the entire basis of the survey at the after-school programs in New York included statements like "I put enough effort into learning science" and "I study hard to learn science." These pre/post surveys proved utterly irrelevant to the students at the learning center, who had not learned science as a separate classroom subject. In Nicaragua, Syndemic would need to introduce students to biology for the first time rather than help increase engagement in science so that students would be more receptive to the science taught in their classrooms. Based on anecdotal evidence, it was apparent that Syndemic was far less effective at engaging children with such limited exposure to science. The kids tended to be noticeably less interested in the game, and the ones who did get into it made comments indicating that they viewed it as a shooting game without grasping its intended purpose.

Moreover, written surveys proved to be an unusable form of data collection. Even after making substantial changes to the surveys to make the statements more relevant to the students' lack of science exposure, we found that their literacy levels were too low to allow them to complete many of the questions. As an alternative, we tried recording children's answers to questions about their experiences with the project using a video camera, but this approach raised a whole different set of concerns. The children strictly defer to the teachers at the center. As a result, we rarely felt as though the children were responding honestly in their answers. It often seemed as though they were hoping to guess the right thing to say that would please the interviewer. For instance, when asking a child what they thought of the activity, such as playing Syndemic, the typical response was "*Si* (Yes)," or "Bien (Good)," even if their behavior during the workshop indicated that they were not engaged or enjoying it. Perhaps we would have found the same answers on written surveys, but the issue seemed compounded when the children were giving their answers looking straight at the teacher holding a video camera. Upon review of the video footage, the only responses that differed from the above phrases involved comments about the game's look. The children seemed particularly impressed by the graphics and often exclaimed, *"Que lindo!* (How pretty!)" Based on this footage, it is hard to reach many data-driven conclusions about the efficacy of Syndemic at the learning center.

I had never attempted to perform research outside the U.S as an undergraduate. Although formal data collection was compromised due to my inexperience, this was a critical stepping-stone in my growth as a scientist. Completing fieldwork in Nicaragua, I often noticed the line between volunteer and researcher becoming blurry. It felt impossible to be present with the children, implement a biology education project, and observe or record useful data. When using Syndemic in workshops in New York, there were always other adults in the room helping to supervise and work individually with the kids. At the learning center, we were often alone with large groups of kids, making the task of assisting children in engaging in an activity and collecting data on that activity feel impossible.

12.6 Concluding Thoughts

Although the conclusions that can accurately be made about the children's learning gains from Syndemic are less than I had hoped, I can see my own learning gains as an undergraduate student. Implementing Syndemic in New York taught me many skills, allowing me to practice the scientific method step by step as I learned how to design and implement a small study. Although working in Nicaragua allowed less practice of the traditional scientific method, the experience gave me invaluable insight into the approach of a researcher trying to adapt their work to diverse settings. I thought that to be consistent, I would need to make as few adaptations to the study of Syndemic as possible. I look back now with embarrassment that I boarded the plane to Nicaragua with 100 copies of the same exact survey I had used in New York, only translated into Spanish, ready to hand to the kids at the learning center. I now know that adapting research to a particular culture and context is not a shortcoming or limitation to my work. It is an integral part of developing research that is accessible and relevant to a diverse range of communities

Future study of Syndemic is critical to completing a more accurate picture of the game's potential. Alex May is currently utilizing the data collected so far to create an updated version of Syndemic that addresses the feedback from students. Changes will include new levels of varying difficulty and opportunities for collaborative play. Long term, May explains,

> My dream version of Syndemic would involve pathogens that digitally 'evolve' alongside your player, so that you must constantly mix up your strategies in order to defeat them. Ideally, there would be no limit to the types of enemies you could encounter, as they would continually adapt to your weapons and tactics. I can also envision a more 3D or perhaps even virtual reality version, where it would be possible to physically explore the human body from the perspective of a microbe. You could examine firsthand how the body gets infected and help the immune system fight it off at the 'ground floor'!

In addition to testing the efficacy of these changes, variations in the study design will also play an important role in future work. A big limitation of the past research conducted with Syndemic has been the short length of each gaming session. Our next step is to assess Syndemic after students have played it for longer intervals over a greater period of time to see if that produces a more significant effect. The impact of doubling the number of game-playing sessions from two to four or six will be compared with the current results to determine the number of times students need to play Syndemic for optimal benefit. Digital games can support science instruction by increasing student engagement and motivation when the games are appropriately designed and tested.

References

Anderson, J. L., & Barnett, M. (2013). Learning physics with digital game simulations in middle school science. *Journal of Science Education and Technology, 22*(6), 914–926. https://doi.org/10.1007/s10956-013-9438-8

Bressler, D. M., & Bodzin, A. M. (2013). A mixed methods assessment of students' flow experiences during a mobile augmented reality science game. *Journal of Computer Assisted Learning, 29*(6), 505–517. https://doi.org/10.1111/jcal.12008

Chen, C.-H., Wang, K.-C., & Lin, Y.-H. (2015). The comparison of solitary and collaborative modes of game-based learning on students' science learning and motivation. *Educational Technology and Society, 18*(2), 237–248. https://eric.ed.gov/?id=EJ1070089

Csikszentmihalyi, M. (1975). *Beyond boredom and anxiety.* Jossey-Bass Publishers.

Glynn, S. M., Brickman, P., Armstrong, N., & Taasoobshirazi, G. (2011). Science motivation questionnaire II: Validation with science majors and nonscience majors. *Journal of Research in Science Teaching, 48*(10), 1159–1176. https://doi.org/10.1002/tea.20442

Guarner, J., & Niño, S. M. (2016). Microbiology learning and education online. *Journal of Clinical Microbiology, 54*(5), 1203–1208. https://doi.org/10.1128/JCM.03176-15

Horton, R. (2020). Offline: COVID-19 is not a pandemic. *Lancet, 396*(10255), 874. https://doi.org/10.1016/S0140-6736(20)32000-6

Inal, Y., & Cagiltay, K. (2007). Flow experiences of children in an interactive social game environment. *British Journal of Educational Technology, 38*(3), 455–464. https://doi.org/10.1111/j.1467-8535.2007.00709.x

Kastberg, D., Chan, J. Y., & Murray, G. (2016). *Performance of US 15-year-old students in science, reading, and mathematics literacy in an international context: First look at PISA 2015.* National Center for Education Statistics, 2017–2028. Retrieved from http://nces.ed.gov/pubsearch

Kelly, D., Nord, C. W., Jenkins, F., Chan, J. Y., & Kastberg, D. (2013). *Performance of US 15-year-old students in mathematics, science, and reading literacy in an international context: First look at PISA 2012.* NCES 2014–024. National Center for Education Statistics. Retrieved from http://nces.ed.gov/pubsearch

Khanna, S., & Tosh, P. K. (2014). *A clinician's primer on the role of the microbiome in human health and disease.* Paper presented at the Mayo Clinic Proceedings. https://doi.org/10.1016/j.mayocp.2013.10.011

Lester, J. C., Spires, H. A., Nietfeld, J. L., Minogue, J., Mott, B. W., & Lobene, E. V. (2014). Designing game-based learning environments for elementary science education: A narrative-centered learning perspective. *Information Sciences, 264*, 4–18. https://doi.org/10.1016/j.ins.2013.09.005

Li, M.-C., & Tsai, C.-C. (2013). Game-based learning in science education: A review of relevant research. *Journal of Science Education and Technology, 22*, 877–898.

Meluso, A., et al. (2012). Enhancing 5th graders' science content knowledge and self-efficacy through game-based learning. *Computers & Education, 59*(2), 497–504.

Miller, L. M., Chang, C. I., Wang, S., Beier, M. E., & Klisch, Y. (2011). Learning and motivational impacts of a multimedia science game. *Computers & Education, 57*(1), 1425–1433. https://doi.org/10.1016/j.compedu.2011.01.016

Osborne, J., Simon, S., & Collins, S. (2003). Attitudes towards science: A review of the literature and its implications. *International Journal of Science Education, 25*(9), 1049–1079. https://doi.org/10.1080/0950069032000032199

Papastergiou, M. (2009). Digital game-based learning in high school computer science education: Impact on educational effectiveness and student motivation. *Computers & Education, 52*(1), 1–12. https://doi.org/10.1016/j.compedu.2008.06.004

Plass, J. L., Homer, B. D., & Kinzer, C. K. (2015). Foundations of game-based learning. *Educational Psychologist, 50*(4), 258–283. https://psycnet.apa.org/doi/10.1080/00461520.2015.1122533

Raymann, K., Shaffer, Z., & Moran, N. A. (2017). Antibiotic exposure perturbs the gut microbiota and elevates mortality in honeybees. *PLoS Biology, 15*(3), e2001861. https://doi.org/10.1371/journal.pbio.2001861

Sender, R., Fuchs, S., & Milo, R. (2016). Revised estimates for the number of human and bacteria cells in the body. *PLoS Biology, 14*(8), e1002533. https://doi.org/10.1371/journal.pbio.1002533

Shaffer, D. W., Squire, K. R., Halverson, R., & Gee, J. P. (2005). Video games and the future of learning. *Phi Delta Kappan, 87*(2), 105–111. https://doi.org/10.1177/003172170508700205

Snow, J. (this volume). What does cell biology have to do with saving pollinators? In M. S. Rivera Maulucci, S. Pfirman, & H. S. Callahan (Eds.), *Transforming sustainability research and teaching: Discourses on justice, inclusion, and authenticity.* Springer.

Snow, C. E., & Dibner, K. A. (2016). *Science literacy: Concepts, contexts, and consequences.* National Academies Press. https://doi.org/10.17226/23595

Tsai, F.-H., Yu, K.-C., & Hsiao, H.-S. (2012). Exploring the factors influencing learning effectiveness in digital game-based learning. *Educational Technology & Society, 15*(3), 240–250. www.jstor.org/stable/jeductechsoci.15.3.240

Emma Ruskin graduated summa cum laude in 2018 from Barnard College with a degree in Cellular and Molecular Biology. In 2022, she began her first year as a Geffen Merit Scholar at the University of California Los Angeles (UCLA) David Geffen School of Medicine. Emma's collaboration with the Danino Laboratory is founded on their shared interest in utilizing art and technology to visually communicate complex scientific concepts.

Tal Danino's research explores the emerging field of microbes and synthetic biology. He primarily focuses on programming bacteria as a cancer therapy. Tal brings this science outside the laboratory as a TED Fellow and through bio-art works. He is currently an Associate Professor of Biomedical Engineering at Columbia University.

Part III
Sustainability and Environmental Justice Perspectives in Undergraduate Science Education

Chapter 13
Teaching Chemistry in Context: Environmental Lead Exposure: Quantification and Interpretation

Rachel Narehood Austin, Ann McDermott, Katrina Korfmacher, Laura Arbelaez, Jamie Bousleiman, Arminda Downey-Mavromatis, Rahma Elsiesy, Sohee Ki, Meena Rao, and Shoshana Williams

13.1 Framing the Issues

The literature on chemistry education over the last few decades suggests two broad themes. First, learning should engage students. Second, students learn more when they care about what they are learning. We begin by explaining our pedagogical goals to improve student engagement and interest. We then briefly describe the science of lead toxicity and its repercussions for environmental justice. This is

R. N. Austin (✉) · M. Rao
Department of Chemistry, Barnard College, New York, NY, USA
e-mail: raustin@barnard.edu

A. McDermott
Chemistry, Biological Sciences, and Chemical Engineering,
Columbia University, New York, NY, USA

K. Korfmacher
Environmental Medicine, University of Rochester, Rochester, NY, USA

L. Arbelaez
Barnard College, New York, NY, USA

J. Bousleiman
Jacobs School of Medicine University at Buffalo, Buffalo, NY, USA

A. Downey-Mavromatis
Chemical & Engineering News, Washington, DC, USA

R. Elsiesy
California Institute of Technology, Pasadena, CA, USA

S. Ki
Keck School of Medicine, University of Southern California, Los Angeles, CA, USA

S. Williams
Stanford University, Stanford, CA, USA

© The Author(s) 2023   227
M. S. Rivera Maulucci et al. (eds.), *Transforming Education for Sustainability*,
Environmental Discourses in Science Education 7,
https://doi.org/10.1007/978-3-031-13536-1_13

followed by the presentation of several detailed examples of teaching lead-chemistry with hands-on activities. We consider the ethical and legal implications of these activities. Finally, we curate a list of similar efforts to help instructors interested in using lead to teach chemistry to find appropriate references.

In providing this overview, we will address one of the book's primary themes: "how to create learning experiences that foster democratic practices in which students are not just following protocols, but have a stake in creative decision-making, collecting and analyzing data, and posing authentic questions." Along the way, we will try to extract from the examples "student perspectives on what it means to engage in environmental research and learning," to provide some guidance on how to create empowering learning experiences. John Dewey (1994, p. 10) describes the importance of placing learners in situations that "give the pupils something to do, not something to learn; and the doing is of such a nature as to demand thinking, or the intentional noting of connections." Under these circumstances "learning naturally results."

Science is intensely creative, but some kinds of science training do little to stimulate creativity and foster thinking "connected with increase of efficiency in action" (Dewey, 1994, p. 9). Science students often blindly follow instructions without asking questions, which may make for an easy-to-teach afternoon in the lab, but can have poor consequences for lab reports when it becomes clear that students had no idea why they took the steps they did. More significantly, it runs the risk of providing "information severed from thoughtful action… [that] simulates knowledge and thereby develops the poison of conceit" (Dewey, 1994, p. 9). Asking students to tackle a question where answers are unknown reinforces the notion that we are all learning. This encourages humility and it reminds us that speedy mastery of already-known chemistry does not necessarily correlate with a significant impact on developing new chemistry.

Knowledge is empowering, and we hope that our annotated bibliography of published examples will provide interested faculty members with the necessary tools to incorporate these ideas into their teaching. We hope to prevent faculty members from "rediscovering the wheel" and free them to learn from our examples. We offer creative ways to empower students to do authentic work: asking and answering questions that people outside of the classroom want answered.

Over the past several decades, scholarship on teaching and learning in the sciences has made it increasingly clear that students learn better when engaged in learning; the more actively, the better. Some students can sit through lectures and still deeply engage in problem-solving outside of class. Many others never manage to switch into a more active mode outside of the classroom; instead, they try to learn chemistry by highlighting sections of book chapters and cramming before each exam.

Many fields of human endeavor rely on authentic experiences in which trainees work alongside skilled tutors. Dancers and musicians learn to dance and play music, respectively, through lessons in which they dance or make music and receive feedback and corrections—and then dance or make music again in an iterative process through which they become better dancers or musicians. Similarly, plumbers and electricians learn via the apprentice model. They connect pipes and wires as they learn. Medical professionals receive extensive clinical training. Active learning strategies in science similarly provide an iterative process in which science students do science and solve problems in the presence of an expert who can coach and guide them. Our efforts in teaching lead chemistry aim to provide an active approach to early college chemistry education. Our approach is designed to improve student engagement and understanding and foster a more inclusive learning environment.

The literature on collaborative learning v. competitive learning is vast and complex and defies simple categorization. In our experience, creating a classroom that fosters democratic practices by encouraging students to work together, discuss problems, and share strengths to solve them creates a more just and inclusive community.

A recent book, *Teaching across cultural strengths: A guide to balancing integrated and individuated cultural frameworks in college teaching* (Chávez & Longerbeam, 2016, p. xx), was designed to serve as "an incremental and pragmatic guide for faculty to transform teaching practices for the purpose of inclusively drawing from a balance of cultural strengths to enhance student learning over time." In it, Chávez and Longerbeam (2016) state:

> What matters most to student learning, success, and increased retention and graduation rates is for faculty to teach from a strengths-based approach. Strengths-based approaches begin with recognizing cultural and other strengths that students bring to higher education and to their own learning. These approaches cultivate and draw on student strengths to engage and facilitate inclusive learning and are derived from positive psychology and social work. (p. xiii)

Asking students to use their burgeoning knowledge of chemistry to think about the stocks and flows of lead in the environment and how lead exposure impacts human health provides opportunities for many different strengths to be validated in the classroom, including problem-solving skills, practical knowledge about how things work, and social justice issues inherent in the historical and current development of urban landscapes.

13.2 The Chemistry of Lead Toxicity and Environmental Justice

Lead has no known beneficial role in humans. Lead substitutes for zinc or calcium in the body, disrupting a host of essential functions. Lead, zinc, and calcium most commonly are found as the +2 ion. All three elements have full electron shells or subshells in their ionic forms and have similar chemistries. Their ionic radii are also similar (Godwin, 2001). Lead can be found in most parts of our environment, including soil, water, air, and homes. Lead enters the environment through industrial emissions, deposition from leaded gasoline, contaminated former lead smelters, paint in older homes, and soils where older homes have been destroyed. Lead exposure also comes from drinking water contaminated by lead corrosion in plumbing.

Lead can enter the body through inhalation, ingestion, or exposure through the skin. Once lead enters the body, it is distributed through the bloodstream and can harm numerous organs. Lead's effects on individuals vary based on amount, duration, and timing of exposure. Lead poisoning can damage the brain and nervous system, impairing learning and memory. Lead exposure is especially harmful to the brains of developing children. Early childhood exposure to lead has been linked to decreased IQ, learning difficulties, behavioral problems, and growth delays. Children and infants experience deleterious effects following lead exposure levels well below those at which adults show effects. However, lead can also damage the cognitive abilities of adults. In extreme cases, lead poisoning can cause seizures or even death. Childhood lead exposure has profound development and learning effects. However, the mechanism(s) by which it exerts these effects on the brain are yet to be fully understood (See Ordemann and Austin (2016) for a discussion of potential mechanisms by which lead can be linked to neurological impacts). There is no known safe level of lead: the current Centers for Disease Control and Prevention (CDC) reference level for blood lead in children is unsafe at 3.5 micrograms per deciliter (ug/dL).

Historically, lead exposure was widespread. In 1970, more than 2000 children in New York City were found to have blood lead levels of 60 ug/dL or higher—severe enough for hospitalization. Advocacy by parents, community organizers, and health care providers led to a wide range of local, state, and federal policies, including banning lead from gasoline (in 1976) and paint (first locally in New York City, then nationally in 1978) and reducing allowable levels of lead in plumbing over time. As a result of these and other changes, including reductions in industrial emissions and controls on consumer products, the proportion of children with elevated blood lead levels declined dramatically. Although these policies are perceived as a public health success, they may also be viewed as an environmental justice failure. Today, low-income and children of color living in pre-1978 housing are most likely to have elevated blood lead levels.

Although most children with significantly elevated blood lead levels are exposed to lead in paint, dust, or soil around their homes, lead in water remains a concern, particularly in historically disadvantaged communities. Flint, Michigan, has been dealing with lead levels above the EPA's action levels since the decision was made in 2014 to switch to a less expensive source of public drinking water. Average blood lead levels there reached about five ug/dL in children younger than 5 years old. In Flint, a city of 98,310, 41.2% of the people live below the poverty line. The median household income is $24,862. 56.6% of the residents are African American. Newark, NJ faces similar problems and again represents a less white and less wealthy demographic than nearby areas in the state where lead levels are much lower.

Childhood lead exposure provides many opportunities to explore the causes and consequences of environmental injustice. Academic institutions increasingly emphasize helping students understand and address social inequities. For example, Barnard College's Diversity Mission Statement recognizes that the success of the institution's mission, which is to "rigorously educate and empower women, providing them with the ability to think, discern, and move effectively in the world … depends on the extent to which our community is diverse, inclusive and equitable (Barnard College, 2017)." In addition, "[o]ur definition of diversity encompasses structural and social differences that form the basis of inequality in our society, including race, ethnicity, gender, sexuality, socioeconomic class, disability, religion, citizenship status and country of origin" (Barnard College, 2017). Teaching chemistry students why and how environmental toxins disproportionately affect historically disadvantaged communities connects science students to our institutional mission, which we hope resonates with many readers.

Just as the brunt of lead poisoning falls on low-income communities in the US (Mielke et al., 1983), in the global context, lead poisoning is most severe in the middle- and low-income countries. In countries with improper waste disposal, such as Senegal, people are at risk of lead poisoning from battery recycling. Artisanal mining activities may expose workers and their families to dangerous lead levels. The Institute for Health Metrics and Evaluation (2019) puts the number of annual deaths worldwide from lead exposure at more than 900,000. Integrating information on the causes and consequences of lead poisoning into a chemistry course allows students to understand the real-world impacts of what they are studying. Not all students report more interest in a scientific topic when they see its applications, but many do, and those that do often value inclusivity. In any case, connecting abstract and concrete topics is a skill all scientists need.

13.3   Detailed Teaching Examples

The most widely-disseminated model for active learning using lead chemistry and the consequences of human lead exposure was funded by the HHMI professorship program and carried out by HHMI professor Hilary Godwin when she was on the chemistry faculty at Northwestern University. Godwin developed a summer course for Northwestern students from groups underrepresented in the sciences. These students worked with her and a postdoctoral fellow to measure lead in Chicago's soil. Her Undergraduate Success in Science (USS) students sampled soils in the Chicago area and did outreach in local communities to inform people about the dangers of lead exposure. Godwin was, at the time, one of the foremost researchers in lead biochemistry, and the close intellectual connection between her research and this project infused it with vitality. (Her work on this project is briefly mentioned in Godwin & Davis, 2005).

13.3.1   Small Project-Based Course for Incoming Students: A Bridge Course

Several years after Godwin's work, Austin emulated it with a similar course at Bates College, working with a small group of students who hailed from groups or backgrounds underrepresented in the sciences (e.g., first-generation college). Students in the course were participants in a bridge program, initially funded by the Howard Hughes Medical Institute (HHMI) to facilitate the transition of students with backgrounds not well-represented in the sciences, into the intellectual and cultural life of the college in the summer before they matriculated. Bates College students in this summer course measured lead levels in the soils in Lewiston, Maine, an old Franco-American mill town with the highest levels of lead poisoning in the state. Austin was inspired by Godwin's work and similar work by the environmental soil scientist, Samantha Langley-Turnbaugh, then at the University of Southern Maine (USM; see annotated bibliography for more information on her work), both of whom had demonstrated the power of this topic to broadly engage students.

Several aspects of the Bates College course were very successful. The students learned how to use ArcGIS and hand-held GIS devices to record locations where they took soil samples. Later, they plotted lead levels on a map of the city and overlaid that with data from the Maine CDC documenting the number of reported cases of elevated blood lead levels by census block. The students learned how to make calibration curves and convert the amount of lead in ppm as measured by inductively coupled plasma-optical emission spectroscopy (ICP-OES) and the percentage of lead by weight in soil. They set up and ran the ICP-OES, gaining exposure to a sophisticated instrument that illustrates basic principles of atomic structure that can be difficult for students to grasp.

A colleague (Holly Ewing) created an excellent exercise for the students on sampling. She described a hypothetical environment—a lot with a house on it—and asked them to develop a sampling strategy to determine the concentration of lead in the soils on the property. Students typically had no prior experience designing sampling strategies. They had to develop a sampling strategy relying only on their intuition and prior experiences about how something like lead might be distributed in the world. After developing a sampling strategy, students were taken into an empty classroom where Ewing had laid out the scenario she described, using loose sand to represent the lead in the soil. It was immediately apparent to the students that the lead was distributed very heterogeneously. Chance would determine whether their sampling strategy would detect the existing lead. This emphasis on uncertainty—an idea undergraduate chemistry students often do not get to think about—ran through the whole class. Students also tested their methodology against an authentic standard, working through the calculations that would enable them to assess how representative their sampling protocol would have been.

The students in the class created a map with approximately 40 soil lead measurements. The map also contained data on soil lead levels from local community gardens previously collected by a Bates student and publicly available data on lead-poisoned children. Students got to appreciate the complexity of policy efforts to mitigate lead. Some students associated with the class volunteered to be part of a neighborhood notification program, warning residents about the dangers of lead exposure during demolitions of old buildings. The city was demolishing abandoned buildings in a revitalization effort but did not have the resources to warn nearby residents of the potential risks that dust from these demolitions might pose. Class members also met with a local landlord, who permitted the class to sample some of his properties. There were no clear patterns in the data. In general, almost all of the samples had lead levels below the EPA action level and lower than those in the more industrial city to our south, Portland, Maine.

At the end of the semester, the class hosted a community discussion about the data. It was not framed as a presentation, because the class members did not feel that they had the answers; rather, community members were invited and shown what had been learned. Among the people who came was a local congressperson who was unfamiliar with the lead problem. She later co-sponsored a bill dealing with lead poisoning and pointed to our discussion as formative.

13.3.2 Large, General Chemistry Courses

Another example provides a contrast to the small, immersion-type courses described above. Recently, Ann McDermott and her teaching assistants implemented an initiative in which all 197 students in a general chemistry lecture course were invited but not required to participate in measuring the lead content in playground soil in the vicinity of Columbia University's Morningside campus. As a class, they would construct a collective dataset and create a larger picture of the lead situation. Students were invited to write one-page proposals about lead in the environment, including health implications, sources, remediation, or detection, including primary references. The topic was introduced in lectures and used to illustrate course material through in-class and homework problems.

The analytical chemistry of lead has strong connections to the content of standard introductory chemistry courses and provides an excellent vehicle for illustrating many of the general principles of a first-year chemistry course. Analyzing the presence of lead in the environment illustrates the atomistic and chemical view of the material world. It makes extensive use of atomic symbols and refers to isotopes in a practical sense because the various isotopes of lead have distinct origins and can be used to trace lead's source. The approaches used to uniquely identify the presence of lead make practical use of line spectra. Atomic absorption spectroscopy can be used for a quantitative determination of the amount of lead. These techniques make a practical connection to more abstract aspects of the general chemistry curriculum, specifically spectroscopy and quantum mechanics. They also reinforce concepts related to the periodic table and periodicity in terms of characteristic properties, such as ionic radii, and preferred charge or oxidation state. When bioavailability of lead is assessed in terms of solubility in acidic buffers, these studies can additionally illustrate kinetics, equilibria, pH, and redox chemistry. Biological aspects of lead can be wonderful vehicles for illustrating coordination chemistry. It also provides numerous opportunities to solve multi-step, quantitative problems.

These activities have the potential to expose students to resources and learning opportunities that exceed those typically offered in the general chemistry curriculum. For example, students are exposed to literature search strategies and structured critical reading of the primary literature. They participate in the creative selection of a problem and its justification in terms of fundamental discovery, planning research, exploring practical impacts, deepening their current understanding, developing testable hypotheses, selecting experimental techniques, implementing a sampling strategy, and using estimates and prior data to select experimental conditions. Students also gain experience with specific research skills and bench techniques and learn more about reproducibility and the importance of controls. They participate in data and error analysis and have an opportunity to consider over-interpretation and inference from data. They learn about strategies for sharing data, scientific writing, and presenting.

McDermott and other members of her teaching team compared their perceptions and intentions with student feedback and student choices. The project goals, as elaborated above, were that these active learning assignments would require student initiative and provide students with the opportunity to select a topic of personal passion and current societal relevance—thus creating a context that was authentic for each student. McDermott hoped that this segment of the course would be goal-oriented and topically focused, in the way that the practice of scientific research is, and provide students an opportunity to see operational applications of and connections among the wide breadth of topics presented in a survey-style introductory chemistry course.

Student response was assessed through participation rates. Both the research paper and the experimental research on environmental lead were optional. They were among several mechanisms for obtaining participation points in the course (a small increment of their total grade). In the context of their academic schedules, these choices competed with other more exam-pertinent activities. Nevertheless, 43% participated by writing a paper regarding lead or collecting data on lead in the local environment. (Dropout of the class overall was less than 5%.)

How did students perceive this activity? What aspects interested the students in this activity? Over 90% of the class participated in a feedback survey about these choices. Were their reasons for participating similar to McDermott's intentions in presenting the option? Most students who participated in the lead research project expressed a passion for the topic, civic commitment, or cited the importance of hands-on learning. Several students explained that they were motivated by the possibility of additional face-to-face contact with the professor. Many who participated in the lead research paper stated an interest in the intellectual challenge of writing about science and using library materials. Many also cited engagement with the specific topic, civic engagement, or connecting the class material to the real world.

The students' experiences and comments echo some of McDermott's original intentions that the experiments and the research would allow students to pursue a passionate interest. Students also saw the research paper as an opportunity to strengthen a broader set of discipline-relevant skills. In addition, student feedback highlighted that these "crowd-sourced" data studies could be socially engaging cooperative endeavors. This finding provides an exciting opportunity for follow-up pedagogical interventions.

What concerns did students have? Those who did not participate in either the research paper or the experiments cited time management as their top reason for avoiding the selections (some commented that they would have been interested in principle). Some cited a perception that the topic would be too advanced, and a handful stated that they had little interest in this specific topic. Indeed, the practical challenges regarding time management and student perceptions of the challenge are essential considerations for course planning and communication. Other vital questions will be valuable to address in future efforts. Is there a correlation between high educational gains and classroom integration of societally relevant chemical studies?

Austin reinforced the notion that it is possible to do project-based activities in large introductory chemistry lecture classes. She taught a semester-long project-based lab developed by her colleague, Tom Wenzel. The lab was part of a two-semester introductory chemistry sequence with an environmental focus that they co-developed. In this class, students planned, carried out, and analyzed the results from an experiment to answer the question: "Does acid rain mobilize lead from soils and lead to its uptake in plants?" Wenzel (2006) describes this project. Introductory chemistry students worked in small groups throughout the semester to do this project. Lab hours were somewhat flexible. Students came to the greenhouse to water their plants when needed, and they spent time planting and harvesting. These were not typical introductory chemistry lab activities but were required to determine if acid rain mobilizes lead from soils and leads to the uptake of lead in plants. The end-of-semester results were often inconclusive and not statistically significant. Wrestling with uncertainty is one of the most fundamental aspects of the practice of science and yet we often try to shield young chemistry students from encountering it. However, by shielding students from uncertainty, we reinforce the notion of an omnipotent perspective and, perhaps unintentionally, make science seem less inclusive than it is.

13.3.3 Non-majors Course Without a Laboratory Component

"Jazz of Chemistry" is a chemistry course offered at Barnard College targeting first-year students. It surveys chemistry in everyday life, from forensics to poisons to food chemistry. This course also satisfies a science requirement in the core curriculum, "Foundations." The course does not include a lab and does not serve as a pre-requisite for other science classes, nor does it fulfill requirements for medical or other health professional schools.

Students' background in this course varied from a post-baccalaureate biochemistry major to a general studies student with a minimal chemistry background. The majority of the students in the class were first-year students fulfilling one of the science requirements. Students were assigned a project on the Flint Water Crisis during the semester. Students could choose which type of project they wanted to do: a video presentation, a play with the topic as the theme, or a small group presentation. Students chose to do small group presentations, focusing on different aspects of the crisis: background information about lead, the Flint Water Crisis, and remediation steps. Each student also submitted a written report with references.

Students researched a wide range of subjects, including government involvement, how the event affected the population of Flint, and the possibility of cures for lead poisoning. Students also researched information about the age of water pipes in other cities and how to avoid this type of crisis in cities with older infrastructures. This project was ideal for first-year students to develop a bigger picture of what goes on in situations like the Flint water crisis. The project also enabled students to be creative. One of the students presented their report on background information with all the characters involved using a Playbill.

The students learned precautions to prevent consuming tap water contaminated with lead, when the water coming from the tap would contain the highest lead levels, and how drinking water is treated before distribution. In summary, this project successfully fulfilled the objectives of a course like Jazz of Chemistry, where the emphasis is helping students see a global picture that links chemistry to significant social issues. The course also emphasized cooperative learning, professional presentation, and scientific research and writing.

13.4 Responding to the Legal and Ethical Implications of Lead Assessment

Because of health, legal, and economic issues associated with lead contamination, students need to be respectful of uncertainty and understand the ethical complexity of the issue. Program designs also need to comply with pertinent laws. Partnering with local community groups, health care providers, and government agencies can help contextualize student findings.

There are a number of particular ethical and legal issues to consider when students engage in assessments of people's homes and community environments. This applies to all assessments—visual, survey, or sampling—particularly when testing for lead. First, students need to appreciate the residents' tenure (renters/owners), as well as their economic, cultural, and political circumstances. Renters are often rightly concerned that any information or action related to housing quality could lead to conflict with their landlord and possible eviction. Although there may be legal protections against retaliatory eviction, in reality, these protections may not hold. The Hippocratic Oath of "first, do no harm" is an important reminder to students and faculty to take every precaution that their involvement with someone's housing situation does not cause unintended adverse effects. Precautions include providing referrals to free legal resources, sharing information in ways residents can understand, supporting follow-up actions or questions, and being clear with residents about potential outcomes of having additional information about their housing quality—before testing. Although these issues are most clear for renters, low-income owner-occupants may have concerns that identifying hazards will obligate them to make repairs they cannot afford. Information about lead hazards may also lower their home's value, cause conflicts with neighbors, or create conflict within families about appropriate courses of action. It is also essential to recognize that lead may come from many sources in addition to paint, dust, and soil around pre-1978 housing. Children may be exposed to lead in water, from their parents' jobs, consumer products, traditional medicines, legacy industrial contamination, or other sources. Any housing assessment or intervention is an opportunity to raise awareness of all these potential sources of exposure.

Second, it is essential to recognize that lead is a highly regulated hazard with unique legal and health implications. For example, the federal disclosure law requires owners who have "knowledge" of lead to disclose or report this information to any future renters or buyers (EPA, n.d.). It has not been clearly established by law whether this applies to a positive reading for lead using a home test kit, but the argument could be made that any evidence of lead triggers the disclosure law, potentially affecting the resale value of the property. Finding lead in a home may encourage residents to test their children for lead; if they are found to have elevated lead, this may empower the family to prevent further exposure. It may also trigger intervention by the local health department. Renters may pursue legal action against the property owner. Finding a potential lead hazard may cause the property owner to do renovation work to address the hazard (e.g., fixing peeling paint), which can reduce future exposures. However, if such work is not done carefully using lead-safe work practices, it is easy to generate lead dust, creating an even more severe hazard. The EPA requires people who are paid to disturb paint in pre-1978 housing (including DIY landlords) to have Renovation, Repair and Painting Program (RRP) and EPA certification precisely to avoid creating lead hazards during renovation (EPA, n.d.).

Third, and perhaps most concerning, is the potential for false negatives. Any test may yield a false positive (indicating lead when there is no lead) or a false negative (indicating there is no lead when lead is present). In the latter case, residents may be falsely reassured that their house is lead-safe, and fail to take precautions such as lead-safe cleaning and work practices during renovation, keeping children from chewing on walls or window sills, covering bare soil, and using walk-off (dust) mats at entrances (See Korfmacher and Dixon (2007) for more information on the reliability of spot tests for lead; also see Waggett et al., n.d.). A false negative for lead may increase the potential for poisoning a child. Clear and thoughtful communication about the many uncertainties involved in lead testing is critical, particularly when community residents or groups are involved.

Fourth, students tend to overreach in the conclusions they draw as they begin to study lead chemistry. They may assume that any child with elevated levels of lead in their blood will have a lower IQ than they would have otherwise and may not recognize the impacts this implication could have on affected families. Instructors must remind students that a statistical association between two variables does not mean we know precisely how lead exposure will affect a given individual. Sharing information about the potential lifelong health and behavioral impacts of lead on children can deeply upset parents and community members. Partnering with local health departments, health care providers, or housing agencies to provide appropriate context, information about uncertainty, and follow-up action can be helpful.

Finally, talking about social justice issues in the classroom requires a skilled facilitator. College science teachers may not be trained or experienced in moderating such difficult conversations. It can be helpful to lay out some guidelines for talking about racism, equity, and justice at the beginning of the class, particularly since their colleagues and community members may have very different life experiences and perspectives. It is important to remind students to remain humble and acknowledge how much they do not know about diverse communities' experiences and multidisciplinary perspectives. It can be challenging to balance this message while encouraging students to become as confident of their basic scientific skills as quickly as possible.

These are just some of the ethical, legal, and economic factors to consider before embarking on environmental assessments, particularly those that involve testing housing or other community environments for lead. Each of these issues may be fodder for rich discussions about how to prevent, mitigate, or address such real-world complexities and how these principles may apply to other kinds of citizen science. Colleagues in social sciences, humanities, or student services can assist in fostering constructive and respectful discussions of such difficult issues.

13.5 Conclusions

Teaching chemistry through the lens of urban lead hazards and children's environmental health provides an opportunity to connect fundamental concepts in chemistry to social and environmental justice issues. This problem-based approach creates multiple opportunities for real-world and community-engaged learning experiences for chemistry students. We hope that the examples provided in this chapter and included in the annotated bibliography will provide foundational information on which interested science educators can build authentic classroom and laboratory experiences that respond to the strengths and needs of their students, as well as their local community contexts and partnerships.

Annotated Bibliography
The annotated bibliography primarily contains examples of published cases that use lead chemistry to engage students. It also includes some valuable readings to help create ethical and authentic classroom and laboratory learning experiences.

Bachofer, S J. (2008). Sampling the soils around a residence containing lead-based paints: an X-ray fluorescence experiment, *Journal of Chemical Education*, 85(7), pp. 980–982, https://doi.org/10.1021/ed085p980

> The goal of Bachofer's experiment on soil was student exposure to sampling and X-ray fluorescence. The students collected samples from the soil around a residence with known exterior lead paint. The students determined exact sampling locations based on the EPA's Lead-Safe Yards information. Lead content was determined by X-ray Fluorescence. They analyzed their samples by averaging four generated spectra. The average lead values from the soil were then juxtaposed with the EPA's standards (preliminary remediation goals) and a background sample. To connect back with the community, stu-

dents drafted letters to the homeowner. The layout of the experiment allowed for the practical application of environmental chemistry, community engagement, student problem solving, and decision making.

Breslin, V. T. & Sañudo-Wilhelmy, S. A. (2001). The lead project: An environmental instrumental analysis case study. *Journal of Chemistry Education, 78*(12), 1647–1651. https://doi.org/10.1021/ed078p1647

> Breslin and Sañudo-Wilhelmy describe a course in environmental instrumental analysis focused on teaching techniques and engaging students in original community work. Students were given preliminary articles on lead and the environment. From this point, students engaged in problem definition, sampling design and statistics, analytical and instrumental principles, sampling protocols and field sampling, sample preparation and analysis, data analysis and interpretation, and final report and presentation. Each step included a lecture and/or active learning portion, in which students were given the tools for the work ahead. The students did fieldwork in a suburban area, testing the lead levels in the soil and lead content in house paint. Analysis revealed comparable lead levels in the soil surrounding older homes in the New York suburbs (Suffolk County, in particular) and in large cities. Finally, students worked one-on-one with professors to write and revise a lab report in manuscript form, exposing them to the realities of the publication process and providing students with a fuller understanding of the research process. Students presented their work to the public, and homeowners who came to the presentation were given information on lead levels and harm prevention. After completing the project, some students credited their enrollment in the course and engagement in the project as a catalyst for being hired for environmental analysis and consulting work.

Cancilla, D.A., (2001). Integration of environmental analytical chemistry with environmental law: the development of a problem-based laboratory, *Journal of Chemical Education, 78*(12), pp. 1652–1660, https://doi.org/10.1021/ed078p1652

> Cancilla outlines the creation of a "problem-based environmental analytical chemistry laboratory and its integration with an undergraduate environmental law course." This combination of environmental law and environmental chemistry posited some unique discussions, like the differences between scientific uncertainty and legal uncertainty. It gave students the vocabulary necessary to integrate the scientific and the legal. The course organized a group of environmental law students and one environmental chemistry student into simulated governmental entities, such as the EPA or the Department of Health. The environmental chemistry students and environmental law students were given the same basic information about a fictional city to introduce them to real-life guidelines for quality control and contextual circumstances. The chemistry students ran studies on water samples (generated by instructors) and gave their results to the law students for their argument. The students then

presented their cases to a board. Chemistry students gained exposure to various tests of toxicity, including microtox experiments, immunoassay experiments, solid-phase extraction, microextraction, and atomic absorption spectroscopy. Due to the fictional legal matter, chemistry students were exposed to the chain of custody forms, and law students could use mishandled chain of custody forms in their arguments. The quality of the data generated was also used in arguments before the board, which encouraged chemistry students to be conscientious of the data generated, as poor or unusable data would impact their group's arguments. Each student was assessed based on a consulting report, modeled on given real-life examples of consulting reports. Similar to students in the Breslin and Sañudo-Wilhelmy (2001) course, the environmental law and the environmental chemistry students found this course beneficial in securing jobs in related fields.

Fillman, K. L. & Palkendo J. A. (2014). Collection, extraction, and analysis of lead in atmospheric particles, *Journal of Chemical Education, 91*(4), pp. 590–592. https://doi.org/10.1021/ed400589r

> Environmental science students sampled ambient air to find whether or not lead levels in Berks County, Pennsylvania were compliant with National Ambient Air Quality Standards. Through this experiment, students were introduced to atomic spectroscopy techniques necessary for metal analysis, learning the importance of the proper technique selection. Students were familiarized with air sampling, quality control parameters, and percent recovery in a similar context to government use. Samples were collected in a glass fiber filter. Students tested a 1 × 2 inch piece of the filter, which was placed in a centrifuge tube with digestion acid. The sample was centrifuged, and the supernatant was collected in two aliquots. A known amount of a lead standard was added to one of the two samples. A blank filter was also prepared to determine an unused filter's lead content for comparison. The samples were then run in a Varian GTA 100 and a Varian 220 FS atomic absorption spectrometer. The paper notes that this analysis could also have been done with inductively coupled plasma-optical emission spectroscopy or mass spectrometry. More advanced students analyzed the samples with anodic stripping voltammetry. The majority of students involved stated that they found the experiment interesting. There was no noted community engagement, and the students followed a pre-existing procedure.

Goldcamp, M. J., Underwood, M. N., Cloud, J. L., Harshman, S. & Ashley, K. (2008). An environmentally friendly, cost-effective determination of lead in environmental samples using anodic stripping voltammetry, *Journal of Chemical Education, 85*(7), 976–979, https://doi.org/10.1021/ed085p976

> Anodic stripping voltammetry can be used to teach students about lead in the environment in a lab-friendly, fieldwork-oriented way. This article is concerned with introducing the technique to students in a cost-effective and time-effective manner. The authors used pencil graphite as electrodes instead of more expensive glassy carbon electrodes. The students collected samples

from waterbeds in Ohio. They sifted, dried, and extracted the samples for voltammetric analysis. This experiment taught electrochemistry and environmental chemistry, emphasizing the standard addition method while allowing students to do fieldwork. After the collection of the samples, students worked from an established protocol. The procedure was time-consuming, though the authors provide ways to run the experiment under time constraints (such as preparing solutions and samples in advance). The four students who participated noted that their fieldwork at the waterbeds increased their interest in the experiment.

Langley-Turnbaugh, S. J. & Belanger, L. G. (2010). Phytoremediation of lead in urban residential soils of Portland, Maine, *Soil Survey Horizons, 51*(4)

This article examined whether phytoremediation could be used to successfully remove lead from urban soil. While planting spinach in lead-contaminated soils decreased lead levels in soils, the decrease was not statistically significant and more alarmingly, the levels of lead in spinach were troubling, pointing to concerns about lead levels in urban gardens. As phytoremediation might be an inexpensive strategy employed in urban settings, this article provides a cautionary note about engaging in this practice without monitoring lead levels in vegetables.

Weidenhamer, J. D. (2007). Circuit board analysis for lead by atomic absorption spectroscopy in a course for nonscience majors, *Journal of Chemical Education, 84*(7), pp. 1165–1166, https://doi.org/10.1021/ed084p1165

A considerable amount of environmental lead contamination today comes from electronic waste. Since electronics, such as phones and laptops, are ubiquitous for many college students, it would be wise to include electronic waste in curricula when teaching environmental chemistry. Circuit boards from old laboratory instruments were tested for lead content using atomic absorption. The samples were digested and prepared for AA in the first lab period. In the second lab period, absorbance readings were analyzed with Excel. The varying colors of the solutions made a point about qualitative variability. The third lab period was a wrap-up of the analysis done on the second day. This experiment was designed and carried out without decision-making on the part of the student, and while it engaged the student body, it did not go beyond the lab itself.

Wenzel, T. J. (2006). General Chemistry: expanding the learning outcomes and promoting interdisciplinary connections through the use of a semester-long project. *CBE Life Sciences Education, 5*(1), pp. 76–84. https://doi.org/10.1187/cbe.05-05-0077.

This article summarizes a semester-long project-based laboratory studying the mobility of lead ions through the soil and into plants as a function of the pH of rain. The lab was developed for a first-semester introductory chemistry class and illustrated that problem solving and experimental design could be taught at the earliest stages of scientific education.

Acknowledgements We would like to acknowledge all of the students we have taught in various courses over the years where lead chemistry and the injustices associated with lead poisoning have been featured, and the staff at our various institutions who have supported our efforts. Initial funding for CHEM 103 "Urban Lead Pollution", taught at Bates College, was provided by the Howard Hughes Medical Institute (HHMI). RNA acknowledges the National Science Foundation (CHE-1151957) for support of her work in this area. Any opinions, findings, and conclusions or recommendations expressed in this material are those of the author(s) and do not necessarily reflect the views of the National Science Foundation. KSK is supported by National Institute of Environmental Health Sciences grant P30 ES001247and developed problem-based teaching resources on lead through the InTeGrate program https://serc.carleton.edu/integrate/index.html.

References

Bachofer, S. J. (2008). Sampling the soils around a residence containing lead-based paints: An X-ray fluorescence experiment. *Journal of Chemical Education, 85*(7), 980–982. https://doi.org/10.1021/ed085p980

Breslin, V. T., & Sañudo-Wilhelmy, S. A. (2001). The lead project: An environmental instrumental analysis case study. *Journal of Chemical Education, 78*(12), 1647–1651. https://doi.org/10.1021/ed078p1647

Cancilla, D. A. (2001). Integration of environmental analytical chemistry with environmental law: The development of a problem-based laboratory. *Journal of Chemical Education, 78*(12), 1652–1660. https://doi.org/10.1021/ed078p1652

Chávez, A. F., & Longerbeam, S. D. (2016). *Teaching across cultural strengths: A guide to balancing integrated and individuated cultural frameworks in college teaching* (1st ed.). Stylus Publishing.

Dewey, J. (1994). Thinking in education. In L. B. Barnest, C. R. Christensen, & A. J. Hansen (Eds.), *Teaching and the case method* (3rd ed., pp. 9–14). Harvard Business School Press.

Diversity, Equity, and Inclusion: DEI Mission. (2017, February 15). Barnard College. https://barnard.edu/dei/mission

Environmental Protection Agency (EPA). (n.d.). https://www.epa.gov/lead/real-estate-disclosures-about-potential-lead-hazards

EPA. (n.d.). *Lead renovation, repair and painting program.* https://www.epa.gov/lead/lead-renovation-repair-and-painting-program

Fillman, K. L., & Palkendo, J. A. (2014). Collection, extraction, and analysis of lead in atmospheric particles. *Journal of Chemical Education, 91*(4), 590–592. https://doi.org/10.1021/ed400589r

Godwin, H. (2001). The biological chemistry of lead. *Current Opinion in Chemical Biology, 4,* 223–227. https://doi.org/10.1016/s1367-5931(00)00194-0

Godwin, H. A., & Davis, B. L. (2005). Teaching undergraduates at the interface of chemistry and biology: Challenges and opportunities. *Nature Chemical Biology, 1,* 176–179. https://doi.org/10.1038/nchembio0905-176

Goldcamp, M. J., Underwood, M. N., Cloud, J. L., Harshman, S., & Ashley, K. (2008). An environmentally friendly, cost-effective determination of lead in environmental samples using anodic stripping voltammetry. *Journal of Chemical Education, 85*(7), 976–979. https://doi.org/10.1021/ed085p976

Korfmacher, K. S., & Dixon, S. (2007). Reliability of spot test kits for detecting lead in household dust. *Environmental Research, 104,* 241–249. https://doi.org/10.1016/j.envres.2007.02.001

Langley-Turnbaugh, S. J., & Belanger, L. G. (2010). Phytoremediation of lead in urban residential soils of Portland, Maine. *Soil Survey Horizons, 51*(4). https://doi.org/10.2136/sh2010.4.0095

Langley-Turnbaugh, S. J., & Hardy, A. (2012). Soil trace element concentrations on a historic Maine island. *Soil Horizons, 53*(5). https://doi.org/10.2136/sh12-02-0001

Mielke, H. W., Anderson, J. C., Berry, K. J., Mielke, P. W., Chaney, R. L., & Leech, M. (1983). Lead concentrations in inner-city soils as a factor in the child lead problem. *American Journal of Public Health, 73*, 1366–1369. https://doi.org/10.2105/2Fajph.73.12.1366

Ordemann, J. M., & Austin, R. N. (2016). Lead neurotoxicity: Exploring the potential impact of lead substitution in zinc-finger proteins on mental health. *Metallomics, 8*, 579–588. https://doi.org/10.1039/C5MT00300H

The Institute for Health Metrics and Evaluation. (2019). *Lead exposure—Level 3 risk*. http://www.healthdata.org/results/gbd_summaries/2019/lead-exposure-level-3-risk

Waggett, C. G. III, Gragg, R. D., Korfmacher, K. S., & Richmond, M. (n.d.). *Lead in the environment*. https://serc.carleton.edu/integrate/teaching_materials/lead/index.html. Accessed 31 Oct 2021 (Short URL https://serc.carleton.edu/187377).

Weidenhamer, J. D. (2007). Circuit board analysis for lead by atomic absorption spectroscopy in a course for nonscience majors. *Journal of Chemical Education, 84*(7), 1165–1166. https://doi.org/10.1021/ed084p1165

Wenzel, T. J. (2006). General chemistry: Expanding the learning outcomes and promoting interdisciplinary connections through the use of a semester-long project. *CBE Life Sciences Education, 5*(1), 76–84. https://doi.org/10.1187/cbe.05-05-0077

Rachel Narehood Austin is the Diana T. and P. Roy Vagelos Professor of Chemistry at Barnard College, Columbia University. Work in her lab focuses on understanding the mechanisms of metalloproteins, especially those important in the global cycling of elements and neurochemistry, and developing and characterizing heterogeneous catalysts that can be used for green chemistry, biofuels upgrading, or environmental remediation.

Ann McDermott is the Esther and Ronald Breslow Professor of Biological Chemistry and Professor of Biological Sciences and Chemical Engineering at Columbia University. An expert in using Nuclear Magnetic Resonance (NMR) spectroscopy to study biomolecules, Professor McDermott's work has been recognized with numerous awards, including acceptance into the National Academy of Sciences.

Katrina Korfmacher is a Professor of Environmental Medicine at the University of Rochester. Dr. Korfmacher has worked with community partnerships related to lead, healthy homes, air quality, and land use for 20 years. She is the author of "Bridging Silos: Collaborating for Environmental Health and Justice in Urban Communities."

Laura Arbelaez was born in Pereira, Colombia and raised in Paterson, New Jersey. She received her BA in Biochemistry at Barnard College of Columbia University. She currently works in the Environmental and Occupational Health Industry in New York City and will be attending medical school in 2022.

Jamie Bousleiman graduated from Barnard College in 2019 with a major in biochemistry. She is first generation Lebanese-American currently attending the Jacobs School of Medicine in Buffalo, class of 2025. At Barnard College, she discovered a passion for connecting science to real world issues, which led to her interest in this project.

Arminda Downey-Mavromatis graduated from Barnard College in 2020 with a degree in biochemistry. She is currently an Assistant Editor at Chemical & Engineering News.

Rahma Elsiesy is a bioengineering Ph.D. student based in California, interested in bioinformatics and medical device development. She is originally from Cairo, Egypt but grew up between NYC, the United Arab Emirates, and Saudi Arabia. Rahma is passionate about global public health and making healthcare accessible.

Sohee Ki is currently a third-year medical student at the Keck School of Medicine of the University of Southern California. She aspires to become a leading clinician committed to medical education and research. She graduated Bates College in 2016 with a double major in chemistry and music.

Meena Rao is the Director of Organic Chemistry Laboratories at Barnard College. She is passionate about teaching and has received prestigious awards for teaching from the university. Dr. Rao also has been at the forefront of developing new courses and is actively involved with Collegiate Science and Technology Entry (CSTEP) and Opportunity Programs.

Shoshana Williams earned her bachelor's in chemistry with a minor in religion at Barnard College, Columbia University. She is now pursuing her Ph.D. in chemistry at Stanford University in the laboratory of Professor Eric Appel. Her current work focuses on the development of novel polymeric materials for drug delivery applications.

Chapter 14
Brownfield Action: A Civic-Oriented, Web-Based, Active Learning Simulation

Peter Bower and Sedelia Rodriguez

14.1 Brownfields

Brownfields are a critical concern for environmental justice and sustainability. They are properties, often abandoned, where the expansion, redevelopment, or reuse of the property may be complicated by the presence or potential presence of hazardous substances, pollutants, or contaminants. According to the EPA, there are presently over half a million brownfields in the United States, but this number only includes sites for which an environmental site assessment (ESA) has been conducted. The actual number of brownfields is certainly in the millions and brownfield remediation and redevelopment constitutes one of the major environmental issues confronting communities today. This importance is magnified by the disproportionate number of brownfields in economically stressed communities where revitalization is critical for jobs and economic growth as well as environmental justice and sustainability. Brownfield remediation and redevelopment have been identified as a priority by federal and state governments and constitute an important job focus for science, technology, engineering, and mathematics (STEM) workers (Fig. 14.1).

14.2 Brownfield Action

Introducing students to the basic concepts of environmental science and justice, groundwater, and environmental contamination, including the authentic methods used to study and inclusive approaches to address environmental contamination, is the main goal of Brownfield Action (BA). Because brownfields are often a source of

P. Bower (✉) · S. Rodriguez
Department of Environmental Science, Barnard College, New York, NY, USA
e-mail: pb119@columbia.edu

© The Author(s) 2023
M. S. Rivera Maulucci et al. (eds.), *Transforming Education for Sustainability*,
Environmental Discourses in Science Education 7,
https://doi.org/10.1007/978-3-031-13536-1_14

Fig. 14.1 The Brownfield Action Logo

environmental toxification, they are also a concern for the environmental health and safety of communities, therefore civic engagement and inclusion are important components of BA.

BA is a web-based, interactive, three-dimensional digital space and learning simulation in which students form geotechnical consulting companies and work collaboratively to explore and solve problems in environmental forensics as they engage in an organic, evolving, semester-long, laboratory exploration of a simulated brownfield and its local community. BA marries a civic-minded, constructivist approach to learning about the environmental, economic, and civic importance of brownfields and the toxification of the environment. Created at Barnard College in collaboration with the Columbia Center for Teaching and Learning, BA has been used for over 15 years at Barnard College for one semester of a two-semester Introduction to Environmental Science course that is taken by more than 100 female, undergraduate, non-science majors each year to satisfy their laboratory science requirement. BA was selected in 2003 as a "national model curriculum" by SENCER (Science Education for New Civic Engagements and Responsibilities), an NSF STEM education initiative. What makes the BA SENCER model curriculum unique is that it includes a significant component of engagement with the civic dimensions of environmental contamination interwoven with the technical investigations being conducted by the students (Bower et al., 2011).

A signature feature of BA is its companion website at (Bower, n.d.) featuring information about BA and a guided walkthrough of the simulation. In the Registered Instructors section, instructors maintain profiles of their experiences with BA and add downloadable original curriculum materials for sharing with new BA users. Interested faculty can consult profiles, find and contact faculty with similar courses and interests, re-use or create derivative materials from the posted documents, and also become part of the collaborative network. A library of documents, maps, and images related to the simulation and its use in the classroom can also be found in the Appendix.

14.3   Pedagogical Rationale

The pedagogical methods and design of the BA model are grounded in substantial research literature focused on the design, use, and effectiveness of games and simulations in education. Benefits of a simulation approach to learning include increased engagement; adaptations for students with high or low prior knowledge; effective replacement for expensive/impractical field trips; control over the pace and direction of learning; increase in student participation over large lecture hall formats; effectiveness in representing complex subject matter; application to realistic situations; and the packaging of complex ideas into a consistent narrative. Much of the literature on computer-based simulations cited below is built upon the legacy of researchers such as Greenblat (1981), Lederman (1984), and Petranek (1992, 1994), who showed how paper-based educational simulations could motivate learners to be active participants in their own learning through individualized activities and immediate feedback. Many researchers are actively engaged in the study of particular teaching and learning strategies that employ a custom-created computer-based simulation or game similar to BA (Barab et al., 2000; Barab & Plucker, 2002, 2005, 2010; Dede & Ketlehut, 2003; Rosenbaum et al., 2007; Kim et al., 2009; Jacobson et al., 2009). A number of researchers have explored the use of simulation technologies in creating virtual field trips as a response to educational, logistical, and economic constraints (Arrowsmith et al., 2005; Whitelock & Jelfs, 2005; Ramasundaram et al., 2005).

Simulations allow the packaging of complex issues into consistent narratives, which can facilitate meaning-making (Bruner, 2002; Weinberger et al., 2005). Simulations have been found to be an effective way to represent complex systems and explore inclusive, multi-faceted socio-technical problems (Gee, unpublished; Squire & Jenkins, 2004; Barab et al., 2005; Rosenbaum et al., 2007; Herrington et al., 2007). Learners are able to understand educational content by exploring it within realistic situations, consistent with the principles of situated learning (Lave & Wenger, 1991; Barsalou, 1999; Pedretti, 1999; Barab & Plucker, 2002; Rosenberg, 2006).

Mayer and Chandler (2001) found that students who had more control over the pace and direction of educational simulations showed better learning outcomes than those who had less, or what Betrancourt (2005) has called the interactivity principle. Accordingly, engaging students with the educational content at hand is key. BA accomplishes this through the gradual unveiling of additional components as students learn more concepts and discover more of the town and its underlying hydrogeologic features and civic infrastructure. Student teams control the pace and direction of their explorations to varying degrees depending on the course level in which BA is implemented (Bower et al., 2011).

Some suggest that using simulations and educational games can be an effective way to engage the "video game generation" (Katz, 2000; Prensky, 2006). Researchers in recent years have also found that the learning and engagement seen in young people's playing of video games, including simulation-based games, can be translated into meaningful educational gains in the classroom (Shaffer & Gee, 2005; Gee, 2007; Rieber, 2001; Rieber et al., 2004; Rieber & Noah, 2008; Prensky, 2006; Herrington et al., 2007; Van Eck, 2007; Chinn & Malhotra, 2002). Through its use of a complex narrative with interactive videos and maps as well as environmental testing tools and other game-based features, BA takes advantage of the motivational features of video games to engage students in the narrative of an environmental contamination scenario that weaves in a knowledge base of scientific skills and concepts.

Simulations also allow educators to move away from large lecture halls, where students are typically passive and increase participation through inquiry-based learning. However, the role of teachers in scripting and facilitating simulation-based activities remains crucial (Barab et al., 2000; Weinberger et al., 2005). While measurable benefits of educational simulations have been shown to vary according to prior knowledge, the medium has been shown to positively impact comprehension, cognitive load, and learning efficiency (Park et al., 2009). BA is inclusive and is adaptable to students with both high and low prior knowledge by modifying the amount of prior information and kinds of assignments given to the students and by tailoring contextual help provided by instructors to the appropriate level of the intended audience. While simulations have many positive aspects, negatives in general include the cost and expertise required for development, maintenance and updating., the time required for students to learn how to use the simulations and "buy into" its virtual reality, the inability of many simulations to deal with the ambiguity of dynamic, real-life situations, and the potential for some loss of control and flexibility on the part of the instructor.

BA is one of a small but growing number of computer simulation-based teaching tools that have been developed to facilitate student learning through interaction and decision-making in a virtual environment. In STEM fields, other examples include CLAIM (Bauchau et al., 1993) for mineral exploration; DRILLBIT (Johnson & Guth, 1997) and MacOil, (Burger, 1989) for oil exploration; BEST SiteSim (Santi & Petrikovitsch, 2001) for hazardous waste and geotechnical investigations; Virtual Volcano (Parham et al., 2009) to investigate volcanic eruptions and associated hazards; and eGEO (Slator et al., 2011) for environmental science education. These virtual simulations give students access to environments and experiences that are too dangerous, cost-prohibitive, or otherwise impractical to explore (Saini-Eidukat et al., 1998). Through directed role-play, they also provide opportunities for social interaction and student inquiry into the human element of technical analysis and decision making (Aide, 2008).

14.4 Simulation Overview

The heart of the BA simulation is a virtual world containing over 2 million hydrogeologic data points in a three-dimensional grid representing roughly 150 acres of land in a virtual town. The grid points contain many different types of natural data including surface elevation, depth to water table and bedrock, soil or sediment type, and vegetation. The town has a fully-realized human infrastructure (buildings, roads, wells, water towers, homes, and businesses) as well as a municipal government. There is a compelling storyline of economic development in the town involving individual people and their particular civic roles and life histories. The story (embedded and to be discovered in the simulation) is one of groundwater contamination caused by underground contaminant plumes stemming from the failure of underground infrastructure (pipes and tanks). As students become familiar with the simulated town and the history of its abandoned plant, the students gradually reconstruct the details of an all-too-familiar narrative. The "Self-Lume, Inc." factory, which until recently manufactured radioluminescent signs, has been badly mismanaged and factory employees have dumped radioactive materials into a septic field. A gas station is also discovered to have a leaking underground storage tank. Both have contaminated the local aquifer and the town well.

At Barnard College, students in the BA laboratory form their own environmental consulting companies and work in teams of two. Each team views a video that sets up the background narrative, in which a developer ("Malls-R-Us") plans to bring back an abandoned industrial brownfield in the town into commercial use. Each company signs a detailed contract with a development corporation to perform testing needed for a Phase I Environmental Site Assessment (ESA) of the abandoned factory and surrounding properties and prepares a report for the developer on the advisability of proceeding with mall construction. The contract summarizes the budget and obligations for each company as well as the goals for the semester-long investigation. Everything the student companies do in BA costs money and each company competes with the other companies to successfully complete its investigation and maximize its profit. While BA is a collaborative project, each student at Barnard must write and produce her own Phase I ESA report using the information acquired by the team. After the Phase I ESA, all student companies are hired by the EPA and work together in the Phase II Environmental Site Investigation, with each student preparing a Phase II report delineating the nature and extent of the contamination. Student companies work together as detectives to develop an understanding of the specific roles of individuals at the abandoned factory site. They assist local prosecutors with forensic evidence to help build both civil and criminal lawsuits against the responsible parties.

The digital space and supporting materials draw the students into a "real" integrated world with several mysteries to be solved. The tools required to solve these mysteries come from an interdisciplinary array spanning geology, environmental science, history, economics, civics, physics, biology, chemistry, and law. Each company must strive to obtain the maximum amount of information at the lowest cost, information that will help them make more important and expensive decisions later in the semester in order to fulfill their contractual obligations. Students may obtain historical and qualitative information by visiting the municipal complex or interviewing community members within the simulation. They can choose from an array of technical tools enabling them to determine surface elevations and construct a topographic map, find underground tanks or pipes using ground-penetrating radar, collect seismic data to find depth to bedrock, or drill to determine the depth to the water table and to take groundwater samples. As an example of authenticity, all tools, for example, ground-penetrating radar, can only be used after passing certification tests. As with any real-world professional investigation, students must keep in mind their budget, integrate a variety of information coming from multiple sources, and think strategically about the best way to run their investigation. As they learn critical concepts of geology, hydrology, and chemistry in lectures, students are required to apply this new knowledge to successfully pursue their investigation and to make sound scientific and budgetary choices. Finally, students research and prepare a Phase III ESA that presents a plan for the remediation of the site. Figure 14.2 shows a screenshot from the Brownfield Action software.

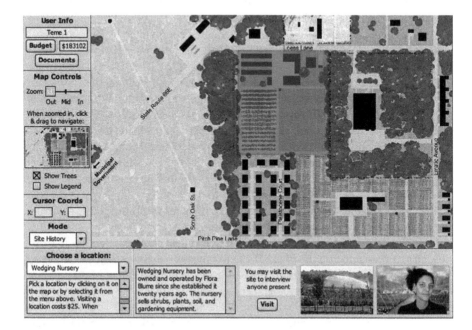

Fig. 14.2 Screenshot of Brownfield Action software in Site History mode

The main learning goals for the use of the BA simulation at BC are:

1. To understand the interdisciplinary nature of the scientific process and the inter-connections between groundwater contamination, toxins, human health, brown-fields, government, economics, and law through a real-life simulation;
2. To work collaboratively in teams as environmental consulting companies and to develop strategic thinking while exploring and solving problems in environmental forensics;
3. To overcome uncertainty and ambiguity by engaging in action planning, net-working, and negotiation; to learn to make choices, learn from the consequences, and develop self-confidence;
4. To develop and encourage civic-mindedness;
5. To collect and analyze data, critically review relevant documents, integrate inter-disciplinary scientific and social information, produce bedrock, topographic, and water table contour maps as well as site maps, and write and communicate their results in three, professional-level Phase 1, 2, and 3 environmental site assessments;
6. To prepare these reports and fulfill contractual obligations while working within a budget and competing to maximize their company's profit; and,
7. To increase student awareness of the constructivist learning environment of which they are a part and to promote "self-renewing intellectual resourcefulness."

14.5   Assessment

Formative assessment strategies for the use of the BA simulation and curriculum at Barnard College were employed using a modified model of Design Research (Bereiter, 2002; Collins, 1992; Edelson, 2002), culminating in a qualitative ethno-graphic approach using monthly interviews to determine the impact of Brownfield Action on the learning process. Results of these ethnographies (see Kelsey, 2003) showed at a high confidence level that the simulation allowed students to apply content knowledge from lectures in a lab setting and to effectively connect disparate topics with both lecture and lab components. Furthermore, it was shown that BA improved student retention and that students made linkages in their reports not likely to be made in a traditional teaching framework. It was also found that, in comparison with their predecessors before the program's adoption, students attained markedly higher levels of precision, depth, sophistication, and authenticity in their analysis of the contamination problem, learning more content and in greater depth. This study also showed that BA supports the growth of each student's relationship to environmental issues and promotes transfer into the student's real-life decision-making and approach to careers, life goals, and science (Bower et al., 2011). A recent Barnard graduate wrote:

> *Brownfields* was one of the most beneficial and realistic courses I took at Barnard/Columbia. The course is an online simulation that takes students through the thinking, writing, inter-viewing, and sampling processes of Phase 1, 2, and 3 Environmental Site Assessments (ESA). *Brownfields* had a major impact on my job search and, ultimately, my career path.

Brownfields gave me the confidence and the necessary terminology and skills to tackle environmental consulting interviews I had along the way. I am now a Field Hydrogeologist/ Environmental Scientist for an environmental consulting firm. Brownfield Action prepared me for the interview process, many of the sampling techniques, and the creative mindset needed for tackling large-scale environmental and hazardous waste projects. Specifically, the interview process at many environmental consulting firms consists of questions regarding relevant courses. Of course, for me, *Brownfields* was the most relevant! During many of my interviews, I was asked to explain how I would go about analyzing water, air, or soil quality, how to balance data or information from multiple sources (i.e. interviews, lab results, documents, etc.), and how to conduct an ESA from beginning to end. Finally, I was asked to provide writing samples. I chose to submit my Phase 1, 2, and 3 ESA Reports written during my *Brownfields* course. I really think these set me apart from other candidates for the position I obtained because I know MANY qualified people that were being interviewed. Being paired up with a lab partner with whom I formed an environmental consulting company and collaborating collectively with the rest of the class has given me the necessary communication and teamwork skills that are imperative in this line of work. Furthermore, the course was taught in a way that gave just enough direction to get us (the students) started but not enough to eliminate the ambiguity of how to approach an ESA at a specific site, find contamination, groundwater contamination plumes and origins or the techniques necessary for analysis and removal of hazardous materials during remediation. It was up to the students to come up with a plan of action and discuss various strategies at each stage.

14.6 Collaborative Network of Users

The BA simulation is also unique in that it has been disseminated to twelve colleges, universities, and high schools where it is currently being used in the classroom and through the development of a collaborative community of users. BA is the only SENCER national model curriculum with a network of faculty collaborating in a community of practice. Moreover, this network has adapted the original simulation and its related products for use with a widening diversity of students, in a variety of classroom settings, and toward an expanding list of pedagogical goals (see Bower et al., 2014).

The collaborative network of users described in Bower et al. (2014) includes educators using the BA simulation to enhance environmental instruction at the high school level, to teach the fundamentals of hydrology and environmental site assessments at an introductory to intermediate undergraduate level, and to train both undergraduate and graduate students in advanced courses in hydrology and environmental remediation. Although many of the applications reported here apply to courses in STEM curricula, BA is not restricted in its utility to teaching students with advanced STEM skills. Rather, BA has proven to be equally effective whether it is used to introduce non-science literate students to basic concepts of environmental science and basic civic issues of environmental contamination (Bower et al., 2011) or to provide advanced training in environmental site assessments and modeling groundwater contamination to future environmental professionals.

BA has increased exposure to STEM education innovations and environmental science to historically underrepresented groups. BA emphasizes brownfields in environmental justice communities, as well as increasing participation of women and underrepresented minorities in STEM. Schools that join the growing BA network also commit to racial and gender diversity for this project. The 112 students who use BA at Barnard College as part of the Introduction to Environmental Science course reflect the overall composition of the college. Except for a few male Columbia students, all are women, 18% are African-American, Latina, or Native American, and 24% are Asian. Over 90% of these students are non-science majors taking the course to fulfill their science requirement. For most of these non-science majors, BA is their last academic contact with science. Several of the collaborators are inner-city institutions with significant minority populations. These include Wayne State in Detroit (nearly 30% Black with a significant Arab minority), California State University East Bay (61% female, 26% Hispanic, 11% Black, 26% Asian/Pacific, and 16% other and international), and City College of New York (with 38% Hispanic, 25% Asian and 15% Black). Other partner institutions include Kansas State University (25% Black, Hispanic, and Asian, including a small native American population, and 50% women), Connecticut College (60% women and 19% Black, Hispanic and Asian), Hofstra University (38% Black, Hispanic, and Asian and 53% women) and Lafayette College (47% women and 15% Black, Hispanic, and Asian).

14.7 Conclusions

Environmental science education plays a vital role in preparing a highly-qualified STEM workforce. Effective environmental science education requires students to collaboratively work together to solve complex problems. BA contributes to the transformation of environmental science education by the use of an adaptive and authentic, technology-rich collaborative learning environment that can enable tailored team-based problem-solving experiences that improve learning and collaborative engagement. As such Brownfield Action successfully incorporates authenticity, active collaborative learning, and justice in its pedagogy, as well as its user network of institutions and students.

Appendix: Brownfield Simulation Materials

The following materials are available in these shared folders:

1. BA Articles: Bower et al. (2011, 2014) and Kelsey Ph.D. Thesis: https://drive. google.com/drive/folders/1LOLwdFbCXuw5aaHR-Gtfp4zgSbOTC2qI?usp=sh aring
2. Scope and Sequence for Twelve BA Laboratories at Barnard: https://drive. google.com/drive/folders/1weQ5gzwE9LReer-GHiRn13ey08wbHiIX?usp=sha ring

3. BA Student Maps: https://drive.google.com/drive/folders/1C4EVlv2lJIA0u
   AW80MydLuBA0bhnZCQg?usp=sharing
4. BA Student Phase 1, 2, & 3 ESAs: https://drive.google.com/drive/folders/1Qw
   1n1Gh6v5uWbmngnmtjsQqDOpzY8D7Y?usp=sharing
5. Documents: https://drive.google.com/drive/folders/1W1HFPBnlKZljdCQZme
   GpjZBHCcCwxvNz?usp=sharing

References

Aide, C. N. (2008). Using online role-playing to demonstrate impracticability of responses to an earthquake scenario. *[Paper] Abstracts with Programs – Geological Society of America, 40*(6), 370.

Arrowsmith, C., Counihan, A., & McGreevy, D. (2005). Development of a multi-scaled virtual field trip for the teaching and learning of geospatial science. *International Journal of Education and Development using ICT [Online], 1*(3). http://ijedict.dec.uwi.edu/viewarticle.php?id=29&layout=html

Barab, S. A., & Plucker, J. A. (2002). Smart people or smart contexts? Cognition, ability, and talent development in an age of situated approaches to knowing and learning. *Educational Psychologist, 37*(3), 165–182. https://doi.org/10.1207/S15326985EP3703_3

Barab, S. A., Hay, K. E., Squire, K., Barnett, M., Schmidt, R., Karrigan, K., Yamagata-Lynch, L., & Johnson, C. (2000). Virtual solar system project: Learning through a technology-rich, inquiry-based, participatory learning environment. *Journal of Science Education and Technology, 9*(1), 7–25. https://doi.org/10.1023/A:1009416822783

Barab, S., Thomas, M., Dodge, T., Carteaux, R., & Tuzun, H. (2005). Making learning fun: Quest Atlantis, a game without guns. *Educational Technology Research and Development, 53*(1), 86–107. https://doi.org/10.1007/BF02504859

Barab, S. A., Gresalfi, M., & Ingram-Goble, A. A. (2010). Transformational play: Using games to position person, content, and context. *Educational Researcher, 39*(7), 525–536. https://doi.org/10.3102/0013189X10386593

Barsalou, L. W. (1999). Language comprehension: Archival memory or preparation for situated action. *Discourse Processes, 28*, 61–80. https://doi.org/10.1080/01638539909545069

Bauchau, C. C., Jaboyedoff, M. M., & Vannier, M. M. (1993). CLAIM: A new personal computer-assisted simulation model for teaching mineral exploration techniques. *Geological Association of Canada Special Paper, 40*, 685–691. https://quanterra.ch/wp-content/uploads/2015/11/claim_eng.pdf

Bereiter, C. (2002). Design research for sustained innovation. *Cognitive Studies: Bulletin of the Japanese Cognitive Science Society, 9*(3), 321–327. https://doi.org/10.11225/jcss.9.321

Betrancourt, M. (2005). The animation and interactivity principles in multimedia learning. In R. E. Mayer (Ed.), *The Cambridge handbook of multimedia learning* (pp. 287–296). Cambridge University Press.

Bower, P. (n.d.). *Brownfield Action.* www.brownfieldaction.org

Bower, P., Kelsey, R., & Moretti, F. (2011). Brownfield Action: An inquiry-based multimedia simulation for teaching and learning environmental science. *Science Education and Civic Engagement, 3*(1), 5–14.

Bower, P., Kelsey, R., Bennington, B., Lemke, L., Liddicoat, J., Miccio, B., Lampousis, A., Thompson, D., Seewald, B., Kney, A., Graham, T., & Datta, S. (2014). Brownfield action: Dissemination of a SENCER model curriculum and the creation of a collaborative STEM education network. *Science Education and Civic Engagement, 6*(1), 5–22. https://academicworks.cuny.edu/cc_pubs/366/

Bruner, J. (2002). *Making stories: Law, literature, life*. Farrar, Straus and Giroux.

Burger, H. R. (1989). MacOil – An oil exploration simulation. *Journal of Geological Education, 37*, 181–186.

Chinn, C. A., & Malhotra, B. (2002). Epistemologically authentic inquiry in schools: A theoretical framework for evaluating inquiry tasks. *Science Education, 86*(2), 175–218. https://doi.org/10.1002/sce.10001

Collins, A. (1992). Toward a design science of education. In E. Scanlon & T. O'Shea (Eds.), *New directions in educational technology* (pp. 15–22). Springer. https://doi.org/10.1007/978-3-642-77750-9_2

Dede, C., & Ketelhut, D. (2003, April). *Designing for motivation and usability in a Web-based multi-user virtual environment*. Paper presented at the Annual Meeting of the American Educational Research Association, Chicago.

Edelson, D. C. (2002). Design research: What we learn when we engage in design. *The Journal of the Learning Sciences, 11*, 105–121. https://doi.org/10.1207/S15327809JLS1101_4

Gee, J. P. (2007). *What video games have to teach us about learning and literacy* (2nd ed.). Palgrave Macmillan.

Gee, J. (unpublished manuscript). *Why are video games good for learning?* Academic Co-Lab. Retrieved from http://www.academiccolab.org/resources/documents/MacArthur.pdf; https://doi.org/10.18261/ISSN1891-943X-2006-03-02

Greenblat, C. S. (1981). Seeing forests and trees: Gaming-simulation and contemporary problems of learning and communication. In C. S. Greenblat & R. D. Duke (Eds.), *Principles and practices of gaming simulation* (pp. 13–18). Sage.

Herrington, J., Reeves, T. C., & Oliver, R. (2007). Immersive learning technologies: Realism and online authentic learning. *Journal of Computing in Higher Education, 19*(1), 80–99. https://eric.ed.gov/?id=EJ830802

Jacobson, A. R., Militello, R., & Baveye, P. C. (2009). Development of computer-assisted virtual field trips to support multidisciplinary learning. *Computers & Education, 52*(3), 571–580. https://doi.org/10.1016/j.compedu.2008.11.007

Johnson, M. C., & Guth, P. L. (1997). A realistic microcomputer exercise designed to teach geologic reasoning. *Journal of Geoscience Education, 45*, 349–353. https://doi.org/10.5408/1089-9995-45.4.349

Katz, J. (2000). *Up, up, down, down*. Slashdot.org. Originally published November 30, 2000. http://slashdot.org/features/00/11/27/1648231.shtml

Kelsey, R. (2003). *Brownfield Action: An education through an environmental science simulation experience for undergraduates*. PhD thesis, Columbia University.

Kim, B., Park, H., & Baek, Y. (2009). Not just fun, but serious strategies: Using meta-cognitive strategies in game-based learning. *Computers & Education, 52*(4), 800–810. https://doi.org/10.1016/j.compedu.2008.12.004

Lave, J., & Wenger, E. (1991). *Situated learning: Legitimate peripheral participation*. Cambridge University Press.

Lederman, L. C. (1984). Debriefing: A critical reexamination of the postexperience analytic process with implications for its effective use. *Simulations and Games, 15*(4), 415–431. https://psycnet.apa.org/doi/10.1177/0037550084154002

Mayer, R. E., & Chandler, P. (2001). When learning is just a click away: Does simple user interaction foster deeper understanding of multimedia messages? *Journal of Educational Psychology, 93*(2), 390–397. https://doi.org/10.1037/0022-0663.93.2.390

Parham, T. R., Boudreaux, H., Bible, P., Stelling, P., Cruz-Niera, C., Cervato, C., & Gallus, W. A. (2009). Journey to the center of a volcano; teaching with a 3-D virtual volcano simulation. *[Paper] Abstracts with Programs – Geological Society of America, 41*(7), 500.

Park, S. I., Lee, G., & Kim, M. (2009). Do students benefit equally from interactive computer simulations regardless of prior knowledge levels? *Computers & Education, 52*(3), 649–655. https://doi.org/10.1016/j.compedu.2008.11.014

Pedretti, E. (1999). Decision making and STS education: Exploring scientific knowledge and social responsibility in schools and science centers through an issues-based approach. *School Science and Mathematics, 99*, 174–181. https://doi.org/10.1111/j.1949-8594.1999.tb17471.x

Petranek, C. (1994). A maturation in experiential learning: Principles of simulation and gaming. *Simulation & Gaming, 25*(4), 513–523. https://doi.org/10.1177/1046878194254008

Petranek, C., Corey, F., & Black, R. (1992). Three levels of learning in simulations: Participating, debriefing, and journal writing. *Simulation & Gaming, 23*, 174–185. https://doi.org/10.1177/1046878192232005

Prensky, M. (2006). *Don't bother me, mom, I'm learning!: How computer and video games are preparing your kids for 21st-century success and how you can help!* Paragon House.

Ramasundaram, V., Grunwald, S., Mangeot, A., Comerford, N. B., & Bliss, C. M. (2005). Development of an environmental virtual field laboratory. *Computers in Education, 45*, 21–34. https://doi.org/10.1016/j.compedu.2004.03.002

Rieber, L.P. (2001, December). *Designing learning environments that excite serious play.* [Paper] Presented at the annual meeting of the Australasian Society for Computers in Learning in Tertiary Education, Melbourne, Australia.

Rieber, L. P., & Noah, D. (2008). Games, simulations, and visual metaphors in education: Antagonism between enjoyment and learning. *Educational Media International, 45*(2), 77–92. https://doi.org/10.1080/09523980802107096

Rieber, L. P., Tzeng, S., & Tribble, K. (2004). Discovery learning, representation, and explanation within a computer-based simulation: Finding the right mix. *Learning and Instruction, 14*(3), 307–323. https://doi.org/10.1016/j.learninstruc.2004.06.008

Rosenbaum, E., Klopfer, E., & Perry, J. (2007). On location learning: Authentic applied science with networked augmented realities. *Journal of Science Education and Technology, 16*(1), 31–45. https://doi.org/10.1007/s10956-006-9036-0

Rosenberg, M. J. (2006). *Beyond e-learning.* Pfeiffer.

Saini-Eidukat, B., Schwert, D. P., & Slator, B. M. (1998). Text-based implementation of the Geology Explorer, a multi-user role-playing virtual world to enhance learning of geological problem-solving. *[Paper] Abstracts with Programs – Geological Society of America, 30*(7), 390–391.

Santi, P. M., & Petrikovitsch, J. F. (2001). Teaching engineering geology by computer simulation of site investigations. *[Paper] Abstracts with Programs – Geological Society of America, 33*(6), 131.

Shaffer, D. W., & Gee, J. P. (2005). *Before every child is left behind: How epistemic games can solve the coming crisis in education* (WCER working paper no. 2005–7). Wisconsin Center for Education Research. School of Education, University of Wisconsin-Madison.

Slator, B. M., Saini-Eidukat, B., Schwert, D. P., Borchert, O., Hokanson, G., Forness, S., & Kariluoma, M. (2011). The eGEO virtual world for environmental science education. *[Paper] Abstracts with Programs – Geological Society of America, 43*(5), 135.

Squire, K., & Jenkins, H. (2004). Harnessing the power of games in education. *Insight, 3*(1), 5–33.

Van Eck, R. (2007). Six ideas in search of a discipline. In B. E. Shelton & D. A. Wiley (Eds.), *The design and use of simulation computer games in education* (pp. 31–60).

Weinberger, A., Ertl, B., Fischer, F., & Mandl, H. (2005). Epistemic and social scripts in computer-supported collaborative learning. *Instructional Science, 33*(1), 1–30. https://doi.org/10.1007/s11251-004-2322-4

Whitelock, D., & Jelfs, A. (2005). Would you rather collect data in the rain or attend a virtual field trip? Findings from a series of virtual science field studies. *International Journal of Continuing Engineering Education and Lifelong Learning, 15*(1/2), 121–131.

Peter Bower, creator of Brownfield Action, is a retired Research Scholar in the Department of Environmental Science (DES) at Barnard College, Columbia University. As its first full-time employee, he built the DES from its beginnings 35 years ago, teaching and mentoring student research until his recent retirement.

Sedelia Rodriguez is a Senior Lecturer in Environmental Science at Barnard College. She is also the Assistant Director/ Academic Coordinator for Barnard's Science Pathways Scholars Program. Her research focuses on the environmental impacts of voluminous lava flows. As a former environmental site assessment professional, she has used Brownfield Action in her classes for over a decade.

Chapter 15
Learning from the Many, Teaching to the Many: Applying Ecojustice Principles to Undergraduate Pedagogy in Environmental Science, Ecology, and Sustainability Classrooms

Mary A. Heskel and Jennings G. A. Mergenthal

15.1 Perspectives in and on Undergraduate Classrooms

This is at least the second rewrite of this chapter – at least. It is nearly unrecognizable from the first. I (Mary) am letting you in on this because, as an educator and a scientist, I view adjusting ideas and practices, welcoming change and diverse perspectives, and seeking out new ideas as part of the job. It is essential to remind ourselves to be adaptive, as the syllabi of our classes, curricula of our departments, and more broadly, our ideas on how students learn and apply knowledge often stagnate and age while we work vigorously to push the research of our labs forward.

Committing your teaching – through practices and content selection – to include authentic connections to Environmental Justice and EcoJustice can broaden the impact of your STEM (i.e., – science, technology, engineering, and mathematics) courses. While we often recognize the importance of human systems and their effects on ecological and environmental processes when teaching – most notably the impacts of climate change – it is less common for professors to expand into the structural inequalities of human systems and how they shape the science we practice and the content we deliver. This chapter encourages scientists, educators, and students to broaden their perspectives, critique historical biases and inequities in ecology and environmental science, and embrace the broad plurality of information available to us, with particular attention to the integration of Indigenous knowledge.

M. A. Heskel (✉) · J. G. A. Mergenthal
Department of Biology, Macalester College, Saint Paul, MN, USA
e-mail: mheskel@macalester.edu

© The Author(s) 2023   261
M. S. Rivera Maulucci et al. (eds.), *Transforming Education for Sustainability*,
Environmental Discourses in Science Education 7,
https://doi.org/10.1007/978-3-031-13536-1_15

15.2   Perspectives Matter – Here Are Ours

You are about to read the outcomes of our research and discussions as a writing team and our individual experiences in different roles in science education and practice. I (Mary) am currently an Assistant Professor of Ecology at a small liberal arts college. My background as an undergraduate and graduate student at large research institutions where innovative teaching was not highly prioritized; as a biology teacher at an under-resourced urban public high school where nurturing connections with students, clear expectations, and active learning were valued and promoted; and as a research scientist learning and navigating explicit and unwritten rules that dictate success, informs my perspectives on teaching and mentoring undergraduate students. However, in my current role, I find my perspective on teaching most informed by the students I teach – what are their goals for the course and for after college as emerging adults faced with challenges that include student debt, climate change, and anxiety over an uncertain future? I aim to develop courses that teach content and transferable skills and provide a space where students feel they are supported to grow and succeed. So, I have been doing a lot of listening and learning (Dewsbury & Brame, 2019), adjusting and adapting, and reading to create inclusive learning environments (Canning et al., 2019). As an ecologist, Environmental and EcoJustice can help with building inclusive learning environments through enacting their principles, informing content, and connecting students to broader impacts.

I (Jennings) have never been comfortable in science classrooms, not in my secondary education or college education. Make no mistake. I am fascinated by science and observing the natural world and always have been. However, the historical lineage of scientists that I have learned about in school does not include anyone who even remotely looks like me. I am also studying history, and while I would say that I enjoy *doing* science more, I am unquestionably more comfortable *being* in history. As a disabled and Indigenous person, I would traditionally be a subject of (or perhaps an inconvenience to) science rather than an active practitioner. Many of the science classrooms I have been in stressed that we (the students) were all fundamentally the same, all equals before the eyes of science. While this rather literal approach does serve to push back against historical 'race science,' it does not consider the different backgrounds of students and what traumas and experiences they bring into the classroom. This one-size-fits-all can alienate marginalized students, especially if the instructor does not share similar marginalized identities. I want to help change this approach and help create a framework for equity in science education.

15.3 A Role for Environmental Justice and EcoJustice in Undergraduate Science Classes

Environmental Justice and EcoJustice (hereafter, EJ) serve as frameworks to assess environmental laws, regulations and their implementation, decision-making, and natural resource use and distribution, with the guiding principle that there be 'fair treatment and meaningful involvement of all people regardless of race, color, national origin, or income' (Environmental Justice Learning Center, 2019). Broader definitions include the identities of gender, sexuality, citizenship, and Indigeneity (Pellow, 2018). Tenets of EJ as listed by delegates to the First National People of Color Environmental Leadership Summit in 1991 (Bullard, 1999) emphasize the rights of all individuals to clean, protected, and accessible natural resources; the political, economic, cultural and environmental self-determination of all people; and sustainable practices to preserve resources for future generations and others. Also amongst the guiding principles of EJ are (1) equity in decision-making, (2) transparency of and access to information, and (3) education based on diverse cultural perspectives (Bullard, 1999). Those values can inform ways to teach EJ and incorporate it into undergraduate science classrooms through shifts in both pedagogy and content. As undergraduate classrooms and campuses embrace sustainability initiatives, they must be reconciled with an EJ perspective. Sustainability on its own does not always explicitly address the need to repair past and current harms of inequity. However, many sustainability tactics borrow (recognized or not) from marginalized cultures and populations to improve resource use.

Exemplary case studies and 'textbook' examples of EJ are challenging to find, if existent, in practice. What gets more often observed, reported on, and studied, are the multitude of environmental and ecological *injustices* that marginalized communities have historically and continue to experience globally. While these injustices are widespread and deeply rooted in structural inequity (Pellow, 2018; Bullard, 2001), they do not often emerge as a focus in undergraduate environmental science, ecology, and sustainability classes. In environmental science, ecology, and sustainability classrooms, case studies, such as those on point source pollution in watersheds or deforestation of tropical forests, may not integrate discussions of the demographics of urban and rural areas where pollution is more common or how titling of land to Indigenous Peoples can serve to protect and conserve tropical rainforest biodiversity (Blackman et al., 2017). Rethinking case studies to include and apply EJ principles can serve as pedagogical innovation and as a guiding topic within environmental science and ecology courses.

One trove of resources that may help professors readily explore EJ principles is the InTeGrate community, hosted and supported by the Science Education Resource Center at Carleton College (SERC, 2021). This sustainability-focused community of educators creates and shares lessons, modules, discussion topics, and other resources that seek to engage undergraduate students in ecology, environmental science, and sustainability classrooms. The goals of this interdisciplinary and cross-institutional network emphasize the need for undergraduates to connect social sciences and natural sciences through a sustainability lens, promote systems thinking, and approach current and future environmental issues with a problem-solving mindset. Resources about InTeGrate can be found online via the SERC website and are all open access.

Science classrooms often espouse the importance of innovation and collaboration to identify new information and concepts within disciplines. As teachers, we also aim to incorporate best practices based on data to push our pedagogy from passive lecture halls to more interactive active learning and inclusive teaching approaches to make content more accessible and center students in their learning (Handelsman et al., 2004). This approach can transcend to activities such as practice reading journal articles (Hoskins et al., 2011) and interpreting graphs in small groups in classes, though recent evidence shows a reluctance amongst students to engage in this format of learning (Deslauriers et al., 2019). Principles of EJ mentioned above – equity, transparency, and inclusion of diverse cultural perspectives – can reinforce many objectives of an authentic, inclusive undergraduate STEM classroom. Equity, as practiced in the classroom, may translate into more student voice on the weight of assignments; democratizing student-led discussions by the random assignment of roles; soliciting anonymous student feedback regularly to identify areas in need of improvement. Even small teaching gestures, such as waiting a few minutes for students to collect their thoughts before calling for answers, enable instructors to be more inclusive and intentional about who contributes to discussions can lead to a more equitable learning environment (Handelsman et al., 2007).

The principles of transparency and diversity of cultural perspectives can broaden whole disciplines when we think about who is historically represented as the practitioners of science and which voices have emerged as dominant in their representation in texts and readings in science classrooms. Discussions on how science is practiced, documented, and taught can lead to new explorations of under-represented voices and perspectives and address implicit and explicit biases that continue to influence the environmental sciences (Ou & Romero-Olivares, 2019). Recently, in the spring of 2019, an essay published in the preeminent, globally-read weekly journal, *Science*, described how natural historians from Europe increased their specimen collections in the Americas and Africa due to their travels on ships transporting enslaved people (Kean, 2019). The stance and tone of the article made assumptions about who could be viewed as scientists, documentors, and keepers of natural history knowledge in the past (Kean, 2019). They strangely posed 'discoveries' as a potential side-benefit to the centuries-long trafficking and suffering of enslaved Africans. A response to the essay (also published in *Science*) by Dr. Rae Wynn-Grant, a carnivore ecologist, detailed the many oversights and biased perspectives of the essay, clarifying how the historical framing of science as for and by a single group of people was and continues to be damaging (Wynn-Grant, 2019). The editorial staff at *Science* would benefit from including and empowering writers, scientists, and editors of diverse backgrounds to question, critique, and confront historical and current biases in the practice of and documentation of science, as well as bring light to and elevate the understudied science and legacies of marginalized people. Democratizing science to reflect the principles of EJ requires action from professors and students and from gatekeepers that include journal editors, conference organizers, funding program officers, and reviewers of grant proposals.

In the undergraduate classroom, confronting historical biases can be discussed and addressed through student-led projects, such as the ongoing initiative to add and edit Wikipedia to include more women scientists (Wade & Zaringhalam, 2018). In this project, students choose or are assigned scientists who do not have a Wikipedia entry; through readings of their work and examinations of their training and background, students create new entries accessible to all with access to the internet. This exercise emphasizes the individuals and teams driving the research. It centers the work, progress, and growth, not just the outcome. Expanding on this existing Wikipedia project to include individual scientists from under-represented identities and groups and annotating Wikipedia entries on environmental science, ecological, and sustainability concepts to include Traditional Ecological Knowledge perspectives, are some ways to uphold and practice EJ principles through innovative student-focused teaching while simultaneously broadening the representation of who scientists are and what is considered science.

Undergraduate environmental science, ecology, and sustainability classrooms will also benefit from integrating topical and recent EJ case studies as a 'hook' for further examination into discipline-based concepts. Once you are attuned to climate justice and EJ issues, it is challenging to get through a day's worth of news without identifying how intersections of race, ethnicity, immigration status, religion, gender, sexuality, class, and other identities play a role in how individuals and communities are differentially impacted by environmental decisions. As I taught an introductory Ecology course during Spring 2019, there were too many current event topics to cover thoroughly in a class that met three times a week! Two major monsoons hit Mozambique, a country that does not contribute any substantial fossil fuel emissions and is increasingly threatened by more extensive and more frequent storms as ocean waters warm due to climate change (Fitchett, 2018). Fires rage in the Brazilian Amazon to clear land for ranching (Fearnside, 2015), soy farms, and mining (Sonter et al., 2017) that in many cases illegally and violently broach Indigenous land reserves (Blackman et al., 2017). Air pollution in the United States is more likely to impact low-income and marginalized populations, despite these groups contributing less demand to the production of fossil-fuel emitting industries (Tessum et al., 2019). Using EJ topics and case studies can engage students motivated by social justice and science, broaden classroom participation, and enrich the perspectives of all students introduced to environmental science, ecology, and sustainability at the undergraduate level.

A major tenet of EJ frameworks is to uphold equity in access, protection, and decision-making across all groups and stakeholders, especially those marginalized by historical and current practices (Bullard, 2001). In this chapter, we encourage the integration of EJ ideals to inform and restructure the undergraduate classroom to promote equity, authenticity, and inclusion amongst students and faculty. We detail the benefits of broadening science pedagogy and pivoting from a practice that emphasizes *single* actors and sources of knowledge to a more authentic, diverse, and democratic representation of science. We focus on the strengths of including multiple perspectives in terms of pedagogy and sources of knowledge.

15.4   Strengthening Teaching and Learning by Including Multiple Perspectives

15.4.1   Courses, Pedagogy, and Classroom Practice

Undergraduate coursework is often predominantly focused on content, not skills. In a world where surges of information flow from every screen in sight (Pennycook & Rand, 2019), it is more important than ever for students to practice one of the most fundamental skills offered by science – evidence-based critical thinking. This skill – basic information literacy – may be the most widely transferable amongst all practiced in undergraduate classrooms and one whose impact is felt everywhere, from elections to vaccination schedules (Scheufele & Krause, 2019). Instructors prioritize and present terms, definitions, processes, case studies, and systems, often at the expense of teaching process skills transferable and applicable to other courses. This content then is packaged into semester-long topics that become increasingly more specific in a hierarchical manner: *Biology* leads to *Ecology,* which leads to *Ecosystem Ecology,* or *Aquatic Ecology.* As a junior or senior undergraduate, students may have access to more interdisciplinary courses within the major – an *Urban Environmental Science* course may weave in sociology, urban planning, and climate science. A *Sustainable Agriculture* course may integrate political science, geography, economics, and plant biology. Placing these integrative courses at the end of a potentially grueling, years-long major sequence seems alienating and disheartening to first-year undergraduates who enter college enthused to learn about the syntheses and integration of ideas. For this reason, many institutions of higher education are re-thinking this building-block approach to better retain and expand the number of students entering STEM majors (Haak et al., 2011). Especially for students motivated by topics in EJ issues, integrating interests in social justice, applied and synthetic topics early in the course sequence may lead to higher retention in STEM majors.

Faculty may argue that the fundamentals are necessary to understand systems fully, and I agree. As a plant physiologist researching respiration, I (Mary) am the first to argue that some jobs require you to know the tri-carboxylic acid (Krebs) cycle! However, the traditional lecture-midterm-final format of many introductory survey courses does little to inspire, let alone teach students skills beyond memorization and high-anxiety studying. Further, many introductory classes are composed of students with a range of high school science backgrounds – and this variation can create uneven and potentially unwelcoming educational environments (Harackiewicz et al., 2014). To build an inclusive classroom that values equity of experience, you need assignments that provide creativity and individual personal perspectives (Tanner, 2013). Instead of content regurgitation, students need room for expansion and application of ideas, independent of background content knowledge (Dewsbury, 2019). We recommend a few of these in the following section. Introductory courses set the tone for a student's undergraduate education. Over-prioritizing memorization and high-stakes, 'make-or-break' exams can alienate students and create classrooms that lack the joy of curiosity that attracts many students to science in the first place. Alternatively, assignments that address original, student-generated questions can elicit curiosity about the content and generate novel explorations of information.

A shift in perspectives informed by principles of EJ can enhance introductory and more discipline-specific courses. Decentralizing the classroom from a teacher-centric model to a student- and team-focused model reinforces the concept that all voices and perspectives are valued. Classroom architecture can often limit or promote this approach – and can be the most challenging element to change; a stadium-style lecture hall emphasizes that everyone should be focused on the board or the professor, not on themselves or each other as students. Bypassing this can be extremely difficult, and it is unlikely that many colleges and universities have the facilities for classrooms to be spatially decentralized. This challenge puts more pressure and more focus on the teaching models – how can faculty reinforce the concept that students themselves are the drivers of change in their studies and the course? How can faculty promote the message that what students bring into the classroom – their experiences, perspectives, interests, and backgrounds – are all assets to learn from and not deficits to work around (Montgomery, 2017)? How can larger introductory classrooms be re-modeled and re-configured to enhance individual learning and build community (Dewsbury & Brame, 2019)? How can undergraduate science classrooms be designed to include marginalized students and faculty authentically? These questions do not have easy answers, though lessons and principles from EJ may serve as a framework to approach thinking about potential solutions.

Active learning is a term that encompasses pedagogy that engages students in their learning (Handelsman et al., 2007). At the heart of active learning is peer collaboration – students working closely with other students to go through problems, questions, and concepts; a practice that mirrors the process of science in the field, lab, at conferences, or in journal clubs. Building self-confidence, relationships, and feelings of belonging in the classroom can be an essential aspect of learning (Dewsbury, 2019) that distinguishes learning in a traditional undergraduate classroom surrounded by peers from an online degree program. EcoJustice espouses inclusion and democratic decision making, transparency in information, and equity amongst resources; in active learning classrooms, information is decentralized, distributed amongst students in small groups where they work together (Haak et al., 2011). Active learning can meet resistance amongst students used to a more passive format but has proven to yield stronger learning outcomes (Deslauriers et al., 2019). Since earlier generations of science classrooms did not apply active learning widely, this newness will be not only for the students but also the professor as well – and the challenges will be felt and approached differently by both.

Individual and small-group active learning may also yield new approaches – by distributing challenges and activities to students and delivering responsibility to them, the diversity of perspectives, experiences, and learning styles can result in new ideas (Tanner, 2013). An example activity could be the development of a metaphor or analogy for an ecological concept like the successional development of an ecosystem. A sole professor might describe and teach the one metaphor that 'clicked' for them, which is unlikely to resonate with students with different cultural experiences and familiarities. Suppose ten small groups of students worked together to develop a working metaphor for succession to explain and teach, using metacognitive practices to justify and explain their reasoning. In that case, the classroom is likely to develop ten distinct, potentially helpful metaphors. When studying succession, there are now many examples to draw from for reinforcement of ideas as a class. This practice further integrates science as a team-based effort that often yields diverse results that reinforce each other (or not!) and lead to future explorations that include more nuance and critical analysis.

15.4.2 Sources of Knowledge

In the classroom, the instructor serves as the legitimizer of knowledge. The perspectives and observations of white men are the sources of knowledge and expertise most legitimized in science classrooms (von Roten, 2011), (classically regarded as the only group capable of doing 'Science') (White, 2016). As historically viewed, science is assimilative, and knowledge sources must be made to fit within the preexisting framework of 'science' (Kimmerer, 2019). However, when considered at its essentials, science is based on observable evidence. Colonial biases only enter when we consider *who* is doing the observation, leading to the creation of the category, 'traditional knowledge,' posited as subjective and fallible compared to 'Western' science (Mistry & Berardi, 2016). I (Jennings) will continue to use the term Indigenous science throughout this article, as it complicates the implicit assumption that science is a purely 'Western' concept. This assumption is to the detriment of science as a field. We can find numerous examples of observations that were previously deemed traditional and unscientific but later found credible and of merit (Bonta et al., 2017). In reality, this dichotomy between Indigenous and 'Western' science is a false one (Kimmerer, 2019); science is a value-neutral term to which no worldview has a monopoly ("Indigenous Science Statement for the March for Science," 2017).

This stance is not, by any means, to advocate the wholescale or uncritical inclusion of 'alternative' sources of knowledge (such as those that have recently manifested in opposition/skepticism towards vaccines and as Bible-based 'creation science'). Rather this greater inclusivity is intended to broaden mental models of who historically has done science (Nicholas, 2018). The lineage of "Western thinkers" have never been the only ones doing science. When science instructors assume a universal and monolithic background of student prior experience and values, they alienate students who are not privy to these values, students who generally tend to be marginalized in any number of ways (Roy, 2018). To present Western science as the only standard of a completely objective, purely empirical tool can further this harm and alienation, especially as it intersects with personal and family histories. We will thus continue to use the term Indigenous Science in lieu of 'Indigenizing', as the latter may subtly imply a distinction between Science and Indigenous Science, which we hope to make clear are not, in fact, distinct.

15.4.3   Token Inclusion of TEK and Indigenous Perspectives

Though discussion of Indigenous science and Indigenous perspectives in the classroom context has increased, such actions are nearly meaningless if Indigenous science is seen as entirely separate and only occasionally complementary (or supplementary) to Western science. Inclusion only matters if perspectives of Indigenous science are fully integrated and respected as legitimate scientific viewpoints. When concepts of Indigenous science are included in curricula, in my (Jennings) experience, it gets relegated to one of two categories:

1. As a framing text for a discussion of science as international or transcultural. These concepts are never brought up again or used to challenge dominant narratives or examine the science in a broader context; or
2. As supplementary materials that support a conclusion that has been drawn by 'Western science.'

Category one approaches lack meaningful or authentic inclusion. For example, opening a curriculum by mentioning an Indigenous, "observation-based tradition" does little to challenge dominant narratives. Instead, these approaches comprise token moves towards innocence, towards ways to claim to value diverse perspectives without actually complicating the worldview of science as a purely Western phenomenon (Smith et al., 2019). Category two approaches are exemplified by coverage of the 'discovery' that hawks in Australia will carry burning branches to start fires. This phenomenon, originally observed by Aboriginal Australians, was only considered credible when the conclusions were separately reached by Western scientists (Bonta et al., 2017). This example speaks to Kimmerer's point about the assimilative nature of science when the observations are remade to fit within the confines of Western science (Kimmerer, 2019). Observation-based traditional knowledge should be recognized under a broader and more holistic definition of science.

Though these two categories have predominated approaches to incorporating Indigenous science, there are ways to include perspectives and promote equity and critical thinking, which can vary significantly based on the scope and content of the curriculum. Examples of Indigenous, observation-based practices that can be included are the domestication of New World crops like maize, pumpkins, and potatoes, or exploring the rich history of Indigenous mariculture (Smith et al., 2019).

To increase the local content, examine the land you are on and its history, both in terms of people and the environmental ecology. Contact Indigenous groups with histories on the land to inform yourself and your students. Do not expect any single source to be the sum of all knowledge or speak for any group as a monolith, but remain open to the resources and histories of local experts.

Some of the foundational narratives to push back on include the view that environments and ecosystems exist independently of humans and that the only legitimate observations are those recorded via Western science methods. However, there are numerous counterexamples to these narratives, including:

- Practices of sea level and ice level observation among Indigenous nations in the Arctic (Igor & Jolly, 2002)
- The shaping of the Amazon rainforest through agroforestry practices (Miller & Nair, 2006)
- Use of controlled burns as a land management practice in North America and Australia (Kimmerer & Kanawha, 2001)
- Timing and extent of mammalian biodiversity declines in Australia (Ziembicki et al., 2013)
- The impacts of Indigenous land stewardship and sovereignty on biodiversity conservation across continents (O'Bryan et al., 2020; Garnett et al., 2018; Schuster et al., 2019)

15.4.4 Assessments That Connect Authenticity with Curiosity, Inclusion, and Identity

Introducing and emphasizing EJ principles and concepts in biology, environmental science, ecology, and sustainability classrooms may be met with some resistance. Traditional undergraduate classroom structures and assessments are often framed around high-stakes exams, with students anxiously studying to provide the single correct answer to each question. Any practitioner of science – especially any science based on field observations – would cringe to think there is a single, consistent, predictable response for many of the variables we measure. If the answers are all available, what is left to explore? Further, this approach leaves little to no room for authentic, individualized explorations of the content that push students and professors to form new questions and ideas.

I (Mary) often fielded questions from students in an introductory ecology course with multiple caveats – does elevated carbon dioxide lead to more significant plant growth? Sure, but more so in plants that employ C3 photosynthesis than C4 photosynthesis – though maybe not over the long term in grasslands. And we need to keep in mind that elevated carbon dioxide drives climate change, which is likely to lead to extended drought periods and increased insect herbivory. And fertilization effects by elevated carbon dioxide may be limited by nitrogen and soil moisture. Also, with soaring rates of deforestation and fires in important ecosystems like the peat forests of Borneo and the Amazon, carbon loss to the atmosphere and removal from forests may negate fertilization effects elsewhere in the world. Oh – you just wanted a yes or no? Acknowledging and addressing unknowns is a challenge for many students in introductory science classrooms and an abrupt change from the formulas and standardized tests that often occupy much of students' time as high schoolers.

Teaching STEM at the K12 and undergraduate levels should be as much or more about learning practices and approaches as learning content. Learning which questions to ask and how to ask them is fundamental to being a scientist, and at the root of these skills are curiosity and creativity. Lecture-heavy, professor-centric undergraduate classrooms encourage passive learning and the messaging that students do not need to push the conversations or question further into unknowns (Deslauriers et al., 2019). Training the next generation of scientists requires students to be comfortable knowing that not all the answers exist yet and to be confident to formulate and follow their own questions. This seemingly small step – to ask a new question and develop a way to address it – is not only at the heart of all science, but can be incredibly empowering to students who have been taught in a way that showcases science as a set of established hard facts and rules to memorize.

Traditional STEM classroom structures and formats do not emphasize or reward curiosity, despite it being so essential to progress in and process of science. Students curious about a topic may be unsatisfied with traditional assessments; their unique interests may not be directly covered in lectures or centered in exams or lab reports. Student curiosity may emerge and develop more in informal discussions, in follow-up questions after class, or be relegated to the summer months or outside of classes when a select few students get to work on research projects or pursue internships. If we value curiosity, and want to teach in a way that allows *all* students to explore their interests and original questions, we need to design assessments that encourage, nourish, and support those traits. Assessments that value individual voice and perspective, firmly place the student in the role of knowledge-creator and scientist, affirming their place in the classroom, especially for students more likely to experience imposter syndrome (Kolligian & Sternberg, 1991). Further, explicitly valuing curiosity – in the syllabus, in assessments, and in-class meetings – provides students with room and freedom to explore with support and may eliminate some anxiety around failure in science.

Traditional secondary education can have a way of making students partition themselves into either humanities or STEM 'types,' and these trajectories often continue into undergraduate education. Unfortunately, students associate creativity more with humanities subjects, but STEM fields such as environmental science, offer so much room for creative thought. New approaches or uses of methods can lead to whole subfields of environmental science. Take, for instance, the use of increased radiocarbon in the atmosphere caused by the proliferation of nuclear weapons testing in the 1960s as a short timescale proxy for dating organic material. This 'bomb spike' is a measurable isotopic signature of the extensive above ground nuclear testing during that period; as bomb-derived atmospheric radiocarbon was assimilated by photosynthetic organisms, it changed the carbon isotope ratios of their organic compounds, which can then be used for short-term dating to track flows of carbon pools through plants and soils (Uno et al., 2013). A recent highlight of these studies examined the age of maple syrup to understand the movement of slow pools of sugars and carbon through trees (Muhr et al., 2016). However, one can not teach about the 'bomb spike' in comfortable scientific isolation – nuclear weapons devastated cities, killed thousands, and decimated tropical, desert, and marine ecosystems. Our job as professors is to highlight creativity and new applications of methods, but we must also provide the whole context of how science emerges. Rather than just covering the shiny advances, we must explore and discuss the difficult and painful histories and current states.

In environmental science and ecology classrooms, assessments that encourage and support creativity, curiosity, individuality, and inclusion can take many forms. Assessments in undergraduate classrooms should be a mix of high-, mid-, and low-stakes, allow students to build up their confidence and understanding through formative assessments, and provide opportunities for students to contribute their voices and perspectives (Handelsman et al., 2007). Assessments can also be areas in the class to emphasize the inclusion of diverse and marginalized perspectives that may not be showcased in textbooks by including readings or topics that integrate EJ concepts and perspectives. Here we provide a variety of examples that apply to introductory and upper-level undergraduate courses.

Mini-Quizzes. Formative assessments are not common in science classrooms – where semester grades are often calculated as the average of a few, large exams. However, research shows that providing 'low-stakes,' formative assessments that test content knowledge and understanding without the added anxiety of grades can improve student outcomes (Handelsman et al., 2007). In the last few semesters, I (Mary) have implemented ungraded quizzes at the beginning of class 1 day a week. Some quizzes assess content and understanding of the previous class meeting, a concept from reading, or a graph to be interpreted. Other questions prompt creative approaches to a concept – how to design an experiment around an idea, create a helpful metaphor for a complex topic, or apply understanding of an idea to a novel system.

Once students have completed the short quizzes (aimed to take 5–10 min), they work in small groups to compare answers and make edits and revisions based on consensus. This collaborative learning helps students practice explaining their understanding, defending their responses, and taking alternative approaches to a prompt. The discussion of answers in small groups can turn into a short 'studying in class' moment, since many of the quiz questions reflect the topics and difficulty of what students will experience in more summative, formalized assessments like exams. After group collaboration, we work as a whole class and share answers and ideas aloud for all to learn. At this point, I aim to clarify questions and concepts, especially when there may be more than one acceptable response. Mini-quizzes can be collected and checked to make sure the completed quizzes contain answers that students can study from; then they are returned.

Knowing what to expect on exams lowers test anxiety, and I hope the collaborative answering supports student interactions that build community. Providing ungraded formative assessments also allows room to 'fail,' space to explore ideas, and can be a place to encourage students to challenge themselves without risk. From a student's perspective (Jennings), mini-quizzes allow students to learn constructively from failure without being penalized. Furthermore, they serve as a way to build classroom community through collaborative learning and ultimately lower test anxiety by increasing familiarity with materials and how the instructor asks questions.

Further, the mini-quizzes can be used as a pedagogical tool to introduce and engage large science classrooms in discussing and applying ecological and environmental science concepts to EJ topics and marginalized perspectives. For instance, we had recently covered remote sensing techniques in a Plant Ecophysiology course. I (Mary) gave students an abstract, and a summary of a remote-sensing paper focused on the natural resource use of forests by adjacent refugee settlements in Rwanda before, during, and after armed conflict in the country (Ordway, 2015). The mini-quiz prompted students to think of other regions and questions where remote-sensing methods could be used to test a hypothesis about forests and land use. The results of that quiz were diverse, creative, and fully integrated the methods while focusing on EJ issues: forest cover in urban vs. suburban regions in the midwest; how a large land-fill site in Mexico impacted adjacent forest cover over time, and who might be most impacted; changes in vegetation and development in North Korea at the DMZ, and palm-oil plantation expansion in Indonesia, among others. While applications of environmental science, ecology, and sustainability can often be apparent to professors, the content covered in class keeps them isolated in idealized, undisturbed (perhaps fantasized) ecosystems distinct from the current 'real world.' This inaccuracy can be discussed and confronted and can empower students to design questions about topics of interest. Removing grading, and the anxiety around it, from this challenge, unleashes more creativity.

'Future of the field' exploratory essay. One way to encourage student perspectives is to create an assessment that values student voice and opinion. Science assessments (essays and exams) are often designed so that there is only one correct answer and approach (or at most, a few). This approach can limit the creativity of the classroom and does not value alternative perspectives, minimizing the potential authentic inclusion of diverse ideas and strategies. For a large introductory Ecology and the Environment course, I (Mary) developed an assessment to invite student opinion on the discipline. As newcomers to Ecology and Environmental Science, it is likely that undergraduate students can be more objective about its shortcomings as a discipline. While I provided base guidelines for the length and the need to include supporting evidence, students could approach the essay in different styles and cover various topics. The only prompt was: "*The future of Ecology is...*".

The responses to this prompt ranged widely, capturing topics from the lack of Black and Indigenous People of Color and their research in textbooks, to the colonialism implied by much land-based conservation research, to how zoos can provide conservation outreach in urban areas, to the need for a re-thinking of agricultural systems, and the oversight of the whole continent of Africa in many 'global' studies on ecosystem responses. In almost every essay, the current and impending threat of climate change was present. Overall, the essays were stimulating, thoughtful, and enjoyable to read while also providing students with the freedom to explore and organize their thoughts on the fields of environmental science, ecology, and sustainability.

Discussions of future developments occur at conferences and in journals - how we as scholars look to the future of the field, where change is most likely to occur, where change is most needed, what ideas are no longer valid, and so on. However, we rarely empower students to take on that forward-looking vantage point, even though they are most likely to lead the next generation of the discipline. Through student-centered assessments like this essay, students can engage in the intellectual criticism of a field, open avenues for new ideas, and provide fresh, needed perspectives without relying on a single viewpoint.

Zines. Zines are informal, cheaply produced portfolios of creative ideas and share perspectives and viewpoints that are less likely to be promoted by mainstream publishing. Zines can be a collection of essays and pictures with many pages or a short eight-panel folded brochure that guides a reader through a topic. In Ecology and Environmental Science classrooms, zines can be used as individual or small-group projects to showcase new perspectives on a topic covered (or not!) in class. Zines emphasize unique and novel perspectives – the goal is to create, not necessarily imitate, what has been published or made previously, and for this reason can take many forms – cartoons, essays, photos, diagrams, character-driven, or a mix of many elements. Christine Liu (see examples at www.twophotonart.com), a science communicator and neuroscientist, has produced multiple zines to communicate science ideas and make them more accessible to a non-science audience. I owe my inspiration to use zines as a low-stakes assessment to her and other science zine makers.

The mix of science, communication, art, and individual perspective that encompass zines can extend into EJ frameworks in undergraduate classrooms. Unlike many more formalized science communication assessments that assume a certain standard that may not be familiar to all in the class – the production of a podcast, website, op-ed article, or oral presentation – zines are grounded in counter-culture and rejecting norms (Duncombe, 1997). Using zines as an assessment to integrate and communicate concepts in science may indirectly signal enhanced freedom of exploration to students with marginalized and underrepresented identities. Zines do not have a predetermined standard or model to which you need to measure against – a major goal of this format is to express your individuality and share amongst the community (Duncombe, 1997).

In ecology, environmental science, and sustainability classrooms, I encourage professors to keep zine guidelines loose to allow greater creativity and expression amongst students. This freedom can be emphasized by grading that acknowledges the flexibility and individuality of this format. Recently, I assigned students to form small groups based on topic interest (theirs to describe and choose, with an association to the content covered in class) and create an 8-panel zine. Topics included why leaves change color, how animals and plants co-evolved, and deciduous and evergreen tree adaptations. I even contributed one, myself, on the use of isotopes in determining the kind of food you eat. The zines emphasize the need for concise, accessible science communication in a format that eschews norms and invites under-represented viewpoints. Also, zines allow students to welcome elements of 'fun' back into the curriculum – something sorely missing from undergraduate STEM curricula. This is a creative and authentic way to highlight and express messages about a concept that students most connect to (Fig. 15.1).

All too often in K-12 education, science is taught as the accomplishments of a singular genius working alone (or two singular geniuses at odds with each other, e.g., Calvin and Benson, Cope and Marsh). The collaborative aspects of science and discovery are rarely acknowledged, which is harmful. When science is taught as the achievements and accomplishments of individuals, it perpetrates a worldview of

Fig. 15.1 Science communication does not need to be based around stale presentations; zines allow students to play with broad audience science communication

individual exceptionalism that has little bearing on how science, especially ecology, is done. Science is done by teams of flawed and limited people doing their best and working together. Challenge your students to push back against this individualistic worldview by examining the collaborative nature of science through

- Encouraging multi-person lab work *and* a collaborative writeup
- Discussion groups about the content/ethical implications of applications of science.
- A group project about scientists who made discoveries as a group (the Human Genome Project team, for example)

15.5 Guiding Questions for Developing Classrooms That Integrate EcoJustice

When developing new or re-visiting the design of existing classes, we encourage faculty to reflect on ways EJ can be integrated into classroom content and practices. The following prompts for reflection may be most effective when considered before and during course development but can also be applied through the semester to re-tool content, pedagogy, and activities to promote an inclusive classroom.

- Do the learning objectives of the course promote the inclusion of all students? Do they encourage collaboration amongst students or only individual efforts? Are the learning objectives of the course grounded in EJ principles of transparency and sharing of information, equitable decision making, and the incorporation of multiple sources of knowledge? How can the guidelines of the course better reflect EJ principles?
- Do the course assessments encourage and require the integration of multiple perspectives, especially those of marginalized communities? How can assessments be designed to place value on the integration of multiple knowledge sources, especially those not represented at the forefront of standard STEM texts and examples?
- Am I designing a class where all students can thrive? Are the assessments and content designed to serve students from non-marginalized communities due to historical biases of science teaching? How can professors design activities, classes, and assessments that benefit the whole student population and promote authentic inclusion and success of all students?
- Does the content delivered in the class reflect implicit biases, or is it harmful to certain communities? Can content be improved to incorporate perspectives and experiences that reflect the student population more? Are there open-ended, low-stakes activities or assessments that allow students to identify and interact personally with content?

We also recognize that integrating EJ concepts into an undergraduate classroom is a challenge many faculty are not prepared for, especially in STEM courses (see Pfirman & Winckler, this volume). However, we encourage faculty to seek guidance from experts at their institutions – potentially faculty and staff housed in their campus Teaching and Learning Centers or Multicultural Centers. They can offer substantive feedback on how to improve their practice to be more inclusive. The benefit of integrating EJ concepts authentically in these courses can be profound. We are not just providing content piece-meal, hoping students will connect environmental and social issues; we are responsible for providing the tools, information, and space to think critically about how many of the world's current issues can be viewed and addressed more effectively when EJ and sustainability are at the forefront. The current generation of college students will be tasked with so much in the next century to address climate change, resource over-extraction, income inequality, and other issues that affect life from the local to global scales. By involving EJ principles in science classes, we can start thinking together about moving toward a more sustainable, equitable future.

References

Blackman, A., Corral, L. S., Lima, E., & Asner, G. P. (2017). Titling indigenous communities protects forests in the Peruvian Amazon. *Proceedings of the National Academy of Sciences, 114*(16), 4123–4128. https://doi.org/10.1073/pnas.1603290114

Bonta, M., Gosford, R., Eussen, D., Ferguson, N., Loveless, E., & Witwer, M. (2017). Intentional fire-spreading by 'Firehawk' raptors in Northern Australia. *Journal of Ethnobiology, 37*(4), 700–718. https://doi.org/10.2993/0278-0771-37.4.700

Bullard, R. D. (1999). Dismantling environmental racism in the USA. *Local Environment, 4*(1), 5–19. https://doi.org/10.1080/13549839908725577

Bullard, R. D. (2001). Environmental justice in the 21st century: Race still matters. *Phylon, 49*(3/4), 151–171. https://doi.org/10.2307/3132626

Canning, E. A., Muenks, K., Green, D. J., & Murphy, M. C. (2019). STEM faculty who believe ability is fixed have larger racial achievement gaps and inspire less student motivation in their classes. *Science Advances, 5*(2), eaau4734. https://doi.org/10.1126/sciadv.aau4734

Deslauriers, L., McCarty, L. S., Miller, K., Callaghan, K., & Kestin, G. (2019). Measuring actual learning versus feeling of learning in response to being actively engaged in the classroom. *Proceedings of the National Academy of Sciences, 116*(39), 19251–19257. https://doi.org/10.1073/pnas.1821936116

Dewsbury, B. (2019). Deep teaching in a college STEM classroom. *Cultural Studies of Education, 15*(2), 169–191. https://doi.org/10.1007/s11422-018-9891-z

Dewsbury, B., & Brame, C. J. (2019). Inclusive teaching. *CBE Life Sciences Education, 18*(2), fe2. https://doi.org/10.1187/cbe.19-01-0021

Duncombe, S. (1997). *Notes from the underground: Zines and the politics of underground culture.* Microcosm Publishing.

Environmental Justice Learning Center. (2019). United States Environmental Protection Agency. https://www.epa.gov/environmentaljustice/learn-about-environmental-justice. Accessed 20 Dec 2021.

Fearnside, P. M. (2015). Deforestation soars in the Amazon. *Nature, 521*(7553), 423–423. https://doi.org/10.1038/521423b

Fitchett, J. M. (2018). Recent emergence of CAT5 tropical cyclones in the South Indian Ocean. *South African Journal of Science, 114*(December), 1–6. https://doi.org/10.17159/sajs.2018/4426

Garnett, S. T., Burgess, N. D., Fa, J. E., Fernandez-Llamazares, A., Molnar, Z., Robinson, C. J., Watson, J. E. M., Zander, K. K., Austin, B., Brondizio, E. S., Collier, N. F., Duncan, T., Ellis, E., Geyle, H., Jackson, M. V., Jonas, H., Malmer, P., McGowan, B., Sivongxay, A., & Lieper, I. (2018). A spatial overview of the global importance of Indigenous lands for conservation. *Nature Sustainability, 1*, 369–374. https://doi.org/10.1038/s41893-0180100-6

Haak, D. C., HilleRisLambers, J., Pitre, E., & Freeman, S. (2011). Increased structure and active learning reduce the achievement gap in introductory biology. *Science, 332*(6034), 1213–1216. https://doi.org/10.1126/science.1204820

Handelsman, J., Ebert-May, D., Beichner, R., Bruns, P., Chang, A., DeHaan, R., Gentile, J., Lauffer, S., Stewart, J., Tilghman, S. A., & Wood, W. B. (2004). Scientific teaching. *Science, 304*(5670), 521–522. https://doi.org/10.1126/science.1096022

Handelsman, J., Miller, S., & Pfund, C. (2007). *Scientific teaching.* W.H. Freeman and Company.

Harackiewicz, J. M., Canning, E. A., Tibbetts, Y., Giffen, C. J., Blair, S. S., Rouse, D. I., & Hyde, J. S. (2014). Closing the social class achievement gap for first-generation students in undergraduate biology. *Journal of Educational Psychology, 106*(2), 375–389. https://doi.org/10.1037/a0034679

Hoskins, S. G., Lopatto, D., & Stevens, L. M. (2011). The C.R.E.A.T.E. approach to primary literature shifts undergraduates' self-assessed ability to read and analyze journal articles, attitudes about science, and epistemological beliefs. *CBE Life Sciences Education, 10*(4), 368–378. https://doi.org/10.1187/cbe.11-03-0027

Igor, K., & Jolly, D. (Eds.). (2002). *The Earth is faster now: Indigenous observations of Arctic environmental change. Frontiers in Polar Social Science.* Arctic Research Consortium of the United States.

Indigenous Science Statement for the March for Science. (2017). *Indigenous Science Letter*, April 18. https://www.esf.edu/indigenous-science-letter/. Accessed 19 Dec 2021.

Kean, S. (2019). Historians expose early scientists' debt to the slave trade. *Science.* https://doi.org/10.1126/science.aax5704

Kimmerer, R. W. (2019). *P-Values and cultural values: Creating symbiosis among indigenous and western knowledge to advance ecological justice*. Ecological Society of America keynote lecture, Louisville, KY, USA. Accessible as video at https://www.youtube.com/watch?v=xKmKFJzviz0

Kimmerer, R. W., & Kanawha Lake, F. (2001). Maintaining the mosaic: The role of indigenous burning in land management. *Journal of Forestry, 99*(11), 36. https://doi.org/10.5751/ES-11945-250411

Kolligian, J., Jr., & Sternberg, R. J. (1991). Perceived fraudulence in young adults: Is there an 'imposter syndrome'? *Journal of Personality Assessment, 56*(2), 308–326. https://doi.org/10.1207/s15327752jpa5602_10

Miller, R. P., & Nair, P. K. R. (2006). Indigenous agroforestry systems in Amazonia: From prehistory to today. *Agroforestry Systems, 66*(2), 151–164. https://doi.org/10.1007/s10457-005-6074-1

Mistry, J., & Berardi, A. (2016). Bridging indigenous and scientific knowledge. *Science, 352*(6291), 1274–1275. https://doi.org/10.1126/science.aaf1160

Montgomery, B. L. (2017). From deficits to possibilities: Mentoring lessons from plants on cultivating individual growth through environmental assessment and optimization. *Public Philosophy Journal, 1*(1). https://doi.org/10.25335/M5/PPJ.1.1-3

Muhr, J., Messier, C., Delagrange, S., Trumbore, S., Xu, X., & Hartmann, H. (2016). How fresh is maple syrup? Sugar maple trees mobilize carbon stored several years previously during early springtime sap-ascent. *New Phytologist, 209*(4), 1410–1416. https://doi.org/10.1111/nph.13782

Nicholas, G. (2018, February 21). It's taken thousands of years, but Western science is finally catching up to traditional knowledge. *The Conversation.*

O'Bryan, C. J., Garnett, S. T., Fa, J. E., Leiper, I., Rehbein, J. A., Fernandez-Llamazares, A., Jackson, M. V., Jonas, H. D., Brondizio, H. D., Burgess, N. D., Robinson, C. J., Zander, K. K., Molnar, Z., Venter, O., & Watson, J. E. M. (2020). The importance of Indigenous Peoples' lands for the conservation of terrestrial mammals. *Conservation Biology, 35*(3), 1002–1008. https://doi.org/10.1111/cobi.13620

Ordway, E. M. (2015). Political shifts and changing forests: Effects of armed conflict on forest conservation in Rwanda. *Global Ecology and Conservation, 3*, 448–460. https://doi.org/10.1016/j.gecco.2015.01.013

Ou, S. X., & Romero-Olivares, A. L. (2019, August 22). Decolonizing ecology for socially just science. *Science Connected Magazine.* https://magazine.scienceconnected.org/2019/08/decolonize-science-with-global-collaboration/

Pellow, D. N. (2018). *What is critical environmental justice.* Policy Press.

Pennycook, G., & Rand, D. G. (2019). Fighting misinformation on social media using crowd-sourced judgments of news source quality. *Proceedings of the National Academy of Sciences, 116*(7), 2521–2526. https://doi.org/10.1073/pnas.1806781116

Pfirman, S., & Winckler, G. (this volume). Perspectives on teaching climate change: Two decades of evolving approaches. In M.S. Rivera Maulucci, S. Pfirman, & H. S. Callahan (Eds.), *Education for sustainability: Discourses on authenticity, inclusion, and justice.* Springer.

Roten, F.C. von. (2011). Gender differences in scientists' public outreach and engagement activities. *Scientific Communication, 33*(1), 52–75. https://doi.org/10.1177/1075547010378658

Roy, R. D. (2018). Decolonise science – Time to end another imperial era. *The Conversation.* https://theconversation.com/decolonise-science-time-to-end-another-imperial-era

Scheufele, D. A., & Krause, N. M. (2019). Science audiences, misinformation, and fake news. *Proceedings of the National Academy of Sciences, 116*(16), 7662–7669. https://doi.org/10.1073/pnas.1805871115

Schuster, R., Germain, R. R., Bennett, J. R., Reo, N. J., & Arcese, P. (2019). Vertebrate biodiversity on indigenous-managed lands in Australia, Brazil, and Canada equals that in protected areas. *Environmental Science & Policy, 101*, 1–6. https://doi.org/10.1016/j.envsci.2019.07.002

Science Education Research Center (SERC). (2021, July). *InTeGrate: Interdisciplinary teaching about Earth and a sustainable future.* https://serc.carleton.edu/integrate/index.html. Accessed 5 Dec 2021.

Smith, N. F., Lepofsky, D., Toniello, G., Holmes, K., Wilson, L., Neudorf, C. M., & Roberts, C. (2019). 3500 years of shellfish mariculture on the northwest coast of North America. *PLoS One, 14*(2), e0211194. https://doi.org/10.1371/journal.pone.0211194

Sonter, L. J., Herrera, D., Barrett, D. J., Galford, G. L., Moran, D. J., & Soares-Filho, B. S. (2017). Mining drives extensive deforestation in the Brazilian Amazon. *Nature Communications, 8*(1), 1013. https://doi.org/10.1038/s41467-017-00557-w

Tanner, K. D. (2013). Structure matters: Twenty-one teaching strategies to promote student engagement and cultivate classroom equity. *CBE Life Sciences Education, 12*(3), 322–331. https://doi.org/10.1187/cbe.13-06-0115

Tessum, C. W., Apte, J. S., Goodkind, A. L., Muller, N. Z., Mullins, K. A., Paolella, D. A., Polasky, S., Springer, N. P., Thakrar, S. K., Marshall, N. D., & Hill, J. D. (2019). Inequity in consumption of goods and services adds to racial-ethnic disparities in air pollution exposure. *Proceedings of the National Academy of Sciences, 116*(13), 6001–6006. https://doi.org/10.1073/pnas.1818859116

Uno, K. T., Quade, J., Fisher, D. C., Wittemyer, G., Douglas-Hamilton, I., Andanje, S., Omondi, P., Litoroh, M., & Cerling, T. E. (2013). Bomb-curve radiocarbon measurement of recent biologic tissues and applications to wildlife forensics and stable isotope (paleo)ecology. *Proceedings of the National Academy of Sciences, 110*(29), 11736–11741. https://doi.org/10.1073/pnas.1302226110

Wade, J., & Zaringhalam, M. (2018). Why we're editing women scientists onto Wikipedia. *Nature*. https://doi.org/10.1038/d41586-018-05947-8

White, P. (2016). The man of science. In B. Lightman (Ed.), *A companion to the history of science* (pp. 153–163). Wiley. https://doi.org/10.1002/9781118620762

Wynn-Grant, R. (2019). On reporting scientific and racial history. Edited by Jennifer Sills. *Science, 365*(6459), 1256–1257. https://doi.org/10.1126/science.aay2459

Ziembicki, M. R., Woinarski, J. C. Z., & Mackey, B. (2013). Evaluating the status of species using Indigenous knowledge: Novel evidence for major native mammal declines in northern Australia. *Biological Conservation, 157*, 78–92. https://doi.org/10.1016/j.biocon.2012.07.004

Mary A. Heskel is an assistant professor of ecology at Macalester College. She examines plant ecological and physiological responses associated with current climate change and teaches undergraduate courses on ecology, data analysis, the Arctic, and plant physiology. She finds teaching an ongoing challenge and inspiration.

Jennings G. A. Mergenthal is a community engagement specialist at the Science Museum of Minnesota. They graduated from Macalester College in 2021 with a B.A., where they studied Biology and American Indian History. They are engaged in activism around equity and community solidarity.

Chapter 16
The UNPAK Project: Much More Than a CURE

Hilary S. Callahan, Michael Wolyniak, Jennifer Jo Thompson, Matthew T. Rutter, Courtney J. Murren, and April Bisner

16.1 Introduction

Back in 2010, a team of faculty researchers at primarily undergraduate institutions (PUI) initiated a program of research in plant science, led by faculty and conducted by undergraduate students, and emphasizing CUREs (course-based undergraduate research experiences).[1] Shortly after garnering a first grant from the National Science Foundation in 2011, the team held a multi-campus acronym contest and adopted our endearing and enduring name: UNPAK, for Undergraduates Phenotyping *Arabidopsis* Knockouts. Here, the "UN" refers to undergraduates engaging in hands-on and authentic research, and the project's development of protocols for amassing the data that students generate across many campuses into a centralized database. The "A" in UNPAK refers to our focal plant species, *Arabidopsis thaliana.*

Outside the lab, *Arabidopsis* is a small, short-lived weed found throughout temperate biomes, related to cabbage and turnips. It is also a human commensal,

---

[1] Auchincloss et al. (2014), Rodenbusch et al. (2016) and Dolan & Weaver (2021).

H. S. Callahan (✉)
Department of Biology, Barnard College, New York, NY, USA
e-mail: hcallaha@barnard.edu

M. Wolyniak
Department of Biology, Hampden-Sydney College, Hampden Sydney, VA, USA

J. J. Thompson
Crop and Soil Sciences, University of Georgia, Athens, GA, USA

M. T. Rutter · C. J. Murren · A. Bisner
Department of Biology, College of Charleston, Charleston, SC, USA

© The Author(s) 2023
M. S. Rivera Maulucci et al. (eds.), *Transforming Education for Sustainability*,
Environmental Discourses in Science Education 7,
https://doi.org/10.1007/978-3-031-13536-1_16

growing as a weed alongside roads, stream banks, fields, and other cultivated land-scapes.[2] In the 1980s and 1990s it emerged as a model system to unite plant research-ers in genetics, physiology, ecology and evolution. The *Arabidopsis* genome project and later spin-offs parallel the human genome project in many ways, including sus-tained focus, massive government grant support, and international scope. These projects enabled many applications of basic science and ancillary technologies.[3]

Knockouts, the "K" in our acronym, are now a key and routine technology for UNPAK and the entire plant genetics community, thanks to pioneering projects in the 2000s that created them by introducing large DNA "tags" into the genomes of plants. These tags simultaneously mutate the DNA-based information in the gene and molecularly tag the gene's physical location. Comprehensive collections of these induced mutations, known as knockout libraries, were created. Our project focuses on one created at the Salk Institute, with coverage of the vast majority of the 25,000+ genes comprising the *Arabidopsis* genome.[4] Similar comprehensive knock-out libraries exist and are widely used by researchers, not only in *Arabidopsis* but also in yeast and many other microbes, and in fruit flies, mice and other animals.[5]

The potential of such libraries relates to the "P" in our acronym—the phenotypes associated with an individual instance of knockout mutation. Every tagged mutant in any knockout library must be confirmed: has the tagged mutation altered the organismal characteristic(s) as predicted or expected based on the gene disrupted, and with existing genetic knowledge? Here, UNPAK innovates by focusing on less investigated and more complex phenotype traits, those relevant to ecology and agronomy. How likely is a mutant seed to germinate? How fast? How slow or how fast do knockout mutants produce flowers? How often do full-size fruits or seeds fail to ripen? UNPAK also focuses not just on specific mutants or types of mutants, but on a large and randomly representative portion of the overall genome (10+%).

UNPAK was gratified when the National Science Foundation's announced in 2016 its "Ten Big Ideas" and included the statement that a "universally recognized biggest gap in biological knowledge is our inability to predict an organism's observable char-acteristics—its phenotype—from what we know about its genetics".[6] This affirmed UNPAK's hard-won phenotyping pipeline: growing plants in standardized conditions, measuring traits reliably and reproducibly, and checking data quality as it compiles into a database shared across UNPAK campuses and beyond UNPAK.[7] With the con-cept of a gene now in its second century, we UNPAKers want young biologists to learn and research genetics and genomics while keeping sight of phenotypes via "eyes on" and "hands on" work, both connected to the "minds on" activities of asking answerable questions, designing experiments, analyzing and interpreting results, and extracting data appropriately from databases to conduct meta-analyses.

---

[2] Lee et al. (2017).

[3] AGI (2000), Koorneef and Meinke (2010) and Buell and Last (2010).

[4] O'Malley and Eckert (2010).

[5] Hillenmeyer et al. (2008) and Rancati et al. (2018).

[6] Gropp (2016).

[7] Full details are in Rutter et al. (2019).

UNPAK's earlier publications have discussed its aims and progress in mentoring and researching plant genomics and evolutionary ecology, and some of the CUREs implemented by the UNPAK team.[8] Members of UNPAK who are social scientists have assessed its impact on students' development of social, cultural, and human capital in the context of networked research, the role of faculty recognition and mentorship in student development, and the impact on faculty career development.[9]

Here, we reflect on UNPAK's emphasis on authenticity and how that intersects with students gaining not just access to STEM learning, but also a sense of inclusion and belonging. The investigators participating in this conversation have a mix of backgrounds. Some work in ecology or environmental biology and others are more focused on molecular biology, or on education and the social sciences. This is also a circle of colleagues who helped UNPAK in its earliest years and at critical turning points in its decade-long existence. We agreed to respond in writing to a series of questions addressing how faculty have balanced their teaching and their research, and how their teaching and research connects to plant and environmental science, and applications of these sciences. We discuss our views of how UNPAK, directly and indirectly, helps diverse students to persist into future education and careers, whether outside of STEM or in STEM, and whether or not they pursue careers directly relevant to "the environment" or "sustainability". We hope it offers some illumination on how STEM faculty and staff in higher education support society-wide efforts to conceive and build a future that is more environmentally sustainable, and also more inclusive and just.

16.2 UNPAK and Its People

Question To introduce UNPAK and our group please share your current position and institutional affiliations, when you joined UNPAK, and any comments about how your job and expertise are interdisciplinary, including any relevance to environmental science, sustainability or conservation.

Matthew T. Rutter I am now a Professor in the Department of Biology at College of Charleston in South Carolina, and also Academic Director of Stono Preserve. I am a regular faculty member in Biology, but also act as director of College activities at our Stono Preserve property, a 1000-acre natural and historic site 17 miles from the main campus. The College of Charleston is a public, primarily undergraduate urban college, with about 10,000 students. My job as a faculty member is primarily STEM, although my role at Stono Preserve has involved working across almost every discipline at the college, and is more project management. I am interested in

---

[8] Rutter et al. (2019, 2020) and Murren et al. (2019).
[9] Thompson et al. (2016), Thompson and Jensen-Ryan (2018) and Jensen-Ryan et al. (2020).

both conservation and education, conducting research in both areas, so I suppose I am interdisciplinary.

I was one of the founding members of UNPAK, so I got in right at the beginning! I was an early Assistant Professor, going through my 3rd year review. I was Principal Investigator for the College of Charleston's original UNPAK awards, and have continued that role as UNPAK transitioned to its current form, where we have a summer undergraduate research experience and CUREs as our main activities.

Courtney J. Murren I am also Professor in the Department of Biology at the College of Charleston. In addition to Biology, I do work in environmental studies and in education. I joined UNPAK before UNPAK was a thing. I helped with analyzing and designing experiments that were the pilot data for the first grant submissions. At that point, I was an Assistant Professor. My involvement moved from being in the greenhouse collecting data, to writing proposals and working with new faculty who have since joined UNPAK, especially after April Bisner joined the team in 2015 as project manager. Now, being tenured and a full professor frees me to add to my research portfolio work on education research, which is not a topic that "counts" towards research productivity at our institution. I do it because it is super valuable work, grant funding agencies value this expansion, and I care about getting opportunities in the hands of students!

Hilary S. Callahan I am a Professor of Biology and department chair at Barnard College. Ours is a distinctive liberal arts college serving about 2700 students in 2021. Barnard does not admit men nor grant graduate degrees, but it has formal links to—and is geographically right across the street from—Columbia University. Since my undergraduate days, I have tried to combine my STEM expertise with my activism on a variety of issues. The urge to "change things" in positive ways is something I admire in Barnard's students.

Courtney Murren welcomed me to join her and collaborators at the College of Charleston in the earliest stages of creating UNPAK. At the time, I was tenured and well versed in working with program directors within the funding agencies. UNPAK became one of the more ambitious, intense and thoughtful projects I have done, and it also helped me to pioneer various remote technologies in the early 2010s. In our early years, team meetings were frequent, and always via conference call. Later, we started to use the original Google Hangout platform and now, like so many others, we rely on Zoom. In a decade, I have traveled just a few times to work in-person with other UNPAK team members, flying twice to the College of Charleston and once taking Amtrak to Hampden-Sydney in Virginia.

Michael Wolyniak I am the McGavacks Associate Professor of Biology and the Director of the Office of Undergraduate Research at Hampden-Sydney College, a small liberal arts college for men located in central Virginia. My training is in genetics and molecular biology. That said, I have had several experiences with working as an interdisciplinary scholar and teacher. I regularly teach bioethics courses that require students to consider issues related to science and society. I am a part of our institution's Core Cultures program that looks at widely read texts about the history of Western civilization to identify how society has evolved. With respect to my scholarly work, I focus on course-based undergraduate research experiences (CUREs) and how they may be best integrated into scientific coursework. This has given me a lot of experience working with education experts. While I have not directly taught environmental science, conservation or sustainability, the issues come up regularly in classes, typically because of the impact that humans and their choices have on their environment. None of this would be possible without an interdisciplinary approach.

I was blessed to be a founding member of UNPAK, joining in 2010 as a new professor. I originally joined UNPAK to gain opportunities that I could use in my classes to expose students to exciting research opportunities. Over time, my involvement with UNPAK has focused on CUREs and finding ways to integrate CUREs into all of my classes.

Jennifer Jo Thompson I am a Director of the Sustainable Food Systems Initiative and an Associate Research Scientist in the Department of Crop & Soil Sciences at University of Georgia. I am an interdisciplinary scholar. My primary discipline is anthropology, but I work primarily on collaborative, inter- and transdisciplinary research focused on sustainable agriculture and food systems. One thrust of this research focuses on the way various publics engage with science, including how to support students to develop various forms of capital in science.[10] I continue to have components of my research, teaching, and service that focus on expanding participation in science.

I joined UNPAK early, in 2012, when I took a postdoctoral position with Erin Dolan. Our early work on the project specifically focused on social capital: whether undergraduates participating in a networked research project would have unique access to additional social resources due to their potential for access to additional faculty and students across institutions. Ultimately, we found that despite valiant efforts to encourage communication and connections across groups (e.g., early efforts with Google Hangout, etc.) undergraduates built social ties primarily within their labs (with their faculty mentors and fellow students). There were exceptions

---

[10] Thompson et al. (2016) and Thompson and Jensen-Ryan (2018).

among those students who had unique opportunities to visit and spend time in other labs, and build connections with other students or faculty. Our data suggests that this is because undergraduates are focused on building skills, identity, and relationships *within* their lab. In 2013, I became a co-PI (with the second UNPAK grant), and our work shifted somewhat to understanding how to expand undergraduate participation among students from underrepresented groups. This work focused on understanding, in greater detail, how undergraduates developed their identities as scientists and how faculty were able to support this identity development by recognizing and affirming their potential.[11]

April Bisner My job and expertise combine STEM and project management. The UNPAK project's research goals are relevant to environmental conservation and sustainability in that we study gene-environment interactions, and this is a model issue relevant to climate change such as increased drought, or flooding. I was hired as the UNPAK project manager in 2014. At the time I had nine years of experience working on a variety of research projects with increasing management and training responsibilities. I have been with UNPAK for nearly seven years. In our early years, prior to 2018, we had more full-time student interns. Today, we do more work with a virtual or in-person summer REEU (Research and Extension Experiences for Undergraduates), which is a program funded by USDA through AFRI-EWR (Agriculture and Food Research Initiative, Education and Workforce Development). With CUREs, our network of partner institutions also continues to grow.

Question If you are willing to share, can you describe your experiences growing up, your family or your early and later schooling? Does that background affect your current work in STEM research, mentoring and teaching?

Courtney J. Murren I was raised in the woods in a family of modest means with a family business. My parents and grandparents had a nature-loving ethic and any down-time we had as a family was at the nearby shore or mountains. Nature was my teacher. I pulled apart tulip poplar inflorescences, replanted shore grasses in little rows, excavated acorns and climbed trees. I read, and read, and read growing up. Today, I encourage my students to look, observe, be curious and read. Having been from modest means, I had to work all the time and had no "backup" resources. I had not thought of graduate school during college until mentors told me that tuition was waived and that there are stipends for working while studying, to help with living expenses and health insurance. Academics sounded more stable than the family business and I loved the explore/research components so I thought I would give it a try. I traveled as an undergraduate to the tropics with a group of other undergraduates on an NSF REU (Research Experiences for Undergraduates). That trip opened my world.

---

[11] Thompson and Jensen-Ryan (2018).

As a faculty member, I'm committed to paying stipends to all students, hiring students through programs such as Federal Work-Study, and as much as possible providing opportunities for students to travel for their work in their early career. I also take time to talk to students about the paths into graduate school as it is quite different from many of their peers who are seeking schooling for health careers.

Hilary S. Callahan I did not have parents or grandparents who were scientists or in STEM professionals. My dad attended a business college, and worked as an auditor for the USDA for his whole career, often with side hustles, as we call them today. My mother pursued her college degree, a teaching career and an M.Ed., all while doing most of the work at home with three daughters. Over time, I have noticed a divide between professors and scientists with backgrounds like mine, and those who have parents or other relatives who are also professors, scientists, or professionals in remunerative fields. I always try to bear in mind that some work much harder than others to develop the know-how and resources for navigating toward opportunity and eventual success in STEM careers or other careers.

Michael Wolyniak I was raised to always pursue excellence, which I think was essential to me getting to where I am in my career arc. I was blessed with a K-12 education that gave me the opportunity to engage in the STEM disciplines and to develop my interests. I often felt a degree of pressure to always succeed, and this pressure influences how I now mentor my students. I do not want my students to feel that they have anybody to please but themselves when it comes to their professional development and considering their career opportunities. I feel as if I have a stronger sense of empathy for my students and a better ability to consider their backgrounds when mentoring them as a result of my upbringing and education.

Matthew T. Rutter I grew up within walking distance from a state park—there is no question that made a tremendous impression on me and gave me a broad interest in biology, ecology, and environmental conservation. One of my parents worked on the non-academic side of a university, and the other was a high school math teacher. They both completely supported me in the idea that I could follow my intellectual interests, and were supportive of an academic career, particularly one that incorporated teaching. I was fortunate to have that kind of support, and try to relay it to my students.

April Bisner Science was a second career for me because I did not have the attention span or grades when I was in high school and was more drawn to the arts. Attending community college as an adult learner was a valuable experience. It allowed me to attend small classes with enthusiastic professors and gain hands-on experience in the lab and field. This prepared me for transferring into a four-year institution to complete my studies.

Jennifer Jo Thompson I do not come from a family of scientists—and I definitely experienced a kind of science aversion in high school that pushed me toward the social sciences and humanities. As a PhD student, I started developing an interest in science as an important domain of culture—a domain that is powerful, but which also receives a great deal of skepticism from the public, especially when science is emergent and uncertain. At this point the issue of when and how people engage with and (dis)trust science across their daily lives drives a good deal of my research.

16.3 UNPAK as Plant, Agricultural and Sustainability Science

Question Have you or your students gained, via UNPAK, a more nuanced understanding of complex issues in plant science, and more broadly in fields relevant to agriculture or sustainability? How strongly motivated are you and your students to engage in UNPAK because it is _environmental_ research and learning? Is this interest primary, secondary, ancillary, non-existent? Do your interests, or students' interests, include environmental activism?

Michael Wolyniak We are addressing the sophistication and nuance of the relationship between genotype and phenotype. I have gained a more nuanced understanding of the effects of mutations on plant biology. UNPAK does help engage students mostly by increasing plant awareness which is something many students lack—even biology students! I would say environmental research and activism are valued moderately by my department and slightly by my institution. Generally, words seem to be prioritized over actions.

Also, my interests in UNPAK revolve around students at Hampden Sydney gaining the scientific experience, which is necessary to develop a spark that encourages them to develop scientific careers. I am not directly attempting to promote environmental research and learning. That said, the work that I have done with UNPAK students has, in several cases, led to students seeking and obtaining additional environmental related research experiences. Thus, UNPAK has served as the springboard for these students to gain environmental research experiences for themselves.

Matthew T. Rutter I would say the environmental research component may be secondary to more general and fundamental questions about biology.

Hilary S. Callahan At Barnard, UNPAK appeals to students interested in spending time in our greenhouse, and working with plants. All biology majors take genetics, so many enjoy an opportunity to connect course material to research. And our greenhouse is wonderful because it is filled with plants from all across the globe.[12] Probing a little more deeply into the science of UNPAK, I often discuss with students how places like the lab or the greenhouse are "real"—meaning empirical and material rather than theoretical—but not "real world" in the same way as a farm or a forest reserve. The UNPAK project has appealed to me because it allows me to authentically pursue my long-term interest in how different environments have different levels of predictability, specifically from the point of view of plants. I'm also interested in how much or how little environments are disturbed, subsidized or controlled by human activities, both inadvertent impacts and deliberate activities like human labor or other inputs.

I also think that students working in the UNPAK network learn that "fast silver bullets" for improving crop phenotypes do not exist. We cannot magically use biotechnologies or other technologies given how drastically the environment impacts phenotypes and the side-effects of these technologies on other organisms, soils, or the atmosphere. Thus, training in UNPAK helps students to be better at understanding and communicating to other scientists and to lay-people the benefits and risks of biotechnologies in agriculture, or in human health for that matter. I think experience in UNPAK can, potentially, help students move past naive or oversimplified narratives about biotechnology, such as the misconceptions that it is easy and magic and profitable, or that it is inherently dangerous or evil. To me, the key is to encourage thinking holistically, whether one is thinking more narrowly about plant-gene interactions or more broadly about any issue in environmental and sustainability science.

Question Are UNPAK students gaining specific insight into the consequences of scientific knowledge about agriculture and crop genetics? Do they gain insight into multiple sectors of society or the economy: basic research vs. applied research, biotechnology vs. agriculture, conservation vs. policy?

Courtney J. Murren We have particular sub-projects centered around crops and agriculture. We have partners who contribute with this lens.

Michael Wolyniak The issues in this question are covered more in my bioethics courses as opposed to my UNPAK courses. It would make sense to try to combine the science and the societal questions in greater detail in my UNPAK coursework; however, time constraints often prevent that from happening.

Matthew T. Rutter The students are definitely learning about the consequences of knowledge. It is probably focused on agriculture, biotechnology, basic research, and conservation rather than on policy. Much of it is treated broadly, but when students focus on the effects of individual genes or environmental treatments it can become quite narrow. I have seen it affect the personal components of students'

---

[12] See Gershberg (this volume).

lives. This is particularly true in CUREs where many students really haven't touched or interacted with plants very much. For lab and summer students, there certainly are professional elements of their lives that are influenced.

Hilary S. Callahan Most of my UNPAK students have had lots of high school biology, such as AP or IB Biology (Advanced Placement, International Baccalaureate). Barnard also requires an intense series of courses during their first two years. They have listened to lectures and memorized diagrams of molecular biology processes like the use of CRISPR-Cas9 to "edit" gene sequences.[13] I am clear to them about how UNPAK is not about doing that first-hand, but instead focusing on another step in the genomics pipeline, the task of phenotyping plants, important in constructing knowledge and databases in plant organismal biology, and also in the context of human health and disease. From time to time, there have been students whom I have mentored in UNPAK who have become enthralled with thinking about the jobs of farmers, including large-scale farmers in the agri-business that dominates the U.S. today. That was fostered by my UNPAK lab being right next-door to the lab of my colleague, Jon Snow,[14] who focuses on honey bees. The relevance of UNPAK's work to agriculture is also evidenced by its recent development of a Summer REEU, supported by the USDA.

16.4   UNPAK Prioritizes the Undergraduate Experience

Question How much do you prioritize and value mentoring and teaching undergraduates who are preparing for their own careers and lives? Is this a job duty or expectation? A personal avocation? A bit of both?

Courtney J. Murren It is a high priority! It is a job duty, expectation, and avocation. At the College of Charleston, this job duty can vary from regularly to infrequently. For me, I cannot remember a semester or summer where I did not work with undergraduate students in 'independent study'. I am passionate about being part of the solution in expanding opportunities, and helping students on their personal journeys in biology or environmental studies and their related interdisciplinary professional lives.

Hilary S. Callahan As a new professor, I received vivid advice from a colleague about developing "a reputation for doing excellent research that is, additionally, really meaningfully involving undergraduates." Over the years, I internalized this advice, and I have it in common with colleagues who helped create UNPAK. Moreover,

---

[13] https://www.nobelprize.org/uploads/2020/10/advanced-chemistryprize2020.pdf
[14] See Snow (this volume).

work with students has always helped me dispel the cynicism and gloom that periodically creep into my career and my life.

Michael Wolyniak I consider [research] mentoring a central part of my career. I feel it is a personal calling that allows me to gain immense satisfaction from helping my students to attain their professional and personal goals. In graduate school, I considered the work I was doing to be largely a means to an end so that I could obtain a position such as the one I now have that focuses on teaching and mentorship. I am certainly expected to be a mentor and a teacher as part of my job description; however, for me it is the favorite part of my job.

Jennifer Jo Thompson My teaching/mentoring is heavily focused on graduate students, with no professional expectation to teach or mentoring undergraduates. I do teach one mixed-level course, and often have at least one undergraduate participating in research in my lab. An exciting twist is that I am part of a team initiating a collaboration between The University of Georgia's Graduate Certificate in Sustainable Food Systems and Spelman College's undergraduate Food Studies program. This collaboration should expand my personal mentorship of undergrads, but also expand culturally appropriate mentorship and build a pathway for more diverse recruitment into our grad programs. From my perspective, all this begins with working with faculty to support *and reward* effective and inclusive mentorship.

Matthew T. Rutter I value mentoring and teaching undergraduates extremely highly, and it is one of my very top priorities. I participate as a personal avocation, although it is valued at my institution so it does fulfill expectations.

April Bisner One of the great pleasures of my job is helping to train and mentor a diverse group of students. Our intern component of the UNPAK project allows me to work closely with students over the course of a few years and watch them grow as scientists and people. I have also kept in contact with some of our alumni and it is always nice to hear where their lives have headed and what influence our project had on them.

Question Is student work within UNPAK different from the types of jobs or projects and the styles of learning that you experienced during your own undergraduate years? Is your work distinctive relative to most of your professional colleagues, either in your institution or in your sub-discipline?

Hilary S. Callahan UNPAK is so completely different from my undergraduate experiences. The vast majority of my learning about biology and genetics was in the classroom, or involved solitary reading and problem sets. One exception was field trips, which were so important to inspiring me to study ecology. I still value hands-on, place-based, and in-person experiences with faculty and students interacting out on islands or boats or in forests. With UNPAK, we are not in the field, but we are getting data from plants grown in chambers or greenhouses. Students seem to self-select into this type of work, and at the same time many do yearn to work in more natural environments.

Also, UNPAK students certainly have "ownership" of their work, while also seeing that *sole* ownership of a project is rare. In joining UNPAK, students are constantly reminded that they have joined something larger than themselves. My students at Barnard always think it is cool to see students on other campuses also doing summer internships, research projects in their classes, and so on. At its best, UNPAK unites several senses of belonging into one over-arching experience. Students who do UNPAK are also quite intentionally choosing to learn about plants, which is already distinctive as compared to so many other students who do cellular or molecular work, or who work with animal models.[15]

Courtney J. Murren UNPAK is distinctive in my research portfolio by not having a field component! Also, we have over-arching long-term goals, and we ensure that students understand this long-term focus. This matches with others in our biology discipline who have long-term goals.

Michael Wolyniak The work I do in UNPAK has become less distinct because CUREs have become more mainstream. For me, the work remains distinct in the plant-based focus that I take with UNPAK materials.

Jennifer Jo Thompson I did not have any experiences like UNPAK as an undergraduate. Working with UNPAK has given me the skill and confidence to continue to collaborate in large-scale, multi-institutional, interdisciplinary research and teaching endeavors. Although more and more anthropologists work in teams, it is still somewhat unusual to work on 'big data' projects like this.

Michael Wolyniak I think the UNPAK project makes our work, in courses and in the lab, integrate thoroughly to draw undergraduates more deeply into authentic research experiences. This is important both at our institution and in my own sub-discipline of genetics. I never worked on a project as an undergraduate that involved discovery contributing to a larger project. I do think it is quite distinctive. While I certainly had an outstanding scientific education as an undergraduate, there was much more separation between teaching and research. UNPAK has shown me how to meld those two things together in the class to give my students the best educational and potential professional development opportunities that they can get.

April Bisner Work with UNPAK is different from the projects that I worked on in my undergraduate years. Students are given the opportunity to design their own research

---

[15] Also see Snow (this volume), Maenza-Gmelch (this volume) and Rhodes et al. (this volume).

project with the help of mentors. They also have an influence on additional traits that we study in group projects. UNPAK succeeds in being hands-on by having all of the phenotyping measured by students. The data is checked by students and then quality-checked by professors and myself before uploading it to the database. In many cases, students plant and maintain experiments from beginning to end. Students are involved in the statistical analyses of results and under certain circumstances the writing of manuscripts based on those results.

Question What are your views on how actively students in UNPAK are analyzing, dialoguing or arguing from data and evidence? Do they do this? Is it important to the students? How? Why?

Courtney J. Murren The final presentations at the last three summer REEUs (2019, 2020, 2021) are evidence that students are substantiating their claims with data and evidence, and also built from the theory of those who came before us. UNPAK students think and work as scientist-learners, as individuals and in teams.

Michael Wolyniak UNPAK students are certainly actively analyzing and arguing the strengths of their predictions and conclusions based on evidence. These are very important skills to develop for a successful career in science. With respect to dialoguing, UNPAK has been hit-or-miss in terms of encouraging students to collaborate in their research. This is something that I have tried to encourage in my own classes, to a limited degree of success. Since collaboration is so important to the way modern science is conducted, it is a skill that I continue to try to promote among my students.

Hilary S. Callahan I think that general critical thinking and learning scientific methods are front and center in UNPAK, but I also see UNPAK as part of the new trend of "big data research" in which it is also important to recognize that new hypotheses and new explanations sometimes emerge from more descriptive data-exploration and data-mining.[16] I personally see big-data biology as a direct descendant of classical botany and natural history, sometimes referred to as digital natural history. I like that our focus on phenotyping keeps a bit of old-fashioned observational natural history in UNPAK. Students spend weeks focusing on a plant that is almost uselessly cute. UNPAK requires students to look at plants daily for many days in a row, with dedication and Thoreau-like patience. They gain an appreciation for more subtle phenotypes of plants, the timing of life-cycle transitions or the difficult life of a seed. A balance of strong observational skills with critical thinking is something I have valued from the very start of UNPAK, and even from my initial plunge into academic biology.

---

[16] Leonelli (2019).

Matthew T. Rutter My students are actively participating in all of those activities. It is more important to some students than others—some students are still adjusting to learning to think instead of just learning facts. I do think this involves the development of critical thinking, even in a very broad sense of using evidence to build a picture of reality, and predicting future realities.

16.5   UNPAK Supports Access and Persistence in STEM Training

Question Do you think UNPAK is a good pathway for supporting students to persevere in completing their degrees, pursuing advanced degrees, or staying within STEM fields, especially environmental sciences or sustainability?

Hilary S. Callahan I situate my thinking and commenting on this issue by reminding myself that Barnard is an expensive private college and it is really hard for students to gain admission and financial aid to attend. That being said, we have extremely high graduation rates. Having UNPAK in my department ensures that there is somewhere for students to go if they want to take a step away from the high-pressure pathway of preparing for a career in human health.[17] In the middle of a huge city, it is nice to be able to have a place where topics like agricultural yields or weed management are part of our classroom and laboratory and casual discussions, and to learn about research opportunities relevant to such questions.

Courtney J. Murren The social-professional network of the lab environment was really essential! It is not the sole path for students with specific STEM goals, but it makes a big contribution.

Michael Wolyniak There is no better way to encourage students to sustain themselves in STEM fields than to give them the chance to engage in science at it is actually done by and with "real scientists". UNPAK does an outstanding job in providing that initial spark necessary to develop a passion for scientific research. It is a critical springboard for helping students obtain the confidence and skills necessary to pursue an advanced degree in any STEM discipline, but especially in the environmental sciences or sustainability.

Question In what ways do you think UNPAK is helpful in broadening students' understanding of the potential breadth and diversity of STEM training options and STEM career options, especially in environmental sciences or sustainability-related fields?

---

[17] See also Snow (this volume) and Rhodes et al. (this volume).

Hilary S. Callahan I think a lot of students find it refreshing to think about the ecology of plants, and crop ecology, mainly because our college has a strong emphasis on pre-medical preparation. Being a bit "off beat" may help students to pause to think about careers other than being an MD. I have watched some of my UNPAK students at Barnard go into STEM teaching, nursing, epidemiology, even law.

Courtney J. Murren In paper discussions, seminars and workshops with faculty from diverse STEM sectors, we are encouraging of students' personal journeys as individual ones. Given my own interests, I often engage in discussing careers in environmental or sustainability fields.

Michael Wolyniak By participating in an UNPAK project, a student is learning how science is done. In turn, the student learns how scientists are trained, and what is required of them in graduate school. In this way, UNPAK is helping students to better understand that there are more options available for life scientists than simply medical school.

Jennifer Jo Thompson Since a lot of students enter biology thinking about careers in medicine, opportunities to learn about careers in research and ecology are important. UNPAK does this.

Matthew T. Rutter Many students haven't thought of the relevance of plants and plant biology to STEM careers or to environmental sciences. Many who are interested in environmental sciences haven't necessarily thought about the role of genetics and genetic variation in natural populations. UNPAK does help to make those links.

Question Have your initial motivations for joining UNPAK been matched by your experiences? Did UNPAK under-deliver or over-deliver? Does this make work with UNPAK distinctive in some way?

Michael Wolyniak My motivations for joining UNPAK were both altruistic and selfish. Altruistic in the sense of providing my students with outstanding research opportunities for their own development, and selfish in the sense that it would look good for my tenure and promotion opportunities. UNPAK has over-delivered in every sense as I have been able to accomplish both of these goals as a result of my participation.

My UNPAK work started out as distinct because of my lack of training as a plant biologist. Over time my UNPAK work has become less distinct, but this is a positive reflection on the influence it has had on the way that I treat all of my science courses and how they are taught. UNPAK has also profoundly changed my teaching philosophy and showed me the value of authentic research experience as a central part of an effective scientific course. Because of UNPAK, I now teach my classes in a more unstructured and research-focused fashion. In short, my other classes came to resemble UNPAK rather than the other way around.

Hilary S. Callahan I thought that UNPAK was an excellent fit for Barnard, historically and today a college for women, and a college that counters minoritizing women in STEM and in many other aspects of the academy and society more broadly. For me, as a tenured professor, accomplishing work that was "just mine" became less of a priority. Additionally, from an NSF ADVANCE grant at my institution, I learned that collaboration is an effective pathway toward productivity and long-term impact.[18] I think UNPAK has done that, and I am optimistic that it will continue to do that through the final decade of my career.

Matthew T. Rutter As one of the UNPAK founders, I was extremely interested in its science: the connection between genotype, environment, and phenotype in plants. I really had no idea how far UNPAK would go, so I have to say it has exceeded my expectations. Although we can always do more, it amazes me to think how many students we have taught through the program, and how we have amassed hundreds of thousands of data points!

Courtney J. Murren I agree that UNPAK did over-deliver, but differently than I initially imagined! UNPAK involves huge amounts of team work. Other projects have smaller groups of scholars. I've often in my career been part of bigger projects and had other projects with fewer collaborators at the same time.

Jennifer Jo Thompson I was motivated based on my interest in understanding how people engage with science and with scientific knowledge, and my interest in expanding access to participation in science. Although my own work has moved away from working directly with UNPAK, it has been exciting to watch it develop and greatly expand opportunities for faculty and students to participate in research over the years. Its reach has expanded far beyond my initial expectations!

In other ways, my work with UNPAK has deeply impacted the inter- and transdisciplinary work I continue to do. It was my first experience working on a collaborative, cross-disciplinary project, and as a social scientist among a team primarily focused on natural science. The other PIs' respect for, and commitment to supporting the social science was deeply affirming for me as a postdoc and it ultimately inspired me to continue to work on inter- and transdisciplinary projects.

---

[18] Lee and Bozeman (2005).

16.6 UNPAK as a Philosophy and a "Way of Being"

Question Is UNPAK more "liberal arts" or more "vocational"? What do you think about this divide? Is it valid or bogus?

Hilary S. Callahan I dislike the idea of putting a college education into cause-and-effect or return-on-investment relationships, especially if that rigidly shapes students' eventual career satisfaction and learning. That being said, employability and career satisfaction are important, especially for students coming from less advantaged backgrounds. I have worked hard to clarify for myself as a supervisor and for my diverse and ambitious students, in UNPAK and elsewhere, that this is a resolvable issue. A paid job in research part-time during college or full-time during the summer is a way to simultaneously support students tangibly, enhance their learning, and furnish a credential. What's more, that research job adds directly to the collective knowledge of science. Keeping that complex messaging in mind is important, and I know that is part of Barnard's campus culture, and I hope elsewhere.

Michael Wolyniak I see UNPAK as more of a liberal arts project. Training students to develop the critical thinking skills necessary to adapt to any vocation as opposed to providing the training tools for one particular vocation. This does not mean that the liberal arts and the vocational training need to be completely separate. I like to think of myself as a "practical liberal artist" in the sense that I encourage a traditional liberal arts approach to scientific research but with an eye towards potential career opportunities that can develop from the work. UNPAK is certainly effective in allowing students to gain this vital critical thinking ability while also considering what direct employment opportunities may be available as a result of the experiences that they gain from the work.

Matthew T. Rutter I would say more "liberal arts". But I am not sure about the validity of this divide. I think life, even centered around a vocation or skill, should hopefully draw from many areas and be lived like one from the "liberal arts"!

Question To what extent is UNPAK democratic and creative? Does it offer opportunities for all participants to be included in decision-making, pursuing funding, or expanding the network?

Hilary S. Callahan Every year, students come onto Barnard's campus in an ever-more urgent hurry to change the world and solve its problems, but they also are there for four years of learning. Whenever I am designing a course or a research project like UNPAK, I am trying to thread the needle between those two extremes. As faculty, I need to provide some top-down and efficient guidance. I also want to avoid extinguishing students' passion and idealism.

Also, after four years at Barnard, they have fifty to sixty or even more years for a career and the rest of their lives. Recognizing that, I try to have my teaching and mentoring be reasonably well-aligned with political principals like participating in democratic decision-making, and learning that it is important to understand both how and when to lead and how and when to select leadership to follow, or leadership to rebel against. I do not think that is explicitly a goal of UNPAK, but it is something that percolates through Barnard, and probably through other liberal arts campuses in the UNPAK network. Many students that I work with in UNPAK and in other projects are working hard on that balance.

Michael Wolyniak UNPAK has been an awesome force for me in democratizing the research process among my students. I recall in my undergraduate years that student research was largely reserved for people who had the classroom achievement or the persistence necessary to attract the attention of an advisor. UNPAK, by taking place in the classroom, allows all students to participate in the research regardless of their previous background and experience. This is a powerful tool that allows me to potentially inspire many more students than I could have before, and to attract people who may have slipped through the cracks under the apprenticeship model of undergraduate research in developing the sparks necessary to gain a passion for STEM research. The democratization of the process can even allow students to participate in formulating the research questions that we will ultimately pursue in a given class

Jennifer Jo Thompson My thoughts are perhaps somewhat dated, but some of our earlier data suggests that there is an ebb and flow in this over the course of the overall UNPAK project and over the course of students' engagement with the project. When UNPAK was first beginning, students were helping develop structured protocols and thus had agency, ownership, and input into the project development. Once those protocols were (necessarily) codified, students coming into UNPAK no longer had that as a major locus for creative input. My impression is that students continue to have creative input around the specific questions they explore, within a defined set of parameters. This seems like a reasonable balance given the competing demands of the project. At the same time, we found that students who remained with the project for several years often developed independent projects, which gave them greater ownership and agency. Is this the same thing as 'creativity' and 'democracy'? I am not quite sure.

Matthew T. Rutter It is semi-democratic. Students, particularly lab and summer students, have a chance to shape questions. Students in a single CURE are typically learning how to measure plants and work with data and as yet have not had as much creative opportunity in my classes—although the approach that we are launching in our most recent NSF-IUSE-funded work aims to change this. The democratic element is important to me primarily for the more experienced students.

April Bisner   UNPAK is democratic and creative in that students are encouraged to think about other phenotypes they may want to measure outside of the core phenotypes and decide with their classmates or research group which of those phenotypes the class or group will measure. I learn from our student interns and incorporate suggestions on how to improve protocols and workflow. I think it is very important for students to think outside of the box and make decisions as a group.

Question   Are there other ways that UNPAK has fostered an improved network of colleagues or even friendships with colleagues? Examples please!

Hilary S. Callahan   Long before 2020 and 2021, our UNPAK members were "hanging out" using Google tools, and that type of experience is now so ingrained in everyone, so it is no longer as special. It is gratifying to see Zoom enabling other scholars and educators to do this work, and to communicate better.[19]

After so many years, when we meet in person at conferences or other events, it is similar to picking up with life-long friends from middle school or high school. We have many shared experiences. We have celebrated each other's promotions; some of us have seen each other's children grow up and so forth. This is a huge benefit of extending research for more than the usual three-year cycle of funding, and students pick up on that. In a meeting this semester with a recent UNPAK student, she described UNPAK's investigators at College of Charleston as "angels" because of their patience and genuine concern for all. I do, very much, feel privileged to be involved in a project with that much authentic commitment to teaching and mentoring, and to collaborating.

We had a few informal commemorations of UNPAK's tenth anniversary this year, and I do remember Matt Rutter referring to UNPAK as a philosophy and even "a way of being" and that really resonated with me, and with this book's spotlights on authenticity in STEM research.

Michael Wolyniak   I cherish the professional and personal relationships that have developed as a result of my affiliation with UNPAK. I still associate with the majority of my colleagues from the beginning of UNPAK to this very day I am still able to send my students to exciting summer research opportunities that are affiliated with UNPAK. Taken together, UNPAK has been one of the most important parts of my professional development. Thank you to the UNPAK PIs for always welcoming me into your labs and groups. I treasure my professional and personal relationships with each of you and they have really shaped my expectations for collaboration. Thank you!

---

[19] Also see Patterson et al. (this volume).

Matthew T. Rutter Perhaps it is from being in UNPAK since the beginning, but I have literally dozens of new professional colleagues and absolutely count many of them as my friends. Frankly, at the start I knew my local colleagues well and our other collaborators ranged from new acquaintances (Mike Wolyniak, Jenn Thompson) to a collaborator I knew but would get to know much more through UNPAK (Hilary Callahan). Now any of them would be instant buddies that I would insist come out for a good drink and a meal if I were so lucky as to encounter them at a meeting or a visit. As UNPAK has expanded, I can sincerely say that each person I have met has been so impressive and fantastic in their line of work—all are the epitome of teacher-scholars. Some we manage to get to Charleston periodically, some I am just starting to get to know, some I feel like I know very well despite never having been in the same room. I enjoy every minute of it.

16.7 Coda

Hilary S. Callahan Among myriad ways to wrap this wide-ranging conversation, one is to pause to reflect on the shifting context within which UNPAK has continued into a second decade. In hurricane-prone Charleston as well as throughout the U.S., storms and floods impacted many UNPAK campuses, even as UNPAK campuses in California and Seattle experienced wildfires as common climate-related events. On and near several UNPAK campuses, shocking violence and loss of life have drawn media coverage, grief, and rage. Campuses are spotlighted as homes for both ongoing protest of and healing from violence and loss of life, even as they persevere as places for learning and thinking with data and scientific evidence, including the tools for tracing the connections of the sustainability crisis to structural racism, to oppression and injustice.

All of this makes it notable to reflect on turns in this conversation that probed activism or politics.[20] All of us acknowledge, and some even emphasize, that UNPAK is inherently intertwined with broader cultural and political currents. Indeed, the grant that initiated the project in April 2011 was delayed for four months because of a debt ceiling crisis and the failure of a polarized Congress to approve a Federal budget. Fast-forwarding past additional shutdowns and impasses during the Obama administration, we arrive in January and April 2017. UNPAK scientists and students watched or participated in prominent public marches to highlight the role of women and science in society more broadly. Then, in January of 2021, we watched Washington D.C. struggle through massive chaos to resume the hard work of democracy.

The many months of the Covid-19 pandemic forced UNPAK and many others to rely heavily on virtual tools to conduct research, teach and learn together. Amidst

---

[20] See Pfirman and Winckler (this volume, Chap. 19).

frustration and fatigue, some have drawn inspiration from the potential benefits of these tools. For UNPAK, this led to a proposal and new award based at the College of Charleston from the National Science Foundation. The next phase of UNPAK— D-CURE, for Digital-CURE—is premised on virtual work as not just productive, but potentially accessible to more students and campuses, and to more diverse students. Optimistic about this next project, UNPAK is proud of its long journey and prospects for continuing along its ever-adaptable and widening path, learning and training in STEM in solidarity.

Acknowledgements Across its initial and subsequent phases, UNPAK has received funding support from the National Science Foundation (IOS-1052262, IOS-1354771, DUE 212145), an AFRI-REEU award from USDA NIFA (20196703329241) and a Presidential Research Award from Barnard College.

References

AGI, the Arabidopsis Genome Initiative. (2000). Analysis of the genome sequence of the flowering plant *Arabidopsis thaliana*. *Nature, 408*, 796–815. https://doi.org/10.1038/35048692

Auchincloss, L. C., Laursen, S. L., Branchaw, J. L., Eagan, K., Graham, M., Hanauer, D. I., Lawrie, G., McLinn, C. M., Pelaez, N., Rowland, S., & Towns, M. (2014). Assessment of course-based undergraduate research experiences: A meeting report. *CBE Life Sciences Education, 13*(1), 29–40.

Buell, C. R., & Last, R. L. (2010). Twenty-first century plant biology: Impacts of the *Arabidopsis* genome on plant biology and agriculture. *Plant Physiology, 154*(2), 497–500. https://doi.org/10.1104/pp.110.159541

Dolan, E., & Weaver, G. (2021). *A guide to course-based undergraduate research.* W. H. Freeman.

Gershberg, N. (this volume). When a titan arum blooms during quarantine. In M. S. Rivera Maulucci, S. Pfirman, & H. S. Callahan (Eds.), *Transforming sustainability research and teaching: Discourses on justice, inclusion, and authenticity.* Springer.

Gropp, R. E. (2016). NSF: Time for big ideas. *Bioscience, 66*(11), 920. https://doi.org/10.1093/biosci/biw125

Hillenmeyer, M. E., Fung, E., Wildenhain, J., Pierce, S. E., Hoon, S., Lee, W., Proctor, M., St. Onge, R. P., Tyers, M., Koller, D., Altman, R. B., Davis, R. W., Nislow, C., & Giaever, G. (2008). The chemical genomic portrait of yeast. *Science, 320*, 363–365. https://doi.org/10.1126/science.1150021

Jensen-Ryan, D., Murren, C. J., Rutter, M. T., & Thompson, J. J. (2020). Advancing science while training undergraduates: Recommendations from a collaborative biology research network. *CBE Life Sciences Education, 19*, 4. https://doi.org/10.1187/cbe.20-05-0090

Koornneef, M., & Meinke, D. (2010). The development of *Arabidopsis* as a model plant. *The Plant Journal, 61*, 909–921. https://doi.org/10.1111/j.1365-313X.2009.04086.x

Lee, S., & Bozeman, B. (2005). The impact of research collaboration on scientific productivity. *Social Studies of Science, 35*(5), 673–702. https://doi.org/10.1177/0306312705052359

Lee, C., Svardal, H., Farlow, A., Exposito-Alonso, M., Ding, W., Novikova, P., Alonso-Blanco, C., Weigel, D., & Nordborg, M. (2017). On the post-glacial spread of the human commensal *Arabidopsis thaliana*. *Nature Communications, 8*, 14458. https://doi.org/10.1038/ncomms14458

Leonelli, S. (2019). The challenges of big data biology. *eLife, 8*, e47381. https://doi.org/10.7554/eLife.47381

Maenza-Gmelch, T. (this volume). Finding the most important places on earth for birds. In M. S. Rivera Maulucci, S. Pfirman, & H. S. Callahan (Eds.), *Transforming sustainability research and reaching: Discourses on justice, inclusion, and authenticity.* Springer.

Murren, C. J., Wolyniak, M. J., Rutter, M. T., Bisner, A. M., Callahan, H. S., Strand, A. E., & Corwin, L. A. (2019). Undergraduates phenotyping *Arabidopsis* knockouts in a course-based undergraduate research experience: Exploring plant fitness and vigor using quantitative phenotyping methods. *Journal of Microbiology & Biology Education, 20*(2). https://doi.org/10.1128/jmbe.v20i2.1650

O'Malley, R. C., & Ecker, J. R. (2010). Linking genotype to phenotype using the Arabidopsis unimutant collection. *The Plant Journal, 61*(6), 928–940. https://doi.org/10.1111/j.1365-313x.2010.04119.x

Patterson, A. E., Williams, T. M., Ramos, J., & Pierre, S. (this volume). Urban-rural sustainability: Intersectional perspectives. In M. S. Rivera Maulucci, S. Pfirman, & H. S. Callahan (Eds.), *Transforming sustainability research and teaching: Discourses on justice, inclusion, and authenticity.* Springer.

Pfirman, S., & Winckler, G. (this volume). Perspectives on teaching climate change: Two decades of evolving approaches. In M. S. Rivera Maulucci, S. Pfirman, & H. S. Callahan (Eds.), *Transforming sustainability research and teaching: Discourses on justice, inclusion, and authenticity.* Springer.

Rancati, G., Moffat, J., Typas, A., & Pavelka, N. (2018). Emerging and evolving concepts in gene essentiality. *Nature Reviews Genetics, 19*, 34–49. https://doi.org/10.1038/nrg.2017.74

Rhodes, M. E., McGuire, K. L., Shek, K. L., & Gopal, T. S. (this volume). Going up: Incorporating the local ecology of New York City green roof infrastructure into biology laboratory courses. In M. S. Rivera Maulucci, S. Pfirman, & H. S. Callahan (Eds.), *Transforming sustainability research and teaching: Discourses on justice, inclusion, and authenticity.* Springer.

Rodenbusch, S. E., Hernandez, P. R., Simmons, S. L., & Dolan, E. L. (2016). Early engagement in course-based research increases graduation rates and completion of science, engineering, and mathematics degrees. *CBE Life Sciences Education, 15*(2), ar20. https://doi.org/10.1187/cbe.16-03-0117

Rutter, M. T., Murren, C. J., Callahan, H. S., Bisner, A. M., Leebens-Mack, J., Wolyniak, M. J., & Strand, A. E. (2019). Distributed phenomics with the unPAK project reveals the effects of mutations. *The Plant Journal, 100*, 199–211. https://doi.org/10.1111/tpj.14427

Rutter, M. T., Bisner, A. M., Kohler, C., Morgan, K., Musselman, O., Pickel, J., Tan, J., Yamasaki, Y., Willson, J., Callahan, H. S., Strand, A. E., & Murren, C. J. (2020). Disrupting the disruptors: The consequences of mutations in mobile elements for ecologically important life history traits. *Evolutionary Ecology, 34*(3), 363–377. https://doi.org/10.1007/s10682-020-10038-0

Snow, J. (this volume). What does cell biology have to do with saving pollinators? In M. S. Rivera Maulucci, S. Pfirman, & H. S. Callahan (Eds.), *Transforming sustainability research and teaching: Discourses on justice, inclusion, and authenticity.* Springer.

Thompson, J. J., & Jensen-Ryan, D. (2018). Becoming a "science person": Faculty recognition and the development of cultural capital in the context of undergraduate biology research. *CBE Life Sciences Education, 17*, 4. https://doi.org/10.1187/cbe.17-11-0229

Thompson, J. J., Conway, E., & Dolan, E. L. (2016). Undergraduate students' development of social, cultural, and human capital in a networked research experience. *Cultural Studies of Science Education, 11*(4), 959–990. https://doi.org/10.1007/s11422-014-9628-6

Hilary S. Callahan earned her Ph.D. at University of Wisconsin, Madison, and worked at University of Tennessee in Knoxville before joining the faculty of Barnard College. She values her campus for teaching intersectional feminism and botany. She is the proud daughter of a teacher.

Michael Wolyniak is a geneticist and molecular biologist, working since 2009 at Hampden-Sydney, a small liberal arts college for men. He is a Councilor with the Biology Division of the Council on Undergraduate Research (CUR), directing a long-term mentoring network for instructors about effectively using CUREs in the classroom.

Jennifer Jo Thompson is a socio-cultural anthropologist embedded in the College of Agriculture & Environmental Sciences at the University of Georgia. Her inter- and trans-disciplinary focus is the social sustainability of agriculture and food systems. She directs UGA's Sustainable Food Systems Initiative and the Interdisciplinary Graduate Certificate in Sustainable Food Systems.

Matthew T. Rutter is a plant ecological geneticist studying mutation, mutualism, phytoremediation, and methods of education. He is happy to work at an institution that values the teacher-scholar role. He is also Academic Director of Stono Preserve at the College of Charleston, a nearly 1000 acre natural and historic property.

Courtney J. Murren is a plant evolutionary ecologist who melds teaching and research in her professional life, and emphasizes the importance of observation of nature. She is proud to teach biology at a public institution, and to collaborate with students in the research laboratory and in the classroom.

April Bisner is project manager for UNPAK. An alumna of University of Maryland, she also earned her Project Management Professional certification from the Project Management Institute. An enthusiastic botanist and photographer, she formerly surveyed vegetation and assisted in research in varied contexts, including NYC Parks and the National Park Service.

Part IV
Climate Change: Engagement, Politics, and Action

Chapter 17
Teaching About Climate Change from an Astronomical Perspective

Laura Kay and Kassandra Fuiten

Kassandra I am here with Professor Laura Kay, Professor of Physics and Astronomy at Barnard College and author of *21st Century Astronomy.*[1] Professor Kay, you teach an introductory astronomy course called *Life in the Universe*. Can you give me a brief description of this course?

Laura This course explores how scientists think about life in the universe. Currently, we do not have data on life anywhere other than on Earth, so this class is more about how scientists approach the question of "life in the universe" than about what is actually out there. Right now, we just do not know. The class is interdisciplinary: it uses astronomy, physics, Earth science, chemistry, and biology to look at both how scientists are working together on this problem and how they are thinking about this question. But there's no clear answer, and I tell that to my students on the first day. We do not know if there's any life out there. *Life in the Universe* is a class that's about the process of how scientists approach this question.

Kassandra I understand that you also talk to your students about climate change within this course. How do you fit discussions about climate change into a course that, while interdisciplinary, does fall within the Astronomy department?

Laura When students come into my classroom, they have already heard a lot about climate change. Some students have studied it in high school, and all of them have

---

[1] Kay et al. (2019).

L. Kay (✉)
Physics & Astronomy, Barnard College, New York City, NY, USA
e-mail: lkay@barnard.edu

K. Fuiten
Barnard College, New York, NY, USA

© The Author(s) 2023
M. S. Rivera Maulucci et al. (eds.), *Transforming Education for Sustainability*,
Environmental Discourses in Science Education 7,
https://doi.org/10.1007/978-3-031-13536-1_17

heard about it on the news or within a political context. I started including it in this course because climate change is an essential part of the story of why human life evolved on Earth instead of elsewhere in our Solar System. But, because climate change is such a politicized and familiar topic to my students, I aim to approach it from a different angle. Instead of focusing solely on the current state of climate change on Earth, I emphasize to my students that climate change is not just about planet Earth and not just about our current time. Earth has evolved and changed as a planet, as have the other planets in our Solar System.

Kassandra How do you expand your student's understanding of climate change beyond our current Earth context?

Laura I discuss with my students the history of the discovery of contemporary climate change on Earth. This includes looking back to the beginnings of the industrial revolution when people first discovered that carbon dioxide emissions could change the atmosphere. Some of the earliest work was done in the mid-1800s by Eunice Foote and John Tydon, who thought about how atmospheres can block radiation. In the 1890s, the Swedish physical chemist Svante Arrhenius calculated how carbon dioxide emissions could lead to global warming. He won the Nobel Prize for his work in 1903. The point is, the idea that carbon dioxide and other greenhouse gases affect our atmosphere is not a new concept.

Beyond Earth, astronomers have studied climate on other planets, specifically Mars and Venus. Studies of these planets have greatly informed our understanding of climate on Earth, making climate change discussions incredibly relevant to an astronomy course curriculum.

Kassandra How could studies of Mars and Venus, planets which differ so dramatically from Earth, provide insight into climate change on our planet?

Laura Mars and Venus are the two planets closest in distance to the Earth, and they give us examples of large-scale global cooling and global warming, respectively. Studies of Mars allowed astronomers to witness global cooling in real-time. Mars has intense dust storms that occur periodically, which astronomers have observed through telescopes for 200 years. When NASA missions began sending space probes to Mars in the 1960s, scientists could, for the first time, measure Mars's temperature during these dust storms. They found that, as dust quickly covered Mars, the planet's temperature decreased.

Mars's global cooling scenario led to two strong scientific theories about the climate on Earth. The first of these theories concerns the death of the dinosaurs. Scientists now think their extinctions were caused by the impact of a huge asteroid, or possibly a volcanic eruption (or a combination of the two), either of which would have sent enormous amounts of dust into Earth's atmosphere. Similar to the dust storms on Mars, Earth's temperature would have decreased as dust blocked the sunlight. This sudden decrease in temperature would have killed off the food chain;

plants would have died, small animals would have died, and then the dinosaurs, which required large amounts of food to support their massive bodies, would have eventually starved. The second theory to result from the studies of Mars, nuclear winter, also became popular in the 1980s. The theory of nuclear winter proposed that if a nuclear war were to occur, the excess of fire and smoke would block the sunlight and lower the temperature on Earth.

On the other hand, studies of Venus provided insight into global warming. The Soviet Union, which focused its space exploration on Venus, began sending space probes to Venus in the 1960s. Prior to these missions, many people had considered Venus a possible 'sister planet' to Earth. Venus is both the closest planet to Earth and the most similar to Earth in size, and science fiction of the time was filled with fantasies about humans or other life there. These fantasies ended as Soviet missions reported that temperatures on Venus were around nine hundred degrees Fahrenheit. (The missions ended quickly because the probes burned up.) At first, Venus' high temperatures puzzled astronomers. Venus is closer to the Sun than Earth, but not close enough to explain the vast difference between Earth's liveable temperatures and the incredible heat of Venus. Researchers in the early 1970s came up with a possible explanation: the Runaway Greenhouse Effect. The Runaway Greenhouse Effect proposed that the abundant carbon dioxide in Venus's atmosphere trapped radiation from the Sun, causing the temperature of the planet to increase. Any water that may have been on Venus's surface eventually evaporated. Then, the water molecules would have been split by ultraviolet radiation from the sun, with the hydrogen escaping to space.

In the 1970s, astronomers at NASA started building computer models to understand what happened on Venus, and they then applied the basics of these models to Earth's climate. These `general circulation models' or GCMs are used to reproduce or explain climate trends we've seen in the past, what is seen now, and possibly predict what will happen in the future. When similar models were applied to Earth, with increased carbon dioxide levels beginning during industrialization in the 1800s, they found that these models could explain the changing temperatures on Earth. Over the years, models of Earth's global warming have become increasingly sophisticated, but the earlier models originated through the study of Venus.

Kassandra If carbon dioxide in Venus's atmosphere caused such a dramatic increase in temperatures, why didn't Earth also experience this runaway greenhouse effect?

Laura In the past, Mars, Venus, and Earth had liquid water on their surfaces, and they all started out with a primarily carbon dioxide atmosphere. But all three planets evolved. Venus is closer to the Sun than Earth, so it was always warmer. On Earth, life dramatically impacted the evolution of Earth's atmosphere and climate. Photosynthesis, first by bacteria and later by plants, fundamentally altered Earth's atmosphere by producing oxygen which, over time, gave us the oxygen atmosphere that we have now. The existence of an oxygen atmosphere then allowed more complex life to form. Life changed Earth's atmosphere, and Earth's atmosphere changed life. Mars and Venus have 95% carbon dioxide atmospheres, whereas Earth's atmosphere is 78% nitrogen, 21% oxygen, and only 0.04% carbon dioxide. In addition, the atmosphere of Venus is about 100 times thicker than Earth's (Mars' atmosphere is 100 times thinner), so the greenhouse effect is much more substantial on Venus.

Kassandra How did early climate change on Earth differ from the current climate changes that we see?

Laura When we look at historical records of carbon dioxide levels, which we can obtain by studying tree rings and ice cores (See Baxi et al., this volume, chapter 25), we see that carbon dioxide levels have varied over time. By studying ice cores in Antarctica, scientists can measure the amount of carbon dioxide going back 800,000 years. By looking at different chemicals and isotopes, they can see that temperature and carbon dioxide levels have fluctuated together. When the temperature was higher, the carbon dioxide was higher, and when the temperature was lower, the carbon dioxide was lower. Multiple factors contribute to these changing levels, including Milankovitch Cycles, which are the cyclical changes in Earth's axial tilt and orbit. These changes in tilt, orbital ellipticity, and precession occur over tens of thousands of years and appear to align fairly well with Earth Ice Ages and the periods in between the Ice Ages.

The ice core measurements, which Al Gore's film, *An Inconvenient Truth* (Paramount Picture Corporation 2006), made famous, shocked many people in the 1990s. They saw that the cycles of changing temperature and carbon dioxide levels go together naturally, but *over tens of thousands of years*, resulting in slow environmental changes. The changes we are seeing on Earth now are over the last 150–200 years since Industrialization. Scientists are nervous about our current rate of climate change because Earth does not have time to respond to these faster changes naturally.

Kassandra Do other planets offer a possible refuge from climate change and environmental destruction on Earth?

Laura Well, I say no. In *Life in the Universe*, I emphasize that humans have evolved on Earth for Earth's specific environment, and Earth has evolved with us. Humans cannot pack up and go to Venus; for example, it is nine hundred degrees Fahrenheit there. While we may no longer have fantasies about going to Venus, there are still plans to build bases on the Moon, which has no atmosphere, has light gravity, and requires humans to bring all supplies with them. And, there are still fantasies about humans escaping Earth and colonizing Mars. On Mars, the Sun would be very dim, and it is quite cold. The atmosphere is very thin and is not breathable, so you would have to wear spacesuits the whole time or live in a pressurized dome. It is unknown if food could be grown under Mars's weak sunlight, which means we do not know whether Mars could sustain humans. The gravity on Mars is somewhat stronger than on the Moon, yet still only forty percent the strength of Earth's gravity, meaning that living on Mars would cause human muscles to weaken dramatically. Even if Mars was somehow able to provide a sustainable environment for humans, it's doubtful that humans would be able to readapt to the higher gravity back on Earth. I do think some people will make short visits to Mars, and maybe small numbers will live there, but I doubt that millions (or billions) of people will relocate from Earth. And, of course, there would be serious questions about *who* gets to escape Earth to live on the Moon or Mars.

Outside of our Solar System, since 1995, astronomers have found thousands of planets orbiting other stars in our Galaxy. However, we have not collected enough data yet to know if any have the kind of oxygen atmosphere free from poisonous gases that human life requires. Many of these planets are likely balls of gas like Jupiter or Saturn, but some are solid like Earth. Most of these planets are not at a distance from their star that suggests they could have temperatures that permit liquid water on their surface, and Earth life needs water. Upcoming space telescopes may identify distant planets with temperatures and atmospheres that are more Earth-like, but at present, we have no technology that will let us travel to them.

As an aside, I will add that culturally, science fiction has influenced many people's perceptions of other planets. Many blockbuster science fiction movies, television shows, and books show people going from planet to planet, where each planet has Earth-like gravity and temperatures and a breathable atmosphere (and everyone speaks English). These media make it seem as though there are a plethora of planets for humans to choose from; so that if Earth 'goes bad,' we can just go somewhere else. As far as astronomers can tell, realistically, that's not what we actually see out there, at least not yet.

Kassandra What do you hope to accomplish by expanding your students' understandings of climate change beyond Earth and beyond our current time period?

Laura Well, professors always aim to get students to think beyond the present moment and think about history. So I do want students to understand that Earth's climate has indeed changed throughout history. We have all seen the 'Ice Age' movies, but the contemporary climate changes are also different in many important ways. Additionally, I hope to get students to think about scientific methodology rather than political frameworks. For example, we talk in class about how we can measure the temperatures of planets in our solar system and make estimates of the temperatures of planets in systems across the Galaxy. Some climate change doubters argue that it is difficult to measure temperatures on Earth, but if we can measure temperatures on Mars, of course, we can measure the temperature of Earth.

I also hope to make my students think about climate change in a universal context rather than getting bogged down in bickering about the details of whether it is real or what to do about it (see Pfirman & Winckler, this volume, chapter 19). Instead, I want them to think about how there are many planets out in space, but it is not known if any of them are suitable for us, given how life has evolved on Earth. On the one hand, studying other planets lets us see the processes that lead to changes in atmospheres, so we can help understand how they happen on Earth. On the other hand, studies of other planets show that there is not a place to relocate to. We need to care about what is happening on Earth.

Kassandra Do you see teaching climate change as a justice issue?

Laura It is certainly an environmental justice issue to keep Earth habitable for humans! There is no way we can relocate everyone to another planet in the foreseeable future. You may have heard the activist slogan, `There is no Planet B.' I don't spend much time in class debating the merits of taxing carbon emissions or carbon capture like an economist might. Instead, I focus on my area: how astronomers think about planets, which extends to how astronomers think about a planet like Earth. General education science classes are possibly the only science class that a college student will take, and *Life in the Universe* may be a student's only opportunity to hear about climate change from a scientific perspective. Climate change discussions should be included in any general science class when possible, and professors of the sciences at Barnard try to put in the most relevant angle to their department. Climate change is a multidisciplinary problem. It needs an interdisciplinary approach, and the more people thinking about the problem, the better.

References

Baxi, et al. (this volume). Perspectives on teaching climate change: Two decades of evolving approaches. In M. S. Rivera Maulucci, S. Pfirman, & H. S. Callahan (Eds.), *Education for sustainability: Discourses on authenticity, inclusion, and justice*. Springer

Kay, L., Palen, S., & Blumenthal, G. (2019). *21st century astronomy* (6 ed.). WW Norton.

Paramount Pictures Corporation. (2006). *An inconvenient truth*. Paramount.

Pfirman, S., & Winckler, G. (this volume). Perspectives on teaching climate change: Two decades of evolving approaches. In M. S. Rivera Maulucci, S. Pfirman, & H. S. Callahan (Eds.), *Education for sustainability: Discourses on authenticity, inclusion, and justice*. Springer

Laura Kay has been a Professor of Physics and Astronomy at Barnard College, since 1991. As a graduate student she spent 13 months at the Amundsen-Scott station at the South Pole in Antarctica. She studies active galactic nuclei. She has taught courses in astronomy, astrobiology, women and science, and polar exploration.

Kassandra Fuiten graduated from Barnard College in 2016 where she studied Elementary Education and Psychology with a special interest in Attention Deficit Hyperactivity Disorder and Autism. Post-graduation, Kassandra earned CELTA certification to teach English as a foreign language and currently works with elementary English language learner students in Madrid, Spain.

Chapter 18
Volcanoes, Climate Change, and Sustainability

Sedelia Rodriguez and Kassandra Fuiten

18.1 Introduction to the Participants

I (Sedelia) was originally trained as a geologist. I study rocks that solidified from magma or lava to decipher how these formed. I think that teaching students about the natural world is especially important because hopefully as they look more at their environment, they will appreciate it more and want to preserve it. Knowing what happens naturally helps plan for a sustainable community. We cannot control everything. What are some things we cannot change? What do we need to work around? When you do not understand the natural world, it is harder to fight for it. Knowledge about the environment becomes a vehicle for activism.

I (Kassandra) am a student. I interviewed Professor Rodriguez about how she integrates her interest in volcanoes into her teaching in a way that is inclusive.

Kassandra What courses do you teach where you talk about volcanoes?

Sedelia Today, I will focus on my teaching of a First-Year Seminar, but I try to integrate volcanoes into all of my teaching, including Introduction to Environmental Science. The First-Year Seminar is not a science course, the focus is on writing and speaking. We want students to gain their voice and speak confidently. So I had to think about how to introduce science in this type of setting when students do not have a strong scientific background or potentially even a scientific interest. Some of the ways I do this is by getting them to relate to some aspect of a volcanic eruption—how volcanoes affect climate and agriculture (and therefore societies), tour-

S. Rodriguez (✉)
Department of Environmental Science, Barnard College, New York City, NY, USA
e-mail: srodrigu@barnard.edu

K. Fuiten
Barnard College, New York, NY, USA

© The Author(s) 2023
M. S. Rivera Maulucci et al. (eds.), *Transforming Education for Sustainability*,
Environmental Discourses in Science Education 7,
https://doi.org/10.1007/978-3-031-13536-1_18

ism, etc. After that, I talk about hazards and volcanic eruptions, and now they are interested because they know the information is relevant. Pop culture helps to make volcanoes relatable and I use that as a tool for student engagement. Also, I do not start lecturing about the science of volcanoes, but rather introduce it slowly throughout the semester. The whole point is to get students interested and for them to see the information as relevant to their own lives.

Kassandra In your experience, students do not initially see volcanoes as relevant to their own lives?

Sedelia Most of my students do not live near volcanoes and, therefore, do not realize the effects that volcanoes can have upon their lives. I had a student come up to me at the beginning of the semester and insist that, because she was not a science major nor did she have any connection to volcanoes, the course was not relevant or interesting to her. I try to show students why they should care. I talk about the effects that volcanoes may have on their lives and the lives of others, in an inclusive way. In some cases, people who live around volcanoes are Indigenous farming populations. I also discuss how air travel disruptions may affect students while traveling even if they do not live near a volcano. I talk about how ash erupted during a volcanic eruption can have health effects on those living near volcanoes, especially those with respiratory problems. And if a Supervolcano were to erupt, such as Yellowstone in the United States, ash could be distributed all over the country, affecting those who live far away from a volcano. We also discuss how crops can be affected by eruptions; nowadays we get fruits and vegetables from all over the world, so an eruption can affect the production of different foods and items. By getting students to connect volcanoes to their own lives, I hope to engage and interest them in a topic they may never have thought about before.

Other students have grown up near areas of volcanic activity; for example, last year I taught students from Ecuador, Mexico, Seattle, and Northern California. The student from Ecuador shared her experiences seeing smoke and ash fall over her city and having to wear a mask to protect against the ash. She also mentioned that people in lower-income areas had fewer resources and were therefore affected more by the eruption. In some cases, stores and businesses were closed for weeks, which directly affected this population. She was able to see the socioeconomic discrepancies and injustices in her country when a natural disaster occurred. Her stories fascinated other students and made them more interested in volcanoes.

I also noticed that the students from volcanic areas developed an authentic eagerness to further investigate their local volcanic activity. I had my students research a volcanic area, and the students who grew up near volcanoes researched their local areas. The student from Mexico had grown up hearing stories about the volcano near her home. After learning more about volcanoes, she wanted to hear all of the old stories from her grandmother and became excited to research her local history. The student from Seattle researched the possible dangers of Mount Rainier and learned about evacuation plans. And, the student from Northern California researched an area near her home where geothermal research was occurring. She had been there before but had never known about the research happening. She was so excited to learn about geothermal energy and vowed to visit the area again when she visits home.

Kassandra We have seen a lot of destruction by volcanoes in the news, with eruptions from Kilauea in Hawaii, beginning in April 2018, and the eruption of Volcán de Fuego in Guatemala on June 3, 2018. How significant of a threat do volcanoes pose to humans?

Sedelia For those who live in the vicinity of a volcano, the destructive impact can be significant. Strato or composite volcanoes such as Volcano del Fuego tend to be very destructive due to the high silica content of the magma, which affects the viscosity, temperature, and amount of dissolved gases. Higher viscosity, lower temperatures, and more gases result in more explosive eruptions. After a large eruption, like the one in Guatemala, a wave of heat, ash, and wind explodes from the volcano with hurricane strength. This is a pyroclastic flow, the most destructive aspect of a volcano, and, if one is near the explosion, there is no way to outrun it. Lahars, which are mudflows generated by a volcano, come very suddenly after an eruption and have buried entire towns without warning. Earthquakes also often accompany an eruption, although they sometimes serve only as precursors, providing warning signs of rising magma. Ash flow, a mixture of gas and fine-grained rock fragments ejected from the volcano, can cause health problems, especially for those with asthma or other breathing difficulties. Volcano ash can also shut down air traffic, leaving people stranded while traveling and causing economic consequences.

For example, in 2010, after the eruption of Eyjafjallajokull in Iceland, there was an air traffic shut down in parts of Europe between April 15 and May 16. It affected areas as far away as the United Kingdom, France, and some of Germany. Lava flows typically cause very few fatalities because they move so slowly, but they can cause great destruction to crops and buildings. The eruption of Volcán de Fuego spewing ash and lava resulted in 200 deaths. In the case of Volcán de Fuego, the eruption came very suddenly, and the public insists that they were not urged to evacuate, although government officials say otherwise. The eruption destroyed corn, bean, and coffee crops, a total of 21,000 acres of crops.[1] This obviously has a negative effect on the nearby population, both as their food resources are destroyed, but also an economic one because they cannot sell their crops. The infrastructure, roads, bridges, etc. were also destroyed, leaving many stranded.

When Kilauea erupted in Hawaii, many people lost their homes and the island suffered significant damage to infrastructure. However, no deaths were reported in Hawaii because residents had adequate time for evacuation. The Hawaiian volcanoes have less explosive eruptions due to their chemistry. Scientists have warning systems that track precursors to eruptions, such as earthquakes. Although it can be difficult to predict the precise day or even month of an eruption, when they notice an increase in earthquakes for example, they put up a warnings and begin evacuations.

---

[1] Fecht 2018.

Sometimes, even when the warning to evacuate does come, people refuse to leave. Some cultures carry beliefs about the godlike nature of volcanoes, believing that eruptions serve as a form of divine retribution. In certain islands of Indonesia, people refused to leave their homes during times of increased volcanic activity, believing that if they prayed and gave offerings, they would be saved. The unwillingness to leave one's home is logical to an extent. To evacuate means to abandon everything they have ever known and often one's livelihood, which can seem too large a price when volcanic eruption predictions are often inaccurate. Even in Hawaii, some residents did not want to leave their homes and were skeptical of the danger. It is necessary that we learn to coexist with volcanoes and continue to improve methods of eruption prediction because, despite the danger and destruction, volcanoes provide tremendously beneficial services for the environment and for humans that live around them.

Volcanic-rich soils allow for crops to flourish, which is one of the reasons why people live close to volcanoes.[2] For example, many crops are grown near the base of Vesuvius in Naples, Italy. Farmers in the area attest to the unique flavor of their grapes, tomatoes, and wine due to the rich volcanic soil. Volcanoes also form new land; the recent eruptions in Hawaii formed an entire mile of new land. Volcanoes provided beneficial gases early in Earth's history to create a greenhouse effect, which warmed the planet Kay & Fuiten (this volume) and added amino acids to the Earth's surface. The release of these gases allowed for life to develop. We need a greenhouse effect to stop too much solar radiation from coming down to the surface of the Earth. This is just one way that volcanism is related to climate change.

Kassandra Do you think students are interested in climate change?

Sedelia Oh definitely, my own son was already being taught about climate change in the first grade. I think that students nowadays are so aware of it—they might not look into it too much, but they know enough to be concerned. They are also aware that certain areas, for example, low-income neighborhoods are affected more by the changes in climate and pollution.

Kassandra What effect do volcanoes currently have on Earth's climate?

Sedelia Historic eruptions by stratovolcanoes have had short-term effects on Earth's climate with effects waning within a few years. These volcanoes typically cause cooling effects. For example, in 1991, the eruption of Pinatubo in the Philippines lowered global temperatures by about one degree Fahrenheit. Within 2 years, the temperature returned to normal. Most volcanic eruptions do not affect the global climate significantly. Generally, an eruption has to be very large to have a global impact. Additional factors contribute to climatic changes, these include, the volcano's location on Earth, where equatorial eruptions have been known to travel across the globe, and plume material has to reach the Earth's stratosphere to effec-

---

[2] Delmelle et al. 2015.

tively disperse sulfur aerosols globally. Most volcanoes do not have any effect on climate, even in the short term.[3]

Kassandra Do volcanoes play any role in climate change? In other words, how valid is the argument that volcanoes contribute as much or more to climate change than human activities like burning fossil fuels or deforestation?

Sedelia Some people will say that volcanoes emit greenhouse gases (CO_2, methane) that could affect our climate. However, the amount of gases emitted by volcanoes on a yearly basis is not nearly enough to cause the climate change that has already happened in recent decades on Earth. Volcanoes that have been monitored recently and in the near past have not shown any levels that would contribute to global warming. Airplanes have more carbon dioxide emissions than volcanoes do today. Prehistoric eruptions, such as flood basalts, on the other hand, have been linked to mass extinction events. During these events, global climate has been affected significantly, many times leading to warming periods. In addition, we have seen ocean acidification conditions and tremendous loss of both flora and fauna.

Kassandra What's the big difference between the volcanic eruptions that occurred early in Earth's history and had such a strong impact on climate versus the more recent volcanic eruptions that have little effect on climate?

Sedelia As I mentioned earlier, in the Earth's past, it has had large igneous provinces or flood basalts eruptions, which are large eruptions of lava in relatively short periods of time, most of which formed large plateaus over much more extensive areas of land as compared to much more localized impacts of contemporary active volcanoes. A great example of these is the Siberian Traps.[4] This continental flood basalt eruption was the largest eruption event in the history of the Earth. Approximately 4 million cubic kilometers of lava erupted within a million years, a short period of time, geologically speaking. Other researchers and I are looking into the effects of the Siberian Traps because, coincidentally, one of the largest extinction events that we've seen occurred around the same time. We haven't seen this type of massive eruption for about 17 million years, since the Columbia River Basalts erupted. This eruption covered parts of Washington, Oregon, Idaho, and Nevada. It would take one of these huge eruptions to significantly impact climate on a global scale. In other words, we can't blame volcanoes for our current global crisis but we can learn about their effects in order to mitigate future disasters.

Kassandra How can volcanoes help cool the climate?

---

[3] Robock and Mao 1995.
[4] Rampino et al. 2017.

Sedelia When volcanoes erupt, they emit plumes of gases, ash, and rock. These gases, if erupted high enough into the stratosphere, can react to form aerosols. Sulfur dioxide (SO_2) in particular can form aerosols that can block some sunlight from reaching the Earth's surface. This creates a cooling effect. Some scientists have proposed monitoring the next volcanic eruption closely to see what happens to SO_2 in the atmosphere before and after it mixes with water vapor to form aerosols. They also have proposed artificially injecting the atmosphere with such gases or other materials that reflect sunlight in order to slow down global warming.[5] But, the idea is very experimental. Most scientists disagree with these methods of geoengineering because we do not know what the side effects would be of doing this.

Kassandra How does this process work, more specifically?

Sedelia What they would do is fly a plane up to the stratosphere and release different particles that can reflect sunlight, just like sulfur aerosols do. These particles would theoretically block sunlight, allowing the Earth to cool. It would obviously have to be a very large area to have a significant impact on the Earth's temperature. This is known as solar geoengineering.[6]

Kassandra What are the ethical considerations concerning solar geoengineering?

Sedelia The big concern is that we don't know what the unintended consequences of releasing these particles would be. Could there be chemical reactions in the atmosphere that we did not anticipate? How would this affect people, animals, and vegetation around the world? Would there be changes in weather patterns, decreased rainfall, or would this process destroy the ozone layer (SO_2 aerosols) and warm the stratosphere? In addition, there could be an unexpected "runaway" effect Kay & Fuiten (this volume) if the process could not be stopped. Even if scientists are careful and conduct small-scale experiments, they cannot account for the long-term, large-scale side effects that could have world-wide consequences. Finally, this process does not reduce CO_2 emissions and some scientists worry that some governments would focus on this method instead of continuing to reduce CO_2 emissions. There are just too many uncertainties at this time.

Kassandra You mentioned in an earlier conversation that you talk about geoengineering with students in your seminar. How can asking students to debate the costs and benefits of geoengineering allow them to connect more deeply with the course material?

Sedelia Geoengineering is something that could definitely affect them since this would be a global event. The students in my class were already concerned about

---

[5] Gibbs 2019.
[6] Kumar and Karmakar 2020.

climate change, therefore contemplating the effects of such an endeavor made them think "wow, this could affect me or my family or my home directly." They were very animated and opinionated about this, and most of their opinions about geoengineering were negative.

Kassandra Since you're teaching a First-Year Seminar, do you ever hear about students using what they learned in later classes or in their communities?

Sedelia The connections that students make between their own experiences and their coursework can have a profound impact, both on the students and their local environments. One particular student, after studying geology, found out that the area where she went camping as a child, has rocks that are ideal for carbon sequestration (Baxi et al., this volume). She went back and studied these particular rocks for her senior thesis project and looked into the possibility of doing some carbon sequestration in that area. After graduating, she moved back to her hometown and is now teaching about the geology of the rocks among other topics at an elementary school in the area. This student's experience and interests prompted her to explore intellectually and do something beneficial for her environment and community. That is the power of tying students' learning to their lived experiences.[7] I want my students to learn for the sake of learning and possess an authentic curiosity about the world, but I also hope that they take this knowledge and do something beneficial with it, especially considering current environmental issues.

Kassandra Why do you think it is important for your students to draw a connection between sustainability, climate change, and volcanoes?

Sedelia In order to truly understand climate change, one needs to understand the Earth. A multitude of factors influences climate change: ocean circulation, solar heat, sun spots, arctic winds, tectonic plate movement, and volcanoes. Some of these factors influence short-term climate change, such as volcanoes, whereas others, such as tectonic plate movement, influence long-term change over millions of years. All of these factors interact with one another in complex ways and the effects of each can be difficult to separate from one another. These complex interactions make it incredibly difficult to create accurate models; when any of the factors fluctuate unexpectedly, it can throw off the model. And, because there are so many natural factors, climate change skeptics can easily blame the influence of natural factors on our current accelerated climate change. A thorough understanding of the Earth provides one with the power to respond to skeptics and recognize the difference between natural and man-made climate influencers. In order to understand the Earth—and how to live sustainably with it—one needs to understand the science and its history, and volcanoes can provide much of that information.

---

[7] Price 2018.

References

Baxi, K., Goldmark, S., Stute, M., Jurich, J., Chen, L., & Wang, O. (this volume). Panel Four: Design decarbonization, March 30, 2021. In M. S. Rivera Maulucci, S. Pfirman, & H. S. Callahan (Eds.), *Transforming sustainability research and teaching: Discourses on justice, inclusion, and authenticity*. Springer.

Delmelle, P., Opfergelt, S., Cornelis, J. T., & Ping, C. L. (2015). Chapter 72: Volcanic soils. In *The encyclopedia of volcanoes* (pp. 1253–1264). https://doi.org/10.1016/B978-0-12-385938-9.00072-9

Fecht, S. (2018, June 26). *Experimental forecasts could help Guatemala recover from volcanic eruption*. https://news.climate.columbia.edu/2018/06/26/volcan-de-fuego-experimental-forecasts/

Gibbs, A. (2019, September 11). An umbrella to combat warming. *The Harvard Gazette*. https://news.harvard.edu/gazette/story/2019/09/harvard-groups-research-planet-cooling-aerosols/

Kay, L., & Fuiten, K. (this volume). Teaching about climate change from an astronomical perspective. In M. S. Rivera Maulucci, S. Pfirman, & H. S. Callahan (Eds.), *Transforming sustainability research and teaching: Discourses on justice, inclusion, and authenticity*. Springer.

Kumar, K. H., & Karmakar, M. K. (2020, January 20). Solar geoengineering. *Encyclopedia of Renewable and Sustainable Materials, 3*, 751–758. https://doi.org/10.1016/B978-0-12-803581-8.11009-4

Price, D. (2018, March 23). *Laziness does not exist*. Medium. https://humanparts.medium.com/laziness-does-not-exist-3af27e312d01

Rampino, M. R., Rodriguez, S., Baransky, E., & Cai, Y. (2017). Global nickel anomaly links Siberian Traps eruption and the latest Permian mass extinction. *Scientific Reports, 7*, 12416. https://doi.org/10.1038/s41598-017-12759-9

Robock, A., & Mao, J. P. (1995). The volcanic signal in surface-temperature observations. *Journal of Climate, 8*(5), 1086–1103. https://doi.org/10.1175/1520-0442(1995)008<1086:TVSIST>2.0.CO;2

Sedelia Rodriguez is a Senior Lecturer in Environmental Science at Barnard College who investigates the connection between the eruption of voluminous lava flows (Large Igneous Provinces) and their connection to mass extinctions events and climate change throughout Earth's history. Dr. Rodriguez is also the Assistant Director/Academic Coordinator for the Science Pathways Scholars Program (SP)².

Kassandra Fuiten graduated from Barnard College in 2016 where she studied Elementary Education and Psychology with a special interest in Attention Deficit Hyperactivity Disorder and Autism. Post-graduation, Kassandra earned CELTA certification to teach English as a foreign language and currently works with elementary English language learner students in Madrid, Spain.

Chapter 19
Perspectives on Teaching Climate Change: Two Decades of Evolving Approaches

Stephanie Pfirman and Gisela Winckler

19.1 Setting the Stage

Since 1996, Columbia University and Barnard College have required that all under-graduate environmental majors take a course called "Earth's Environmental Systems: Climate." (EES: Climate). Over time, the EES: Climate course content has changed, reflecting changes in the science as well as shifting perspectives of both faculty and students on critical aspects of climate systems. As one of the longest, continuously running, required higher education courses on climate, mapping these changes within the larger context of the evolving climate change issues is valuable. To set the stage for understanding our changing perspectives we first step back to take a broader look at the climate change issue.

Over the past two decades there has been a major shift in the academic community's approach to teaching about climate. At first, since most scholars working on climate came from the physical and earth sciences, we focused on the Earth's natural energy balance, then increasingly we addressed how human activities are altering that balance (See Fig. 19.1). From there the academic community began to explore the impacts of climate change, which raised questions of justice: emissions are primarily from developed countries, but developing countries will face major impacts without resources to manage them (i.e., Rosenzweig & Parry, 1994). That analysis naturally leads into actions that could be taken to reduce warming. Actions imply behavior change, on the part of individuals, governments, businesses, etc.

S. Pfirman (✉)
College of Global Futures, Arizona State University, Tempe, Arizona, USA
e-mail: spfirman@asu.edu

G. Winckler
Lamont-Doherty Earth Observatory, and Department of Earth and Environmental Sciences, Columbia University, New York, NY, USA

© The Author(s) 2023
M. S. Rivera Maulucci et al. (eds.), *Transforming Education for Sustainability*, Environmental Discourses in Science Education 7,
https://doi.org/10.1007/978-3-031-13536-1_19

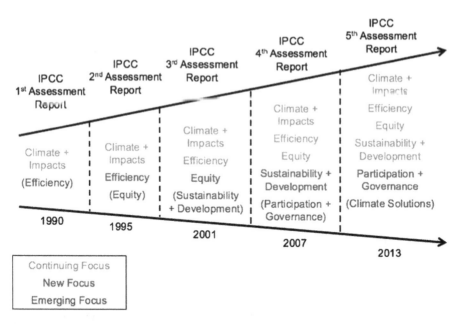

Fig. 19.1 Evolution of the perspective of the academic community on climate change, as reflected in the assessment reports of the Intergovernmental Panel on Climate Change (IPCC); modified and updated from Tariq. (Banuri et al., 2001)

Each of these changes in framing required broadening of the disciplines involved, for example, biology is needed to understand impacts, and the social sciences and humanities are required to consider the values that are behind the choices that lead to decision-making and action. Because all of the people who taught the EES: Climate class were trained in the natural sciences, we had to decide what was authentic to us in terms of presenting and discussing material beyond our professional training (see also Chap. 6 Rivera Maulucci, Pathways).

At the same time that perspectives in the scientific community were changing, so were public perspectives, but in a different way: towards politicization of the issue. In the mid 1990's when we first started teaching the class, surveys indicate that about 40–50% of those surveyed thought that scientists believed that global warming was occurring (Fig. 19.2). While there was a divide along party lines even then, this polarization has grown over time to 87% of Democrats and only 44% of Republicans now thinking that global warming is occurring (Gallup, 2021). Over time, expressing concern about the climate issue effectively became seen as expressing a Democratic political view (Dunlap & McCright, 2008; McCright & Dunlap, 2011). "… [P]articularly in 1997 when the United States signed (but did not ratify) the Kyoto Protocol on reducing carbon dioxide emissions, conservatives began to critique not only the proposals for reducing carbon emissions but also the evidence

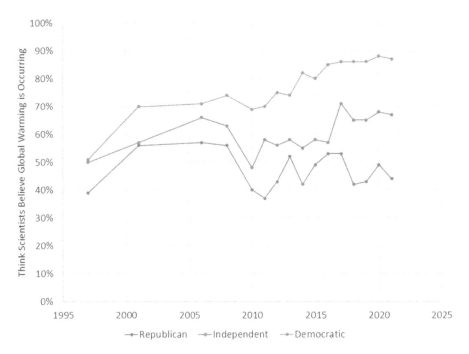

Fig. 19.2 Percent of the public that thinks scientists believe that global warming is occurring. Note increasing consensus until 2008 after which politicization takes place and the curves separate along party lines. Gallup data reproduced with permission (10/29/2021)

for global warming itself" (Dunlap & McCright, 2008, p. 1). Furthermore, Lawrence Hamilton (2008) found "… that ideology can be a powerful filter. More educated people have greater confidence that they understand climate change science, and that this science supports their political predispositions." (p. 677).

This politicization and polarization is felt by teachers in classrooms at all levels, from primary and secondary schools through colleges and universities. Responding to an online poll by the National Science Teachers Association, 82% of science educators answered that they had faced skepticism about climate change and climate change education from a student (NSTA, 2011).

It is a longstanding norm that scientists avoid advocacy in both research and teaching. Edwin Seligman et al. (1915), in the Report of the Committee on the American Association of University Professors on Academic Freedom and Tenure, stated that

> The teacher should be especially on his guard against taking unfair advantage of the student's immaturity by indoctrinating him with the teacher's own opinions before the student has an opportunity to fairly examine other opinions upon the matters in question. (p. 117)

For many of us teaching about climate change, starting around 2007, we realized that we needed to change from explaining how and what we understand about what is happening, to addressing what we can do about it. We were hearing from students that they wanted to know more about actions, and we were asked to provide action-able advice related to our research as well. While most of us did not realize it at the time, this represented a shift from our scholar role to a scholar-practitioner role. For example, a report on the legacy of the 2007–2009 International Polar Year (IPY), stated,

> While the community survey run by this committee revealed concerns when there is a 'mix-ing of advocacy with science' there is a growing sense in the polar research and education community that IPY has made 'knowledge to action' a proper domain for scientists—in the past it was often disregarded as 'activism.' (National Research Council, 2012, p. 102)

However, as seen in these quotes, while 'knowledge to action' and practice grew in community acceptance, advocacy and activism still raise concerns.

Beyond the individual, activism is an issue of concern for the entire scholarly community. Government funding agencies, such as the National Science Foundation, are prohibited from using federal funds for lobbying. However, during the study period back in 2012, according to Brett M. Baker, the NSF Assistant Inspector General for Audit, who analyzed potential advocacy in climate education proposals,

> there are no such restrictions pertaining to the use of federal funds for public policy advo-cacy that falls short of affirmative efforts aimed at influencing legislation. As discussed below, NSF does not have any Foundation-wide restrictions pertaining to public policy advocacy. (p. 1)

Thus, Baker (2012) found "a lack of policy, guidance and criteria on this issue" (p. 1). Furthermore, concerning the specific topic of climate education, Baker noted, "it is not clear how one would know the difference between delving into advocacy for a particular response and presenting evidence so that an informed decision can be made" (p. 2). Thus, at the time, the climate education and research community was left without clear-cut guidance for the difference between knowledge for action, informed decision-making, and advocacy.

While scientists teaching about evolution have long had to face these issues, the rest of the scientific community was largely distanced from being confronted with questions about their opinions or positions versus facts or a scientific consensus. For those teaching about climate, this was an unusual and sometimes uncomfortable situation. We realized that an inclusive classroom entails openness to diverse per-spectives and values, which means that we would have to engage in negotiating conflict, if not within the classroom, then with respect to the larger politicized context for the issue. As articulated by Jonathan Haidt (in Sara Wanous, 2019), people on the political left value issues such as intergenerational equity and are concerned about the vulnerability of the global poor, while those on the right resonate more with stewardship and patriotism. Those of us working on climate change needed to learn how to address these differences with authenticity—congruent with who we are and our own values—both in the classroom and beyond. Here we explore how faculty who taught EES: Climate over the years responded to this shifting landscape of both external context and internal responses.

19.2 History of Earth's Environmental Systems: Climate

In 1994, Columbia University faculty from the Department of Earth and Environmental Sciences joined with faculty from Barnard College's Department of Environmental Science to create three new integrated Earth System introductory courses: EES: Climate, EES: Solid Earth, and EES: Life. With support from a grant from the National Science Foundation and the Columbia Vice Provost at the time, Michael Crow (now President of Arizona State University), a team of faculty brainstormed ideas for the content and approach for the three classes. This process included adapting an online global data access tool, originally developed for research (Hays et al. 1998, 2000), to form the backbone of the corequisite 3 hour labs.

All Columbia and Barnard environmental majors are required to take these classes and the labs. In addition to majors, some students take the classes to fulfill the minor concentration, their general education requirements, or for personal interest. Enrollment increased gradually in 2006–08, and more dramatically in 2009, from about 35–50 to more than 80 students per year (Fig. 19.3). While we are not exactly sure what caused this increase, in 2004, Columbia University introduced a new core requirement called "Frontiers of Science." Part of this one-semester class addressed the issue of climate change, and all students were required to take an additional science class as well. Once students were introduced to the subject of climate change through Frontiers, they may have chosen to fulfill the second part of

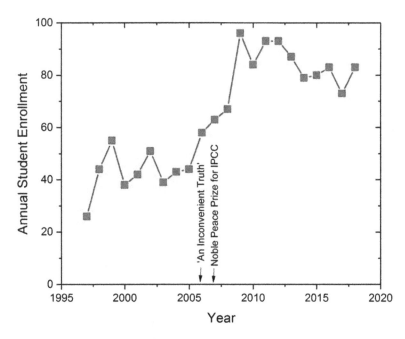

Fig. 19.3 Annual enrollment in Columbia/Barnard's Earth's Environmental Systems: Climate course from initiation to 2018

their science requirement through the EES: Climate class. Other factors that might have contributed are the 2006 release of Al Gore's "An Inconvenient Truth" and the dramatic Arctic sea ice decline in summer 2007, as well as the awarding of the 2007 Nobel Peace Prize to the Intergovernmental Panel on Climate Change, which caught the attention of scientists and students on our campuses and elsewhere.

The goals of the class are to provide the fundamentals needed for understanding the Earth's climate and the climate change issue. We focus on providing physical explanations rather than mathematical derivations of the climate system laws. There is a three-part structure to the class, where first students explore the role of the atmosphere, then they explore the role of the ocean, and then in the last section, they develop an integrated understanding of climate with a focus on climate change (Fig. 19.4). It is this third part, the last section, which has experienced the most change over time. In 1997, the last 2 weeks of the course started with teaching about greenhouse gases and ended with modeled projections of the impacts of future climate change. By 2011, these topics doubled to 4 weeks, including impacts on food and health and mitigation as well as adaptation. As of spring 2018, these issues were covered over 5 weeks, representing 38% of the course (Fig. 19.4).

The increasing emphasis in this third section of the course on future projections and adaptation, as well as choices regarding mitigation, reflects what is felt to be important by the climate community (Fig. 19.1). Also, there was pressure from students to connect the class more with actions. For example, a student responding to a spring 2006 course evaluation noted:

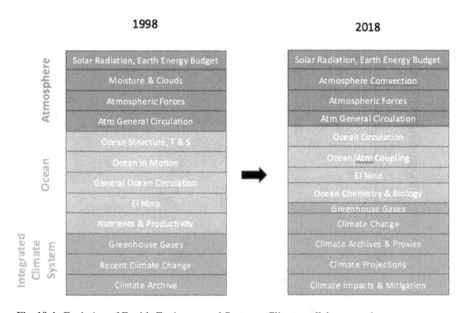

Fig. 19.4 Evolution of Earth's Environmental Systems: Climate syllabus over time

> It would be nice if the atmosphere and ocean portions of the class include[d] a bit more information on what type of legislation and what is being done to protect these systems to keep them working in the future.

As we discuss below, and as the NSF Inspector General noted above, because these topics connect with choices through policies and actions, they have the potential to be perceived as political advocacy or activist.

19.3 Student Perspectives

From the students' perspective, this class has stood the test of time. During senior exit interviews, when we ask our undergraduate environmental majors, "what was the most valuable class you took as a major?" EES: Climate is consistently the course that comes to mind. When asked why it was the most valuable class, students state that it has a good balance of breadth and depth and lays a strong foundation for their future studies.

Analysis of 845 student evaluations from the 28 semesters spanning Spring 1999–2017 (data are missing for nine semesters) indicates similarly positive student experiences. From the early days of the class up to 2017, they note in course evaluations that they appreciate applying what they learned in this class both to their other coursework and to their daily lives. For example, a student wrote, "I loved applying principles of science to the real world, and learning all about the climate." (Fall 1998, Casey, pseudonyms used for students throughout). Another student explained:

> In this course, I was able to learn a lot about how Earth's Climate System works, and how there are many factors and phenomena that play into effect. I was also able to become very familiar with basic physics and chemistry concepts for me to better understand the class. I am now able to tell others and teach others about why we experience different climatic events and weather patterns. (Spring 2016, Blair)

Both of these responses show how students were making connections between the course content, the real world, and their daily lives, even to the point of teaching others about "climatic events and weather patterns."

For students, the last section of the course on climate change, global warming, impacts, and mitigation seems to have the most relevance. For example, many students noted that the best part of the class was "the last third—greenhouse gas mitigation and climate change" (Spring 2006, Sidney). One student explained that they appreciated "The more discussion-oriented lectures at the end of the course [and] the subjects that applied to human impacts and future predictions" (Fall 2007, Adan). Part of the interest and relevance for students was that this section of the course focused more specifically on real-world issues. In addition, students appreciated learning about "how and why" actions on climate change were important:

> I learned a lot about how and why climate change is important as well as the different mechanisms which go into climate change. I gained the insight into how and why it is important to mitigate carbon emissions and protect the atmosphere and ocean. (Fall 2015, Lee)

Beginning in Spring 2006 and continuing, we see a few comments (8 out of 845) from students asking for more information about actions. Surprisingly, there were only two student comments regarding politicization or a lack of inclusiveness. Both of these comments were from the 2007 to 2008 transition period described above, where politics began to be increasingly associated with the issue of climate change.

> One thing that I thought of throughout the semester was that … [the] lectures treat climate change global warming as something that will most definitely happen and … [don't] give a lot of time to explain the positions of reputable scientists including people who participated in the IPCC who have different ideas about what the future world will look like. I do not think that … gives a balanced perspective on the topic of climate change. I personally believe that climate change is real but I'm not as alarmist as the IPCC. In … explanations we are made to believe that the IPCC is the sole authority on projecting climate change and that their reports present the definitive idea of what the world will look like in the future. However, there are many other reports by institutions and individual scientists who have different perspectives about climate change. The IPCC itself has a lot of institutional prob- lems that we never discussed, which could have a profound impact on the reports that it issues. … never gave much credence to alternative opinions about climate change and I think that … [they] should because like all other important matters today there is not only one perspective on this issue. I'm not saying that … [they have] to endorse them—only that … [they] should mention that there are lots of people out there who do not fully agree with what the IPCC is saying or who do not agree with the way that the IPCC has worked to publish its reports—including several scientists who have become so frustrated with the single-mindedness of the IPCC that they have quit. All I ask is that … lectures [present] information that is not solely based on the IPCC reports. I understand that they are impor- tant documents and present really important information about science but I think it is unfair and unwise to base so much of the information we learn in this class on the results of a single reporting group. (Spring 2007, Riley)

Regarding our use of the Intergovernmental Panel on Climate Change report as a teaching foundation, while the Third Assessment Report (2001) largely addressed climate change impacts and adaptations, the Fourth Assessment Report of (2007) was associated with taking action to mitigate climate change for development and sustainability (Fig. 19.1). Mitigation is always more politically charged than adapta- tion as it requires being proactive rather than reactive. Another factor is that because of the long-standing media emphasis on balance (Boykoff & Boykoff 2007), what students were hearing outside of the classroom distorted the extent of disagreement among scientists, leading many to believe that there was stronger scientific evidence against projected warming than there actually was. As Boykoff & Boykoff (2007) stated:

> However, one underconsidered factor—the very norms that guide journalistic decision- making—plays a crucial role in the failure of the central messages in the generally-agreed- upon scientific discourse to transmit successfully into US-backed international policy to combat global warming. (p. 12–13)

The same student (Riley, Spring 2007) continued:

> I also think that it is inappropriate to bring up recent political events such as Al Gore's movie without presenting the other side of the argument or presenting the opposition that many scientists have had to both his science and his tactics. Whenever politics was mentioned in this class it was quite obvious that the professor[s] felt a certain way and wished to inform the class about that viewpoint alone. Just because most of the class may happen to agree with you doesn't mean that you can only sing the praises of Al Gore and his personal environmental crusade. Something that I think would be much more memorable, effective and relevant to students' lives would be to analyze the science he presented and then discuss from there without indicating agreement or disagreement with anything that he says. I know that this is a politically charged issue, and it's an issue that is important to bring up but I think that the way it was done this semester was inappropriate because the other side was never presented or given a chance to speak. (Spring 2007, Riley)

Similarly, Jesse Spring 2008, noted:

> The only thing I didn't like … is that … lectures are consistently politically charged and … clearly pushing … [their] own agenda in the class rather than focusing on the actual science. (Spring 2008, Jesse)

These two students were clearly uncomfortable with what they saw as politics entering the classroom. At the time, we—the professors—did not see Al Gore's film as political, but rather as a high-profile attempt to engage the public with the science. Through its inclusion, we thought we were making the class more relevant to the students. This input from students led to discussions and reconsiderations among the EES: Climate faculty about how to handle issues such as the perceived credibility of sources and different perspectives held by different scientists—including us!—as well as our students.

By 2011, the following comment from a student recognizes that the faculty have opinions, and the student does not feel that they are inappropriate to represent but rather wants to convey that we should consider how time in class is used.

> Sometimes … [they] really inserted a lot of … [their] own opinions on the lectures, but not in jarring ways. (Spring 2011, Cody)

The only other comments regarding political aspects of the course that we saw in our review of the 845 student evaluations were in 2016, and they also refer to balance. But this student says that we can move on from making the argument that humans are causing climate change:

> I also felt that the balance of lecture topics was a little off. We spent the entire second half of the course on showing that anthropogenic climate change is happening and that it is caused by humans—while I appreciated the detail, I don't think anyone in the class was in any doubt that climate change is real, and I would have preferred to go into more depth about other climate topics. (Spring 2016, Taylor)

This comment from the same semester is noteworthy in that the student is more interested in the fundamental science, where we started with the class back in 1996, rather than the projected impacts and potential actions that we focused on increasingly over the years.

"The second half of the course is extremely dry. The very fundamental understanding of the climate system we acquire in the first half does not allow for a smooth translation into the largely policy- and forecast-based lectures we had in the second half. The material in this class is obviously hugely important and could inform students' major declarations—and while I understand this kind of policy maneuvering and forecasting is what climate scientists spend a lot of time doing, I would suggest deemphasizing the bureaucratic image. It's a tiresome introduction to a vital subject. (Spring 2016, Shawn)

19.4 Professor Perspectives

Originally Earth's Environmental Systems: Climate was taught by 3–4 professors, with periodic rotation in and out of co-teachers. Thus, as of spring 2018, a total of 32 different professors had co-taught the class. The course is now co-taught in the fall and the spring semesters by two professors each semester. From 1996 to 2008, Stephanie Pfirman taught the critical third part of the syllabus. In 2009, Gisela Winckler took over teaching it and continued through the spring 2018 study period.

19.4.1 Pfirman Perspectives

I (Stephanie) came to this class with a background in climate science through my Arctic research and my experience with public-oriented climate change education. I was co-PI of the first major (>$1 million) National Science Foundation climate education grant, in 1991, for an exhibition on climate change: "Global Warming: Understanding the Forecast" jointly produced by the American Museum of Natural History and the Environmental Defense Fund. The exhibition largely followed the first report of the Intergovernmental Panel on Climate Change (IPCC 1990) and dealt with historic climate change, causes, and impacts of climate change, ending with a section called "What Can be Done." Both Al Gore and Charlton Heston spoke at the exhibition's opening in 1992, reflecting the lack of politics around this issue at the time (Fig. 19.5).

After the exhibition opened, I moved to Barnard and Columbia. One of the first things that I did was to work with Jim Hays at Columbia to develop the three Earth's Environmental Systems courses as a foundation for environmental majors (Hays et al., 1998, 2000). As the leader of the spring EES: Climate class, when the IPCC reports began to emphasize future scenarios, as well as adaptation and mitigation, we expanded their scope in the course (Fig. 19.4).

I also remember students increasingly asking about actions. I have a vivid memory of teaching about projected climate impacts and saying something along the lines that "For the first time in human history, we have a pretty good idea of how our future will play out." Students were completely quiet—and then asked me directly, "What are you doing about it?" My response was to get back into public education—trying to change the public discourse towards action. This public outreach

Fig. 19.5 Charlton Heston and Stephanie Pfirman at the opening of "Global Warming: Understanding the Forecast," American Museum of Natural History, New York, New York, 1992

coincided with the 2007 Arctic sea ice loss. My research is on Arctic sea ice, and when I saw that the September minimum was 23% below the average, I thought to myself, "Oh no! It's going." With such dramatic ice loss, I knew that the Arctic sea ice was likely to be more vulnerable to future warming and that the whole region was in danger much sooner than we had thought it would be.

I recall being shocked to see the two student comments in 2007 and 2008 course evaluations (quoted above) that criticized the political content of the course. As the lead professor, I was responsible for the overall student experience. The co-teachers had agreed that the course needed to address reducing warming through actions, such as switching to non-fossil fuel energy sources. This content must have been what the students referred to as pushing a political agenda. The comments from students really had a defining impact on me as a teacher and made me think about the difference between informing decisions versus advocating. At the same time, I was upset that a co-teacher, when asked what should be done about global warming, said that they were not a politician and so they would leave it up to politicians to make that decision. It felt to me that this stance was abdicating our responsibility as scientists (See also Chap. 6 Rivera Maulucci, Pathways, this volume). Would an infectious disease researcher who was asked what should be done in the midst of the COVID-19 pandemic not talk about vaccines?

When I moved on to teach other classes, these mixed feelings stayed with me. I felt that it was wrong to be political in the classroom, and I wanted to be inclusive, acknowledging diversity in student perspectives. At one point, I had a student in another class who came from a very conservative, fundamentalist religious background. I found myself choosing my words and becoming tense and vigilant not to sound political. I watched all the students carefully to see their reactions. At the same time, to be authentic and true to the science, I felt that I needed to be strong about articulating actions that would reduce impacts.

Reading back through the EES: Climate evaluations in writing this chapter, I was surprised to see that in the entire history of teaching the class, these two comments in 2007 and 2008 were the only ones of the total of 845 responses regarding politicization and lack of inclusiveness. This fact made me wonder if the comments were representative of the changing public discourse or because of the way the class was taught? The comments came just after the release of Inconvenient Truth (May 2006), and when public opinions about climate change also became more politicized (Fig. 19.2). The reference to Gore in the 2007 comment is perhaps the key. This reference was made right during that transition when concern for the climate was just beginning to be thought of as synonymous with being a Democrat. Some students may have thought that expressing the need to mitigate warming was the same as advocating for the Democratic agenda. Perhaps students taking the class after this time of transition accepted a political framing for the issue. Also, given the growing evidence regarding warming, they may have expected when they enrolled in a class about climate to encounter professors referring to the scientific consensus on the issue coupled with the need to take action.

19.4.2 Winckler Perspectives

I started teaching Climate System in the spring of 2009. I had no prior experience teaching undergraduate students other than occasional guest lectures. Like many of my co-teachers in the class, I have a background in science (physics in my case) and no formal education in public climate change education or policy.

When I started teaching the class, the main challenge was introducing a robust and thorough understanding of the climate system and climate change, with minimal math, as the class is open to non-science majors. Though initially challenging, I quickly started to embrace that concept and appreciate how this very idea of a systems class opens up the dialogue within the class and between the students and the professors.

Over the years, we have increasingly tried to incorporate modules in the lectures and lab that encourage exchange and discussion between the students. For instance, I have the class do a homework exercise based on the Princeton Carbon Mitigation Initiative: Stabilization Wedges (Princeton University, 2021). Students pick their high-priority strategies to mitigate climate change and write about the motivation behind their strategies. I compile the data from their responses and use a graphical illustration of their choices to trigger discussions in the 'Climate Mitigation and Adaptation' class. This graphical representation leads to lively and inspiring discussions in the classroom that often extend after the end of class. It also translates the material to the students as relevant. I find these discussions the most interesting part of teaching the class. In the words of one of the students, "I especially enjoyed

Gisela's lectures on mitigation strategies and global health impacts. She effectively got the class to participate in a discussion of different techniques and made this course all the more applicable to the real world."

Reading back through the evaluations, I realize that my perspective has changed. An increasing number of students express interest and demand information beyond 'just' learning about the scientific basics of the physical climate system. They regard climate change as a given and are more interested in comprehending potential solutions concerning policymaking and technological options. This demand has led me to define my teaching role to improve basic scientific literacy and a deeper general understanding of the challenges related to sustainability and potential solutions. It is also my realization (or hope) that while not every student taking EES: Climate may opt to major in Earth or Environmental Science (and in fact, most do not), those who do not may be inspired to turn their attention towards developing decision-making and/or technologies required to mitigate the societal impacts of a rapidly changing climate.

19.5   Other EES: Climate Professor and Teaching Assistant Perspectives

To expand beyond our own experiences as professors in the Climate class, we conducted a Survey Monkey poll in spring 2018 of all professors who taught the class (besides ourselves), plus six Teaching Assistants (TAs), for a total of 36 invitations. We received 20 responses representing a 56% response rate. We asked six content-based questions, and due to several non-answers to specific questions, we received 112 responses. We found a surprisingly wide range of perspectives on what people thought was appropriate to teach in this introductory class. To represent this range, we selected responses categorized on three themes: integration of so-called climate "deniers" perspectives, consideration of issues relating to justice, and how to handle teaching about actions. We present the selected responses followed by a section that summarizes our views about their responses.

19.5.1   Integration of "Denier" Perspectives

Regarding the issue of whether or not to tackle climate "denier" perspectives head-on, some professors and TAs decided to focus on the science, while others addressed this through a consideration of varying values. Most of the comments below are in response to the question, "Is there anything that you would like to share with us about teaching climate change related to the issue of inclusion—for example, reference to "denier" perspectives?"

19.5.1.1   Focus on the Science

- It is worthwhile to discuss why the deniers' claims are unfounded when they are and what issues they raise that are worth investigating.
- I will, where relevant, bring up common denier talking points and explain why they are incorrect or incomplete. The students really like this, I think because they feel ready to engage with friends and family who do have this perspective.
- Every year I conduct a poll (by using clickers) to probe the students' views on climate change before the class. There are always some who are unconvinced that climate change is an issue and that it belongs as a topic in an Earth Science degree. They tend to have a more informed view by the end of the class and enjoy learning about the topic.
- I didn't want to suggest that scientists were "split" on the climate issue like some anti-climate-change advocates try to claim. So I wasn't worried about being too political because it was important for me to show that the science is robust regardless of what is said by politicians.

19.5.1.2   Consideration of Values

- The opposition of the deniers is rarely rooted in a scientific disagreement, but more so in their basic value system—that is where I try to lead the discussion. With respect to deniers who may speak up in a group, I try to discover whether their stance is based on mistrust of information vs. ideology. If it is mistrust of modeling, for example, I encourage them to skip modeling altogether, and just review the historical data in order to draw their own conclusions. If it's ideology, I try (briefly) to see if there's a possibility to find common ground based on values. If it doesn't look like it, I'll note that such a stance has more place in a policy discussion and move on.

19.5.2   Considering Justice

With respect to justice, again, there was a range of responses, with some professors feeling that this was outside of their area of expertise or outside of the appropriate scope of the class. In contrast, others addressed the social justice aspects of impacts and responses, and still others addressed the complexity of the issue. Most of the comments below are in response to the question, "Is there anything that you would like to share with us about teaching climate change related to the issue of justice—for example, differential impacts on vulnerable populations?"

19.5.2.1   Outside class scope or personal expertise

- No, in my day we did not deal with this in the climate class. We just hoped to help the students understand the natural climate system and the possible impacts man could have on it.
- I don't really discuss this much. Most of my classes focus on the scientific issues since that is my area of expertise.
- Important topic, but not one I would consider myself to be an expert in.

19.5.2.2   Social Justice of Impacts and Responses

- I think that quantifying the impacts of global climate change is the most important thing for the issue of justice—for example, quantifying the people who will die from flooding or famine, etc. can put the social impacts into better perspective.
- I mentioned disparate impacts as a social justice issue, but beyond that tried to stay away from anything political.
- That never came up in my part of the course. Certainly, the idea of differential impacts of climate change on vulnerable populations is an important one. But it is only fair to also recognize that responses to climate change will also have differential impacts on different populations, and I think this is one reason for some of the resistance to doing anything about the problem. I'd like to think we can design climate mitigation policies to account for the different degrees of hardship such policies will impose on different segments of the population. But I don't have much confidence that our and other governments will actually take such things into account.
- Engineering fixes might be good for some regions, bad for others. Countries that can control the climate might do so for their own good, not the good of the whole planet. This is a great challenge.
- Should have more weight in the class. I addressed this more explicitly in [an upper-level class].

19.5.2.3   Cultural/Intergenerational Complexity

- …justice is a complex concept that is maybe one that goes beyond this class. Just think of justice in a country, between generations, or globally between regions of different development. And the issue of justice also has adaptation and mitigation dimensions...
- I think it's important not to generalize. All regions, countries, and cultures have different values and combinations of factors that matter with respect to environmental stress, security, and justice. Climate change can be a very personal issue, I feel.

- I'm not teaching in a formal classroom setting now, but when I talk to groups of students, I try to make the issue relatable on a more personal level through a discussion of financial resources. I invite them to think about impacts they've heard about—"slow-motion" (like a drought) vs. catastrophic (like a heavy rainfall event and flooding). How do people caught up in the impact cope, and what is the personal cost? What should people be expected to handle on their own, and when is there a line crossed to where community and/or government help is required? Because it's a discussion, people have the chance to share their opinions, and perhaps think through their stances more with additional input from their colleagues.

19.5.3   Teaching about Actions

We also asked professors and TAs about the degree to which they incorporated actions in their teaching. Most mentioned that while they initially focused on science, increasingly, they taught students to understand actions. One noted that "some students like to talk after class about climate change issues and how we can respond to that." As the issue is developing into a crisis, many have now moved on to also recommending actions. Most people teaching the class are also climate researchers and perceive the issue as important to address. Milman et al.'s (2017) survey of the motivations of interdisciplinary climate change (IDCC) scholars found that "Respondents' decisions to conduct IDCC research are driven by personal motivations, including personal interest, the importance of IDCC research to society, and enjoyment of interdisciplinary collaborations" (no p. #). In fact, "92% conduct IDCC research to address issues of societal importance" (no p. #). Most of the comments below are in response to the questions:

- Has your approach to talking about actions changed since when you first started teaching about climate change in general, not just in this class?"
- In teaching the EES: Climate class, did you encounter any issues regarding teaching about mitigation actions? Did you feel that it was appropriate or inappropriate for you to discuss actions? For example, did you have concerns about being "political" or being perceived as an advocate? Did you consider being inclusive to different perspectives on climate action?"
- Is there anything else that you would like to share with us about teaching climate change in general?"

19.5.3.1   Focus on Understanding Climate Science

- I began my quest to understand the climate system because I found it interesting. Now it's obvious that it is critical in developing a sustainable environment. This makes the understanding an urgent matter...

- When I taught this subject in [another class] I got positive feedback from the students for making it all about the science and leaving out the politics.
- No [teaching has not changed], but this may be because I still feel like a bit of an outsider to the field, so I wouldn't feel comfortable promoting actions as an "expert".
- I would not say that I have concerns about being "political" or perceived as an advocate. I would instead say that it is my strong preference not to use the position of instructor to try to shape political views in a class about science. In my section of the class, I repeatedly emphasize my belief that the students can form and hold whatever opinion they prefer on the decisions that our government and society should make, but that their opinions about climate should be grounded in scientific understanding.

19.5.3.2 Understanding the Science of Actions

- If one sticks to the natural science facts (which this class should) there is enormous consensus as articulated by the IPCC process. If one talks about adaptation and mitigation measures one can show the natural science dimension of these ... again not so controversial. But when going into recommendations for political actions I would think one should be more inclusive ...
- Yes. I have been dismayed at how the perception of the science of climate change has been contaminated by people's views about whether something should be done about it. I try to separate the science from the possible policy response. And for the latter, although I strongly favor action, I am willing to accept the view of others that they may be the losers in any action that is taken to mitigate climate change, as long as they are willing to consider the possible ramifications of not doing anything and to realize that uncertainty about the magnitude of climate change goes in both the "not as bad as predicted" and "worse than predicted" directions.
- I found that the students were very interested in the discussion of future climate change and adaptation vs. mitigation, to the point that some of them expressed the wish that I'd spent more time on that topic. I tried to steer clear of political discussions, and instead relied on information ... to allow the students to come to their own conclusions about what the effects of continued increasing CO_2 emissions might be. I was also careful to note that there is a difference between what we can do to adapt/mitigate climate change, and what communities decide to do, based on risk assessments, financial abilities, other social priorities—where there is certainly plenty of room for discussion.
- Certainly, the specific agreements that I discuss have changed. For example, Paris accords didn't exist. Also, the landscape for renewables (wind and solar) has changed significantly over the years, and my teaching has reflected this.

- Yes, my approach has evolved to focus much more strongly on mitigation and resilience rather than on prevention.
- Yes, I would now much more deeply consider avoiding direct advocacy. Talking about advocacy is something I would consider but I would stop short of advocating for my position.

19.5.3.3 Recommending Action

- I believe the response side should be (more) prominent...maybe by now, it is...
- I feel it is appropriate to discuss mitigation. I presented a range of proposed ideas, including those I didn't agree with. I did try to be cautious about presenting my own opinion but did express my opinion when asked.
- I think of myself as welcoming towards all perspectives on climate *action* and I consider the science to be very clear about the consequences of *inaction*. I no longer think of things as political on that level.
- The impact of anthropogenic climate change is of increasing concern. We have to do something.
- Yes—my focus has certainly shifted somewhat towards solutions, and I don't hesitate to provide my viewpoint after discussing the various aspects of a potential remedy.
- Yes, I had a paper published that got a lot of attention and forced me to more prominently voice my opinions, and this affected my approach.

19.6 Outlook

We found that Columbia and Barnard professors have a wide range of perspectives on teaching about climate change. They are thoughtful in crafting their classroom strategies to navigate a commitment to students and society on this critical issue. We also found that students—mostly—accept this range of engagement and approaches from professors. Why might this be the case? Potential explanations are (1) that the political values of the students largely align with those of the professors or fellow students, (2) students respect the authority and/or authenticity of the professors, and (3) the facilitation style of the professors. Regarding political alignment, April Kelly-Woessneor and Matthew Woessner (2006) found that almost 90% of students believe that they know their professors' political leanings. As the climate became politicized over the years, students who do not align with the largely Democratic association with the issue may not become environmental majors. As noted by Diana Hess (2018), "… my students suggested ... [t]hey had conservative friends who simply avoided courses with an environmental or climate-related theme. ... some also did not want to be in courses that they perceived to have a liberal agenda."

The facilitation style of the EES: Climate professors varied widely, ranging from intentional integration of controversial material to avoidance. Several of the Climate class professors were explicit about leaving politics and political action out of the class. One professor considered varied student responses as a normal part of being a professor, noting, "I receive mixed reviews every year about every aspect of my teaching, including how students feel about these issues." Hess and Paula McAvoy (2009), in their study of high school teachers and students, similarly found a continuum of disclosure of opinions on controversial subjects on the part of teachers, ranging from pure neutrality to sharing views when asked, to explicit disclosures. Alan McCully's (2006) study of community educators reported that such disclosures could come from a desire for authenticity and avoid risking their students' trust.

Several professors commented that they felt that issues like climate justice and recommendations for action were outside their areas of expertise. However, many respondents described an evolving approach to teaching about climate. Sometimes, they learned from experiences gained through their research, other classes, or public speaking that they could handle conflicting responses and become more comfortable talking about other aspects of this issue. As Sabine Roeser (2010) observed, leaving out ethical issues leads to "complexity neglect." Roeser (2012, p. 4) argues that to motivate action, educators and communicators should incorporate "appeals to emotions such as feelings of justice and sympathy for victims of climate change, in the present and future generations."

In 2020, many colleges and universities provided forums to address systemic racism. Workshops and other professional development helped train people to deal with conflict, emotions, and discomfort and improve classroom discussions on values and justice. Perhaps climate educators can learn from these experiences and push themselves and their colleagues to move beyond their comfort zones. If we do not grapple more directly with teaching the complexities of the climate issue, who will? In the words of the EES: Climate course professors:

The issue became more urgent.
… the science more definite.
… the need for action became more pressing.
… we have to do something.

Acknowledgments We acknowledge the faculty, students, and TAs involved in the class over the years, and we especially appreciate the faculty who took the time to respond thoughtfully to the survey. We also thank Julianna Gwiszcz and Daniel Fischer, Arizona State University, for their constructive reviews of the draft manuscript.

Original NSF grant: A New Approach to Earth and Environmental Science Undergraduate Instruction, James Hays (Principal Investigator) and Stephanie Pfirman (Co-Principal Investigator), DUE, #9455688, 1995–1998. Columbia University also provided support.

References

Baker, B. (2012). *Closure of the audit of NSF controls for ensuring that climate change education program funds are not used improperly for advocacy.* Assistant Inspector General for Audit. https://www.nsf.gov/oig/_pdf/Controlsoveradvocacy.pdf

Banuri, T., Weyant, J., Akumu, G., Najam, A., Pinguelli-Rosa, L., Rayner, S., ... & Toth, F. (2001). 1. Setting the stage: Climate change and sustainable Development. Climate Change 2001: Mitigation.[np].

Boykoff, M. T., & Boykoff, J. M. (2007). Climate change and journalistic norms: A case-study of US mass-media coverage. *Geoforum, 38*(6), 1190–1204. https://doi.org/10.1016/j.geoforum.2007.01.008

Dunlap, R. E., & McCright, A. M. (2008). A widening gap: Republican and democratic views on climate change. *Environment: Science and Policy for Sustainable Development, 50*(5), 26–35. https://doi.org/10.3200/ENVT.50.5.26-35

Gallup, Inc. (2021). Percent of the public that thinks scientists believe that global warming is occurring. https://www.gallup.com/home.aspx

Hamilton, L. C. (2008). Who cares about polar regions? Results from a survey of US public opinion. *Arctic Antarctic and Alpine Research, 40*(4), 671–678. https://doi.org/10.1657/1523-0430(07-105)[HAMILTON]2.0.CO;2

Hays, J., & Pfirman, S. (1998). Kites in the gale: Columbia's new environmental education curricula. *Earth Matters, Fall, 1998,* 16–17.

Hays, J. D., Pfirman, S., Blumenthal, B., Kastens, K., & Menke, W. (2000). Earth science instruction with digital data. *Computer Geosciences, 26*(6), 657–668. https://doi.org/10.1016/S0098-3004(99)00101-6

Hess, D.J. (2018). We aren't teaching what students need to know about climate science. *The Chronicle of Higher Education, Commentary,* June 08, 2018. https://www.chronicle.com/article/We-Aren-t-Teaching-What/243623

Hess, D. E., & McAvoy, P. (2009). To disclose or not to disclose: A controversial choice for teachers. In D. E. Hess (Ed.), *Controversy in the classroom: The democratic power of discussion* (pp. 97–110). Routledge. https://doi.org/10.4324/9780203878880

Intergovernmental Panel on Climate Change by Working Group I. (1990). In J. T. Houghton, G. J. Jenkins, & J. J. Ephraums (Eds.), *Climate change: The IPCC scientific assessment* (p. 410). Cambridge University Press.

Kelly-Woessner, A., & Woessner, M. C. (2006). My professor is a partisan hack: How perceptions of a professor's political views affect student course evaluations. *PS: Political Science & Politics, 39*(3), 495–501. https://doi.org/10.1017/S104909650606080X

McCright, A. M., & Dunlap, R. E. (2011). The politicization of climate change and polarization in the American public's views of global warming, 2001–2010. *The Sociological Quarterly, 52*(2), 155–194. https://doi.org/10.1111/j.1533-8525.2011.01198.x

McCully, A. (2006). Practitioner perceptions of their role in facilitating the handling of controversial issues in contested societies: A northern Irish experience. *Educational Review, 58*(1), 51–65. https://doi.org/10.1080/00131910500352671

Milman, A., Marston, J. M., Godsey, S. E., Bolson, J., Jones, H. P., & Weiler, C. S. (2017). Scholarly motivations to conduct interdisciplinary climate change research. *Journal of Environmental Studies and Sciences, 7*(2), 239–250. https://doi.org/10.1007/s13412-015-0307-z

National Research Council. (2012). *Lessons and legacies of international polar year 2007–2008.* National Academies Press. https://doi.org/10.17226/13321

NSTA (National Association of Science Educators). (2011). *NSTA poll on climate change education.* November 10, 2011, https://ncse.ngo/nsta-poll-climate-change-education

Princeton University. (2021). *Princeton carbon mitigation initiative:* Stabilization wedges. http://cmi.princeton.edu/wedges

Roeser, S. (2010). Emotional reflection about risks. In *Emotions and risky technologies* (pp. 231–244). Springer. https://doi.org/10.1007/978-90-481-8647-1_14

Roeser, S. (2012). Risk communication, public engagement, and climate change: A role for emotions. *Risk Analysis: An International Journal, 32*(6), 1033–1040. https://doi.org/10.1111/j.1539-6924.2012.01812.x

Rosenzweig, C., & Parry, M. L. (1994). Potential impact of climate change on world food supply. *Nature, 367*(6459), 133–138. https://doi.org/10.1038/367133a0

Seligman, et al. (1915). American Association of University. Professors, General report of the committee on academic freedom and academic tenure, *Bulletin of the American Association of University Professors*. Accessed January 10, 2022 at http://ilj.law.indiana.edu/articles/AAUP-1915-Statement.pdf

Wanous, S. (2019). Dr. Jonathan Haidt on how moral psychology can inform climate advocacy. *Citizens Climate Lobby*. Retrieved from https://citizensclimatelobby.org/dr-jonathan-haidt-on-how-moral-psychology-can-inform-climate-advocacy/

Stephanie Pfirman is Foundation Professor, College of Global Futures, and Senior Sustainability Scientist in the Global Institute of Sustainability and Innovation at Arizona State University. Until 2018 she co-chaired Barnard College's Department of Environmental Science, with a joint appointment at Columbia University on the faculties of the Earth Institute and Department of Earth and Environmental Sciences.

Gisela Winckler is a Lamont Research Professor at the Lamont-Doherty Earth Observatory of Columbia University and the Associate Director of the Geochemistry Division. She teaches in the Department of Earth and Environmental Sciences at Columbia University. She investigates climate change in the oceans and on continents, on various timescales, and aims to translate her research experience into the classroom.

Chapter 20
Building a Circular Campus: Consumption, Net Zero Emissions, and Environmental Justice at Barnard College

Sandra Goldmark, Leslie Raucher, and Ana Cardenas

20.1 Welcome to Your First Year Seminar: Please Wear Gloves

In the spring of 2018, a group of students gathered around a table in a small classroom tucked behind a stairwell in Milbank Hall. A blue tarp was spread out on the table, and three bags of garbage were spilled out onto the tarp. Many of the students, all first-years, were still close enough to their high school selves that they indulged in exclamations like "eew, gross!" Equipped with gloves and safety goggles, however, even the most squeamish eventually got to work methodically sorting and weighing the contents of the bags, which had been destined for landfill. They determined that 50% of the contents of the 67 pounds of waste could have been recycled, and 21% could have been composted. Many students seemed dismayed at the lack of basic waste sorting. One pulled a Poland Spring water bottle out of a mound of trash and exclaimed, "but this is, like, recycling 101!"

This waste audit was part of a first-year seminar called "Things and Stuff," an examination of the origins and impacts of modern American patterns of consumption. The structure of the waste audit had been adapted from a Waste Management course taught for many years by Professor Peter Bower in Barnard's Environmental Science department. Professor Bower's course examined waste systems on campus and in New York City, and involved, quite literally, digging deep into the trash at Barnard.

S. Goldmark (✉)
Department of Theatre, Barnard College, New York, NY, USA
e-mail: sgoldmar@barnard.edu

L. Raucher
Sustainability, Barnard College, New York, NY, USA

A. Cardenas
Barnard College, New York, NY, USA

© The Author(s) 2023
M. S. Rivera Maulucci et al. (eds.), *Transforming Education for Sustainability*,
Environmental Discourses in Science Education 7,
https://doi.org/10.1007/978-3-031-13536-1_20

347

The results reported by Professor Bower's students corroborated the slightly more creative interpretations of the first-year seminar students, putting specific numbers to what can be casually observed with a glance in almost any waste receptacle on campus: it is a poor showing, with the basic tenets of recycling 101 regularly ignored. Monthly reports from our waste hauler confirm; our average diversion rate hovers at around 21%. By comparison, zero-waste plans typically aim for a 90% diversion rate.[1]

Why were our diversion rates at Barnard so consistently low? Why were our trash cans filled with compostable food and recyclable plastics, despite a campaign to redesign signs, waste-sorting tutorials offered to students, faculty, and staff, and compost bins placed in every academic building and hand-delivered to 26 offices and 3 dorms? Why did our students consistently observe members of the community walk casually up to a waste sorting station and dump an entire tray full of food, plates, and service items, unsorted, into the trash? And, as one student noted, if we can't even recycle properly at this point in the game, what hope do we have of ever coming together to form a collective, coherent response to the full range of climate challenges?

While waste is a common entry point into sustainability learning and planning, it is also often just that – a starting point. As the students enrolled in "Things and Stuff" learned, and as demonstrated by Barnard alumna Annie Leonard in *The Story of Stuff,* once you start to really look at your trash you wind up pulling on a thread that quickly becomes a web – an interconnected nexus of consumption that will snare you in questions of human rights, global manufacturing, product design, human psychology, global inequality, and modern economics. As the first-year seminar students learned, our trash can speak. The students, in addition to picking through banana peels and name tag lanyards, read *A History of the World in 100 Objects,* in which Neil MacGregor argues that the objects that surround us tell an eloquent story of "the world for which they were made," a story that is especially important because histories told only in words and texts are inherently less equitable, inevitably privileging literate societies (MacGregor, 2011, pp. xv–xvi). Even in the highly literate and often over-sharing world of a college campus, we can find truths in the trash that would otherwise go unsaid. When our students started examining the objects in our garbage bags, they opened a conversation not only about what we toss, but what we buy, what we allow ourselves to see and what we hide, and ultimately, what we value. As another of their seminar readings pointed out, trash provides "potent, unique clues about the inner workings of society and country that could be found nowhere else" (Humes, 2012, p. 155). For, as the students quickly saw, the problem in our trash wasn't just a failure of "recycling 101." Yes, we should all stop and recycle properly – but it's not what we need to focus on. As Leonard argues, "Recycling is easy: it can be done without ever raising questions about the inherent problems with current systems of production and consumption, about the long-term sustainability of a growth-obsessed economic model, or about

---

[1] "Since the mid 1990's, international and US organizations have used 90% or more diversion from incineration and landfill as the equivalent of Zero Waste. This is considered Zero Waste even if some materials – 10% or less – would still be wasted; even as the Zero Waste concept includes a no burn no bury epithet" (Zero Waste International Alliance, n.d.).

equitable distribution of the planet's resources." (Leonard, 2010, p. 233). Good design, reuse, and repair are all steps to not just better sort our waste, but to avoid creating it, and the associated emissions, in the first place. But the first step is to understand how the objects we toss – or don't toss – connect us to each other and to the rest of the world.

When we pull on the threads in our waste, we quickly see how our consumption ties us together across the city and the globe. If we truly strive to build a campus that champions diversity and inclusion, we cannot ignore the impacts of consumption, large and small. To cite just three examples of how consumption intersects with equity and inclusion, both near and far: In a 2017 community meeting, Barnard Facilities staff reported their frustration at how people thoughtlessly tossed full plastic cups of iced coffee in the trash – the bags would often break, spilling liquid all over their feet and ankles, and creating additional work; New York City's waste transfer stations are disproportionately located in low-income communities[2] – the more waste we create on campus, the more the impacts of sorting that waste falls on neighboring communities; the low prices American consumers, including on our campus, have come to expect are made possible by low wages paid in manufacturing communities around the world, largely women and people of color in the Global South.[3]

These intersecting impacts and inequities are revealed by our trash, and both in the first-year seminar and in our sustainability programming on campus, we strive to look for an intersectional response, one that will lead to better recycling rates, yes, but that will also support a healthy ecosystem, support our students' access to affordable supplies, save the college money, share resources with our neighbors here in New York City, and reduce emissions of greenhouse gases. For when we look at the *source* of all that trash on the blue tarp, we find that the actions and behavior patterns they represent are a significant source of Barnard's overall emissions. One Poland Spring bottle is not that big a deal. But the indirect emissions from our food, our travel, our purchases, and our waste, collectively referred to as Scope 3 emissions (as distinct from Scope 1, emissions from fuel combusted on campus or Scope 2, emissions from purchased electricity), represent as much as two-thirds of Barnard's total emissions, as we will discuss in more detail in a later section. The charts in Fig. 20.1 illustrate how significant Scope 3 emissions can be

---

[2] Several communities in New York City (most notably North Brooklyn, the South Bronx, Sunset Park and Southeast Queens) have become saturated with privately owned and operated waste transfer stations where waste is shifted from collection vehicles to long haul trucks, bringing thousands of heavy diesel trucks through these communities each day, communities with some of the highest asthma rates in the country. Since the closing of the Fresh Kills landfill (which re-routed most of NYC's residential waste to these same communities), over 75% of the City's entire solid waste stream is now processed in a handful of communities throughout New York City (New York City Environmental Justice Alliance, n.d.).

[3] For an example from just one industry, the fashion industry, which generated $2.5 trillion in global annual revenues before the pandemic (Amed et al., 2021), employs about 75 million people, 80% of whom are women (Fashion Revolution, n.d.). The people who make our clothes usually earn less than a third of what they need to meet their basic needs (Labour Behind The Label, n.d.).

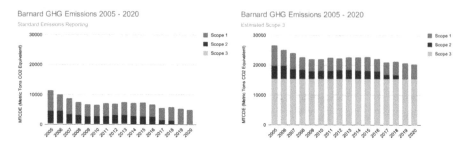

Fig. 20.1 Barnard Emission Sources, 2005–2020, with an estimated contribution of Scope 3 emissions at right

on a college campus. At left are the emissions as reported in a typical protocol, for example to the Mayor's Carbon Challenge in New York City, where Barnard has reported for several years. Emissions from waste are the only Scope 3 category included. At right are Barnard's total emissions with a much more comprehensive estimate of Scope 3 emissions, including travel, food, and purchased goods. These estimates are being developed with consultants from Energy Strategies, as part of Barnard's overall pathway to net zero emissions. The Scope 3 emissions are clearly significant; more importantly, the behaviors associated with them are at the core of how we live as a community, how we spend our money, and what we value.

Given the significant impact, both local and global, of Scope 3 emissions and the social ramifications of addressing the associated behaviors, the questions posed by the garbage spilled over our seminar classroom table pointed us to a response based on principles of justice and circularity. This chapter will explore Barnard's efforts to build a circular campus, an effort spearheaded by the Office of Sustainability and Climate Action in collaboration with multiple campus partners, including student groups, the Office of Diversity, Equity and Inclusion, the Office of Community Engagement and Inclusion, Access Barnard, Finance, Academic departments, Facilities, and students. It will also discuss our work with external partners like Bloxhub, a Danish innovation hub specializing in urban sustainability, the Ellen MacArthur Foundation, a world leader in circularity, and Rheaply, a circularity startup. With these partners, Barnard has developed a Circular Campus framework with five main areas of focus:

1. Reuse and Sustainable Purchasing
2. Design, Construction, and Deconstruction
3. Waste
4. Food and Dining
5. Green Spaces

This chapter will discuss the reasoning behind our approach, share specific case studies from our campus and from other institutions, and make the case for circularity as a dual strategy to reduce environmental impact and support environmental and social justice by increasing equity, affordability, and access for students and neighboring communities. It should be noted that, at Barnard, we are just starting out on our journey towards a circular campus. We do not have all the answers on how to achieve our vision. Our goal in writing this chapter is to share our progress so far and to help other institutions realize their own goals. As such, we describe our process and provide as much "how-to" information as we can, so that others can iterate and build upon our work. At the same time, we write to learn as much as to teach. Circularity, by definition, benefits from diverse inputs, resource sharing, and community.

20.2 Why Circularity?

Barnard's Circular Campus is a holistic, systems-based framework designed to reduce emissions, waste, and costs, transform consumption patterns on campus, increase access and affordability for students, and support the transition to a just, sustainable economy. This framework is helpful in connecting the dots between hundreds of seemingly disparate to-do items in a sustainability action plan, from improving diversion rates to creating healthy green spaces to supporting sustainable and ethical procurement. The framework can be integrated into and support a broader pathway to net zero emissions or carbon neutrality. It's a way to bring together stakeholders who may operate in separate silos but whose work, in fact, is interconnected and interdependent, especially when trying to move sustainable changes through an organization. On a deeper level, building a circular campus is an opportunity to ask ourselves what we really value, and then build a pattern of consumption that reflects those values.

A circular campus is one where, yes, waste is properly sorted – and greatly reduced. But it is also a campus where resources – from textbooks to food to green spaces – are shared, within the campus and with local neighbors. It is a campus where access to those resources is as evenly distributed as possible, by leveraging social interconnectedness. It is a campus where we feel the threads that connect us, through our objects and our actions. It is a system where resources and benefits flow easily to where they are needed, and where "recycling 101" is no longer disregarded, or better yet no longer needed, because habits, norms, and cultural patterns support and reinforce sustainable and ethical purchasing, sharing, reuse, repair, and renewal.

Beyond the campus, circularity has been identified as a critical component of a global climate response. As defined by the Ellen MacArthur Foundation (EMF), a circular economy is a systemic approach "that tackles global challenges like climate change, biodiversity loss, waste, and pollution. It is based on three principles, driven by design: eliminate waste and pollution, circulate products and materials (at their highest value), and regenerate nature." In contrast to the "take-make-waste" linear model, a circular economy is regenerative by design and aims to gradually decouple growth from the consumption of finite resources (EMF, n.d.-a).

The problems circularity strives to address are clear: natural resource extraction and processing make up approximately 50% of total global greenhouse gas emissions (United Nations Environmental Programme (UNEP), 2020). Household consumption accounts for between 50% and 80% of total land, material, and water use (Ivanova et al., 2016). And yet consumption continues to grow, especially in the United States, home to 5% of the world population and responsible for 25% of global consumption (Scientific American, 2012). Globally, since 1970, the population has doubled while the use of resources has tripled, and it could double again by 2050 if business continues as usual (UNEP, 2020). This pattern of production and consumption not only creates carbon emissions, but it also contributes to habitat loss and reduced biodiversity. As the 2021 Biodiversity and Climate Change Workshop Report co-sponsored by the Intergovernmental Science-Policy Platform on Biodiversity and Ecosystem Services (IPBES) and the Intergovernmental Panel on Climate Change (IPCC) argues, the problems of climate, biodiversity, and human behavior – i.e., consumption – cannot be separated and must be tackled holistically:

> Only by considering climate and biodiversity as parts of the same complex problem, which also includes the actions and motivations and aspirations of people, can solutions be developed that avoid maladaptation and maximize the beneficial outcomes. (IPBES and IPCC, 2021)

Faced with these intertwined challenges, the EMF report "Completing the Picture" argues that circularity offers intertwined benefits, both reducing emissions and "meeting other goals such as creating more liveable cities, distributing value more widely in the economy, and spurring innovation" (EMF and Material Economics, 2019).

This concept of multi-layered, shared benefits can be deepened by an explicit and conscious utilization of an environmental justice lens as indispensable to any circular approach. Some critics have argued that the circular economy, especially as conceptualized by the global business community, has not done enough to consider inclusivity, social justice, and antiracism. Raz Godelnik, Professor of Strategic Design and Management at Parsons School of Design – The New School, writes:

> In the past this deficiency could be dismissed with arguments like 'the circular economy would improve air quality and reduce water contamination, therefore it has an implicit social value', but not anymore. In 2020, implicit intentions are not enough anymore – if you believe in something you need to be explicit about it on every level, from design to execution. (Godelnik, 2020)

This important critique invites us to remember that circular principles are nothing new, and can be found far beyond business models, policy think tanks, and economists' conceptions of growth; nature has practiced circularity for thousands of years, as have human beings around the world.

In nature, waste does not really exist. Material by-products from one system are raw materials for another, creating mutually adaptive and regenerative systems that have proven successful for thousands of years. The water cycle is a large-scale example of circularity in nature: in the hydrologic cycle, the sun evaporates the water from the surface of the earth, which then rises as vapor. As the vapor continues to rise in the air, it gets cold and changes back into liquid, forming clouds, which then falls down to earth as precipitation, and is stored in oceans, lakes, glaciers, snow caps, rivers, and below the ground in aquifers. The hydrologic cycle is a continuous cycle that has no beginning or end; it allows the planet's ecosystems to thrive and can serve as a model for human-built systems including water or resource management. On a smaller scale, biomimicry can also offer circular solutions for product design, from waterproof clothing to aerodynamics to packaging. Whether taking inspiration from the water cycle for urban planning to creating food packaging inspired by silkworms,[4] nature can point us to circular, innovative, and collaborative solutions in the design of both products and systems.

We can also learn from cultures that have historically proven to be good stewards of nature. Robin Wall Kimmerer, professor and enrolled member of the Citizen Potawatomi Nation, writes of the power of reciprocity and exchange in "The Gift of Strawberries," demonstrating that a culture of exchange not only helps to distribute resources and share abundance but also creates and strengthens networks and relationships (Kimmerer, 2013). Based on these principles, building a circular campus becomes about more than recycling, waste, or even greenhouse gas emissions; it is about creating a more equitable and just community.

The need for this type of equitable approach is just as clear as the need for global emissions reduction. The links between environmental harm, social inequity, and extractive, linear systems have been well documented. Lisa Woynarksi (2020, p. 56), citing Theresa J. May, writes of the "lockstep between systemic social injustice and environmental lunacy. The so-called natural disaster has exposed the unnatural systems of domination buried under decades of business as usual, unearthing a white supremacist patriarchy through which racism, poverty, and environmental degradation are inseparably institutionalized." But, just as the harms of environmental degradation are linked with inequity, so can the solutions be linked with justice:

> For the greater part of human history, and in places in the world today, common resources were the rule...The market economy story has spread like wildfire, with uneven results for human well-being and devastation for the natural world. But it is just a story we have told ourselves and we are free to tell another, to reclaim the old one... When all the world is a gift in motion, how wealthy we become. (Kimmerer, 2013, p. 31)

The principles of circularity, in very concrete ways, can help distribute benefits and resources more equitably. Garry Cooper, the founder of Rheaply, a reuse platform that Barnard recently adopted, writes that "aside from its environmental benefits, the circular economy promotes resource sharing that will make it easier (and cheaper) for people of all socioeconomic groups to obtain necessary items" (Cooper, 2020). From very different vantage points, Kimmerer and Cooper identify circularity and sharing as a way to distribute resources where they are most needed, one tangible path towards a more equitable and just pattern of consumption.

---

[4] Mori, a Boston-based technology company, develops a protective silk coating product that extends the shelf life of perishable foods to prevent food waste (Mori, n.d.). The edible silk-based coating was developed by MIT Assistant Professor Benedetto Marelli (Apte, 2020).

20.3   Why Higher Education?

According to the Ellen MacArthur Foundation, "if US higher education expenditure were a country, it would be the 21st largest economy in the world" (EMF, 2020); an industry-wide transition to circular systems would have a significant impact in and of itself. But a single campus is much smaller – and a contained economy can also be incredibly valuable as a type of petri dish. A campus is in many ways a microcosm of the larger world, with formal and informal systems for work, living, dining, purchasing, green spaces, waste, and decision making. But unlike larger communities, colleges and universities have significant capacity to influence behaviors through teaching, research, management, and purchasing power. In addition, research and educational institutions have the capacity to experiment, including a built-in corps of people interested in pushing the boundaries of knowledge. This type of small-scale testing ground is well suited to supporting a larger transition, as the circular economy still has plenty of room to grow: the 2021 Circularity Gap Report notes that "100 billion tonnes of materials enter the global economy every year and only 8.6% is cycled back into the economy" (Circle Economy, 2021). Figures like this make Barnard's 24% diversion rate look less discouraging.

In addition to their capacity to develop and test circular approaches as part of sustainability management, college campuses over the previous decades have been a testing ground and incubator of policies designed to support diversity, equity, and inclusion. Taffye Benson Clayton, inaugural VP and Associate Provost for Inclusion and Diversity at Auburn University, writes that "addressing educational inequities has been the focus of institutional diversity, equity, and inclusion (DEI) operations for years. However, the recent triple crisis – the COVID-19 pandemic, the systemic racism in this country, and racial inequities in higher education – has prompted a clarion call for more effective strategies that will result in more equitable outcomes for underrepresented populations by placing DEI at the core of our institutional practice" (Clayton, 2021). The climate crisis presents challenges that are equally daunting and intimately linked with those cited by Clayton. As such, campuses are well positioned to link their work on DEI and climate action. The synergies of climate action and DEI, plus the capacity to pilot innovative educational and operational models make campuses an ideal testing ground for circular systems rooted in justice and inclusion.

20.4   Why Barnard?

So, who were those first-year students picking through the trash, and how might the particularities of the college they chose support or inhibit a circular model? Circularity as a theoretical framework may be very appealing, but as in nature, the specifics of the actual ecosystem will determine real needs and outcomes.

Barnard College is a private women's liberal arts college located on the Upper West Side of Manhattan between 116th and 120th Streets along the west side of Broadway, beside the Hudson River and a few blocks from Central Park, on the traditional land of the Lenape people. Barnard has a total undergraduate enrollment of approximately 2700; nearly 95% of students live on campus (pre-pandemic). The College consists of 16 buildings totaling approximately 1.4 million square feet on four acres, with 251 faculty and 639 staff. Barnard is a self-sustaining entity under the Columbia University umbrella; Barnard is one of the University's four colleges, but it is largely autonomous, with its own leadership, finances, and administration, and its own distinctive approach to historically women's education. Barnard students can take classes, play sports and join organizations at Columbia just across the street, and Columbia students have the same opportunities at Barnard.

Within the context of American higher education, Barnard's location and history make the college a strong candidate for taking a leading role in addressing Scope 3 emissions – the "everything else" after fuel combustion and electricity – and adopting a circular approach. As an urban campus, Barnard has close ties with Columbia University and dozens of other institutions, from near neighbors like Union Theological Seminary, Jewish Theological Seminary, and Manhattan School of Music to those slightly farther afield, like Fordham, the New School, NYU, Brooklyn College, FIT, the CUNY system, plus many more organizations, from public schools to museums to community organizations. This type of rich network is ideal for building robust circular patterns – as in a thriving ecosystem, abundant resources can be moved through the urban system.

In addition, Barnard's location in New York in some ways limits the College's capacity to creatively and aggressively address Scope 1 and 2 emissions; Barnard is tackling these sources, but the College is not in a position to, for example, erect a large solar array on campus. Along with our neighbors, our path to electrification and net zero will be closely tied to developments in city and state energy policies. This does not mean that we will not put resources into tackling Scopes 1 and 2 emissions, but it does mean that we can be most creative and impactful, both in terms of our own emissions and in terms of advancing research and shared knowledge, by taking an innovative and aggressive approach to Scope 3 emissions and circularity.

Finally, Barnard's history as a women's college may provide a special logic for pursuing the historically undercounted and too often overlooked data behind Scope 3 emissions. Since the history of emissions reporting lies in many ways with cities, it is understandable that sources related to the built environment, like electricity and fuel, might be counted first. But this approach may have been misleading when we consider a community more comprehensively; carbon footprint analyses undertaken at Barnard indicate that Scope 3 emissions account for as much as two-thirds of our total footprint (see Fig. 20.1). This significant source of emissions is rooted in consumption-related behaviors – food, purchasing, travel, and other lifestyle-related activities – that are often associated with the "feminine" sphere. These emissions are, perhaps not coincidentally, often dismissed as less important in terms of climate action, despite representing a significant contribution to overall emissions. Feminist activists have criticized overly narrow climate responses as inadequate to the complexity of the problem. Bianca Pitt, the co-founder of She Changes Climate, called explicitly for more balanced gender representation at the recent COP26 conference: "Agriculture, land use, fashion and carbon removal are not on the COP agenda… If you are going to work internationally to find climate change solutions and to pressure countries for better NDCs [Nationally Determined Contributions], it's so important that women are there to provide more lateral thinking" (Palese, 2021).

The intertwined and complex nature of Scope 3 emissions has led to a perception that solutions are difficult to undertake. Indeed, as Barnard's case demonstrates, simply measuring these emissions and associated behaviors is challenging. Addressing these complex Scope 3 emissions and associated behaviors necessarily involve both technological solutions and collaborative, community-based approaches. As a women's college with a strong emphasis on STEM and the humanities, Barnard is well-suited to deal with the complexity of the interplay between technological and social solutions to issues of climate, emissions, and circularity. An intersectional and complex problem rooted in the daily behaviors of an entire community cannot be solved with a top-down technocratic solution. Technology will play a role, but policies and solutions must also address behavioral and culture change. It might be possible, for example, to address Scope 1 and 2 emissions almost entirely "behind the scenes" at Barnard; the administration could theoretically change out all of our boilers, switch to renewable energy, and replace all of our light bulbs with LEDs, and most of the students and faculty would not even notice. A consistent and effective approach to questions of dining, purchasing, and travel, however, will require a blend of technological and policy-based solutions and a coordinated and socially inclusive process. A school like Barnard, with strong science and design programs, a strong commitment to social justice, and a tight-knit community, is well suited to build a comprehensive and inclusive approach to tackling Scope 3 emissions – and building the circular systems to do so.

Increasingly, campus climate action leaders are undertaking a more thorough approach to Scope 3 emissions. For example, to date, Columbia has reported only selected Scope 3 emissions. However, the University's recently released Plan 2030 commits to a much more comprehensive assessment and reduction of Scope 3 emissions: "It is Columbia's intention to set targets for all material sources of Scope 3 emissions that are in line with the science-based concepts and targets described above for Scope 1 and Scope 2 targets" (Columbia University, n.d.). With the climate change clock ticking loudly, we no longer have the option of undercounting a whole range of behaviors and associated emissions.

20.5   What, Exactly, Is a Circular Campus?

Given the specific ecosystem that is Barnard, located in New York City, what would the circular campus "holistic framework" actually look and feel like? How might it work? What will a healthy, just, and equitable circular campus be like 5 or 10 years from now?

In some ways, to imagine a circular campus, we do not have to look very far, nor do we have to imagine something entirely new. Instead, we can look to existing circular systems and imagine scaling and expanding them. The longest-standing and most familiar example is, of course, a cornerstone of educational institutions: the library. A library represents the original campus circular economy. Valuable resources are carefully stewarded and equitably shared with all members of the community. Inputs into the system (new books and tools) are regular and carefully chosen to diversify the total resource pool and enrich the intellectual life of the community. Books and equipment are repaired and maintained; Barnard repairs approximately 100 books in-house every year, and regularly services AV equipment and technology. Outputs are also made thoughtfully – in Barnard's case, used books are sent to Better World Books or recycled whenever possible.

In addition to these existing and already largely sustainable practices, in an ideal circular campus library of the future, all inputs of new books and equipment would be sustainably and ethically sourced, and outputs carefully tracked for re-homing with local communities, repairing, or other means of maintaining value. These principles, already familiar in the case of the library, might be expanded to include construction and renovation projects which use existing materials from elsewhere on campus or nearby community projects, energy systems that create renewable energy and send surplus back to the grid, dining halls that produce zero waste and where no students go hungry, grounds and buildings designed to capture storm water, even recapturing human waste for energy or fertilizer in green spaces filled with plants that support pollinators. Technology is shared, repaired, and when eventually obsolete, sent back into a remanufacturing process. Resources are shared among students, employees, and community neighbors, including supplies, access to green spaces, and food. A circular campus in the future is one where students, faculty, and staff are taking active climate action, and sustainability, circularity, and justice are built into courses across the curriculum and the college's systems and operations.

One salient characteristic might be things that are *not* there: perhaps a circular campus simply does not have the trash cans that dot our current campuses at every entry point, every lobby, every hallway. Waste haulers are no longer paid to remove a loosely differentiated mass of materials by the ton. Instead, outside vendors are contracted to redistribute a much smaller amount of materials and goods – only those that can no longer be used and reused within the campus. Purchases of new goods are extremely rare, and when undertaken, are carefully thought through in terms of sustainability and ethical production. Reuse and repair are the status quo, supported by a rich range of logistical and economic systems. Perhaps the Barnard Store no longer sells new teddy bears and mugs, but the Barnard Reuse and Repair Center restores teddy bears or resells gently used mugs for each generation of students. The vision might seem too utopian, but the models already exist all around us, in natural systems, in human social systems; we will discuss several specific campus examples later in the chapter. These circular systems do not need to be invented – they need to be seen, named, resourced, and scaled. As with our current libraries, the defining characteristic of a circular campus is based on stewarding abundant resources and distributing them widely and equitably.

20.6   Barnard's Circular Campus Framework: Beyond the Trash

In some ways, our circular journey started in the trash, with our students. It was greatly accelerated by two studies indicating that as much as two-thirds of our total campus emissions came from Scope 3-related behaviors. Through student surveys, we knew that students were struggling to get access to materials for their classes and even dropping out of their majors for not being able to afford specific materials. And every day, the newspapers told us about the inequitable impacts of consumption and climate change on women and people of color around the world. We had seen the importance of grass-roots circular systems in addressing student needs or operational challenges and began to take steps to formalize those systems and expand our understanding of the potential areas of impact.

In 2021, we partnered with Bloxhub, a Danish urban sustainability hub, to further develop the circular principles we had been working on since 2017. Bloxhub members (usually design or architecture firms) and Barnard faculty and staff came together in a series of workshops to identify the challenges, opportunities, and guiding principles for our circular campus, defined as having five main areas of focus: Reuse and Sustainable Purchasing; Design, Construction, and Deconstruction; Waste; Food and Dining; and Green Spaces. The workshops focused on the first three main areas and identified a series of guiding principles and next steps for each area. Creating these principles allowed us to think through, articulate, and discuss internally our current priorities and how the college can achieve them. It allowed us to identify our pain points and translate them into feasible initiatives that foster circularity, equity, and access.

As we zoomed in and out of these areas of focus, we became more aware of the connections and cycles that emerge from these systems. We started to look for what Donella Meadows (1999) calls leverage points or "places within a complex system where a small shift in one thing can produce big changes in everything." For example, centralizing the college's purchases, which Barnard did many years ago, emerged as one potential strategy for supporting sustainable and ethical purchasing. The transition to Workday will streamline purchasing and expense reports, and workshop participants felt that, alongside this new management system, returning to a more centralized process would allow the College to steer users to preferred vendors, like BIPOC-owned local businesses, or reuse on a platform called Rheaply, a partnership we'll discuss in depth later in the chapter. In addition, consolidating purchases significantly reduces costs, waste, and emissions. Other ideas that emerged from the Bloxhub/Barnard workshops also pointed towards supporting more thoughtful end-to-end decision-making, from providing contractors with sustainability guidelines before projects begin, to starting plants for the grounds from seeds grown in our greenhouse, rather than purchasing seedlings from an external nursery. Over and over again, workshop participants sought to find ways to curb the flow of unnecessary consumption on campus and create systems that would support a more equitable and circular culture of decision-making. A series of three case studies illustrate the intersections between circularity, emissions reduction, and equity.

20.7   Three Barnard Case Studies: Reuse, Renovation, and Reallocation

Like many institutions, Barnard has long had an informal circular economy for office supplies, furniture, textbooks, mini-fridges, and all the various accoutrements of college life. From cardboard boxes in the hall labeled "free books" to a lively student marketplace on Facebook to a FLIP library dedicated to supporting low-income students, sharing was everywhere – but still somehow invisible. The years-long efforts of a determined staff member in the purchasing office to "rehome" furniture probably saved the College thousands of dollars and thousands of pounds of emissions. But this type of work was neither accounted for, nor resourced. To elevate and support this invisible ecosystem, in 2020 Barnard began piloting with Rheaply, a digital peer-to-peer internal reuse platform, and officially launched on campus in the fall of 2021.

The Sustainability Office, established in 2017, had been advocating for adopting the Rheaply platform for years with little success (see barriers section below). In the spring of 2019, a student conducted a survey that indicated that access to affordable supplies was a significant barrier to entry for many students considering majors in the arts and architecture. At the same time, many arts departments on campus were concerned about the environmental impacts of materials commonly used in the studio classes. This coincidence of environmental and social concerns helped overcome the hesitance to adopt Rheaply, and, in 2020 the Office of Diversity, Equity, and Inclusion, Access Barnard, and Barnard Sustainability launched a pilot program. Rheaply provides a means for faculty, students, and staff to exchange supplies and materials, and allows the College to track metrics on these exchanges, making the invisible circular ecosystem more measurable. In addition, the reuse platform helps address some of the physical barriers of circularity by obviating the need for centralized storage; users simply hold on to an item until it is claimed. A report published by the EMF stated that for the Massachusetts Institute of Technology (MIT), "using Rheaply for just 6 months resulted in nearly 500 items being shared across campus laboratories and departments, and saved $50,000 in avoided purchasing costs" (EMF, 2020). In the first month of Rheaply's campus platform in the fall semester of 2021, the number of users has increased sixfold, from 100 initial users onboarded during the pilot period to over 600. In the first 8 weeks of the 2021 fall semester, Barnard's Rheaply site hosted over 300 exchanges between community members; the total value of these exchanges was $15,631, and 1.1 tons of waste were diverted from landfills.

The value of these exchanges extends beyond dollars and pounds; sharing builds a harder-to-quantify but equally important element – community. Robust systems of sharing and mutual support have been shown to increase community connections, connections which are a key driver of resilience in the face of natural disasters and social upheaval. The argument becomes circular (pun intended): circularity practices are effective in fighting climate change, and are also effective in building the resilience communities will need in order to cope with climate change.

> By leveraging all the knowledge, power and influence we have to push forward on these priorities we can build a circular economy. It will take deep collaboration between business, government and civil society, but the rewards will be well worth it; a stronger ecosystem that will be resilient for the decades to come, and a world where people and nature can live together in harmony. (Ishii & van Houten, 2020)

These benefits are relevant not only within the campus but in neighboring communities as well. Again, it's a feedback loop; circular systems work better with a rich network of engaged users. Rich engagement supports community connections, builds resilience, and fosters yet more mutual support. Cognizant of these extended benefits, Barnard and Rheaply are actively working to extend the Barnard exchange ecosystem to include our neighbors, including Columbia University, local community organizations, and local arts organizations.

The benefits of circularity extend beyond trash, or even course supplies, to some of the largest "objects" on a campus. In 2020–2021, Barnard launched a $150 million renovation of Altschul Hall, Barnard's 126,000 square-foot science building. Constructed in 1969, the building as it currently stands is a significant energy hog and therefore presents strong potential for electrification: the building is currently served by a dual-fuel system of oil and natural gas that not only heats and cools the building itself but also serves other buildings on campus. Redesigning the Altschul plant will be a significant step towards decarbonizing campus energy systems and is a necessary step towards net zero emissions. However, in addition to the Scope 1 and 2 emissions reduction potential of the building redesign, a circular approach will also significantly reduce the impact of the renovation itself. First and foremost, the decision to renovate and add to the building, rather than tear it down and build from scratch is a strong starting point: according to the Institute for Market Transformation, retrofitting can save up to 75% of a construction project's embodied carbon (Duncan, 2019). A project the scale of Altschul, built with all-new materials, could represent as much as 52,000 tons of carbon consumption (Heath, 2020).

Given the scale of the project and the potential savings across all three Scopes, Barnard developed a set of "Design, Construction, and Deconstruction Guiding Principles" to serve as a sustainability guideline for the project. Key principles include prioritizing reclaimed materials, designing for repair, longevity, maintenance, and upgrade, and providing documentation of circular and sustainable design and material choices. For example, metrics required of architects and project managers include tracking the percentage of used or reclaimed materials vs. new materials, percentage of certified sustainably and ethically produced new materials, and percentage of demolition materials recycled or rehomed.

Altschul Hall represents a large-scale example of the power of a circular approach in reducing emissions from a construction project. A much smaller case study, from Barnard's Department of Theatre, illustrates how this decision-making framework, in addition to addressing emissions and material impacts, can also support greater equity. A 2017 study commissioned by the Barnard College Department of Theatre demonstrates that simply by increasing the amount of reclaimed materials used in design and construction, a given production can significantly reduce emissions:

> One show built with roughly 50% used materials, with a $7,000 materials budget, was responsible for the equivalent of emissions from one American home for 6 months. Building the show entirely out of used materials would cut emissions by as much as 75%, and building 'all new' would have doubled emissions. (Gotham 360, 2017)

Following this study, the department designed and implemented policies to track and incentivize reuse, including a new budget structure, publicizing reuse statistics in programs, and introducing sustainability and circular goals during the hiring process for guest directors and designers. As the department began increasing the percentage of reused materials over time (from about 50% in 2016 to 68% in 2019), the intersections between environmental benefits and questions of human sustainability and equity became tangible: when Barnard Theatre began spending less on new materials, the department reallocated that money to more fairly compensate those undertaking the labor needed to source used items. Barnard was able to spend this newfound money on local artisans skilled in sourcing, refinishing, and refining used items. Since 2016 the department has raised design fees by 58% and prop artisan compensation by about 70%. The reallocation of funds formerly spent on new materials constituted 60% of the increase. Once the production team began 'spending money on people, not stuff,' advocating for increased fees and prioritizing the artists involved in circular practices, it became easier to reallocate departmental funding for artist compensation, a major equity issue in the American theatre (Goldmark & Purdum, 2021).

These three examples – reuse with Rheaply, the Altschul Hall renovation, and the reallocation of material budgets towards artist fees in theatre – represent circularity in different arenas and at different scales. Together, they illustrate how a circular framework can be a powerful tool for reducing waste and emissions, and for addressing issues of community, equity, and justice.

It is important to acknowledge that other higher education institutions across the globe are working on developing and implementing their own circular frameworks. For example, MIT has taken "a complex, multi-solution approach" to waste and circularity that includes analyzing "inflows," "stock," and "outputs" across campus, promoting reuse, creating better signs, introducing centralized sorting stations and food collection, and community engagement programs (EMF, 2020). The MIT case study underscores the need for a complex strategy that blends technological and social approaches, a theme that applies to circularity programs designed to address a range of issues. The University of Manchester created a Living Campus Plan which includes high-quality, greenspace with room to learn, think and connect with people; preventative and restorative health benefits; natural resilience to climate change; support for biodiversity; and space, habitats, and corridors for wildlife (The University of Manchester, n.d.). The University of Portsmouth is doing numerous things at a campus level to reduce waste and embed circular practices. Examples include composting of all food waste at the university; the reuse of glass milk bottles for any event catering; cooking oil (a vegetable oil) turned into biofuel by a charity organization (Yateley), which is then bought back by the university to power its catering vans; and the re-use of broken slates that had been removed as part of refurbishment from the university's roof and turned into trays to be used for catering events (EMF, n.d.-b). Falmouth University includes activities such as sustainable printing, campus tree planting, and vegan and vegetarian catering. Falmouth's wildflower areas on campus have increased by 60% (4.5 acres), supporting native flora and fauna regeneration. In addition, they provide discounts on food and drink when students bring their own reusable containers to campus outlets, plastic bottles are no longer offered in vending machines, and food prepared on-site is packaged in biodegradable and compostable packaging (EMF, n.d.-c). As part of Stellenbosch University's ecological sustainability strategy, the South African institution aims to rehabilitate 90% of eroded areas by increasing local, endemic species by 20% and increasing the living microorganisms in the soil of garden areas (EMF, n.d.-d).

One common characteristic in many of these plans, including Barnard's, is a multi-faceted approach to circularity solutions. Some strategies might be top-down, like replacing a boiler or instituting a purchasing policy. Some require technology, like Rheaply. Some solutions are more bottom-up, like cultivating a culture shift towards reuse. By leveraging the power of both toolsets in tandem – the technological and the social – we can plan for and ensure a circular, equitable campus future.

20.8   Challenges to Circularity

Barnard's location in New York may make our campus an ideal testing ground for circular systems, but the city also complicates certain aspects of college life, perhaps best exemplified in the annual high-drama spectacle that is move-out. This springtime ritual represents an opportunity to actively support equity and circularity and also exemplifies many of the challenges typical to making the transition to new patterns of behavior and operations.

Barnard does not store any student belongings over the summer, so each May births a chaotic, frenzied swirl of students packing up clothes, books, bedding, cookware, mini-fridges, electronics, and all the other accoutrements of dorm life, piling them high into wheeled hampers and wobbling down the sidewalks while coordinating pickups with harried parents, who are themselves battling the unsavory realities of New York City traffic and parking rules. This massive displacement of students and their "stuff" results in an enormous amount of discarded goods common to American residential colleges. Campuses all over the country have developed versions of what Barnard and Columbia call "Give and Go Green." In our case, until 2017, students would "drop off" (a better word might be dump) unneeded items in various lounges or other spaces in the residential buildings. Other students often combed through the piles, taking whatever they needed. Of the large quantities of goods that remained, the majority was donated to Goodwill or similar partner organizations. A very small fraction was stored – usually in a small departmental storage space managed by a friendly administrator – and sold in the fall at a Green Sale. Originally managed entirely by students, in 2017 the program was in danger of being canceled due to general mess, stress on the facilities team, and an incident of a student's property getting erroneously absorbed into the "Give and Go" vortex and disappearing. Give and Go Green was an attempt to address a larger problem: an enormous overflow of goods, a grotesque amount of waste, and the pressure to compress the move-out process for 2700 students into a very short period of time. But the program was quickly on its way to being perceived *as* the problem.

A number of campus partners, led by the Sustainability team and Student Government representatives, began trying to find a way to salvage and hopefully improve the program. Our goals were to better manage the flow of goods so that the Facilities team was not overwhelmed with mounds of abandoned items, create a more orderly system for dropoff, and find a way to store a much larger number of items over the summer so that students did not have to buy as much new stuff in the fall. We had some successes, some failures, and many ongoing challenges. We moved the program out of the dorms, though it was always difficult to secure space since move-out happens right around commencement when large spaces are often used for events. We secured offsite storage for the summer so that we could sell more used goods on campus, as opposed to donating to already overwhelmed second-hand partner organizations. In doing so, we increased the revenues from the Green Sale from $600 to $6000. We reserved hours for Barnard staff members to select items to take home; most of the staff who came to "shop" work in Facilities and Dining Service, two of the lowest paid departments on campus. By increasing student staffing, limiting donation hours, and communicating more clearly with the student body, we began to find ways to operate the program that worked better, for students, facilities, residential life, and events management.[5]

Despite these successes, the program remained challenging on many fronts. First and foremost, it's important to remember that Give and Go Green is, at best, a necessarily limited response to a much bigger problem: a system related to on-campus factors like the tight move-out schedule as well as global realities like the low price of new goods, all of which encourage fast consumption and disposal of material goods. Nonetheless, the challenges that besieged our Give and Go Green program are emblematic of the barriers endemic to making sustainability changes, and especially building circular systems. These include (but are certainly not limited to):

---

[5] The Covid-19 pandemic, of course, put this process on hold. We're still not sure how the program will look post-pandemic.

- Space: Our urban campus is limited in the amount of space available to host large events, store items for the summer, or set up a workshop to repair items for reuse.
- Funding: Give and Go Green, when managed by the students, was run entirely by volunteers, who were given meals and a housing extension. The Office of Sustainability is committed to paying student workers a fair wage, so when the program became administered by staff, we worked to provide monetary compensation as well. This meant using our limited budget, securing grants, and increasing the revenues from the sale – while keeping prices low enough to be accessible to students.
- Bandwidth: For all involved, Give and Go Green comes at a challenging time of year. Students are finishing finals and rushing to pack up, staff are preparing for commencement, the Facilities team is clearing out the dorms, and everyone is stretched thin.
- Inertia: This type of program requires collaboration with several other departments who are often very supportive of the idea but understandably hesitant to commit to learning and implementing new systems when they are already stretched to capacity. Habitual patterns are easy to repeat, especially when under pressure.
- Low prioritization: The Give and Go Green program coincides with the high-profile commencement ceremonies, and the fall Green Sale is just one event in a week jam-packed with new student orientation activities. These busy times of year leave little extra capacity for initiatives that are not necessarily perceived as high priorities.
- Cost (or perception of cost): As a relatively resource-strapped college, the potential cost of expanding or building new programs is of course a factor. However, the savings from reuse are rarely factored into decision-making, sometimes leaving cost discussions imbalanced.
- Buy-in: The benefits of a circular economy model are not always immediately obvious to all campus partners. To some, these events are messy and more work than they are worth.
- Inclusion: Many students need similar items, but not all of them have the same opportunities to obtain them. What started as an environmentally-friendly initiative expanded into an attempt to provide necessary items to as many students as possible. This can sometimes create tensions, for example between the need to pay student workers fairly and the need to keep prices low for students.
- Distributed Accountability: Reducing emissions and changing ingrained patterns cuts across operational and disciplinary divisions; assigning accountability for College-wide goals to the appropriate departments can be challenging, especially when accountability is shared.

These challenges are not unique to Give and Go Green, or to Barnard. The University of Manchester identifies similar barriers, and translates them to the organizational scale:

> Lack of awareness and understanding of the CE (Circular Economy) concept, principles, benefits and applicability to the university context; suitable analytical frameworks, data

gathering systems and KPI's [key performance indicators] to identify, evaluate, prioritise, implement, monitor and manage CE solutions; and leadership teams, allocation of responsibilities, stakeholder engagement and effective policies targeting CE as an instrumental strategy for a long-term sustainability. (Mendoza et al., 2019, p. 838)

These barriers to circularity are similar to many typical barriers to sustainability management – often with the added dimension that partners may not perceive the connections between circularity, climate action, and environmental justice.

However, when you begin to think not only about reducing waste and making move-out less of a colossal headache but begin to consciously consider inclusion and justice as well, some of the ways we think about these common challenges can shift. For example, one core principle of environmental justice is that environmental burdens and benefits should be fairly distributed. With a student body with a wide range of socioeconomic backgrounds, the incredible waste of move-out starts to look even more grotesque. Some students are able to abandon room furnishings, duvet sets, and winter coats, while other students struggle to pay for dorm supplies.

In the context of these economic differences in our community, move-out and green sale programs become an opportunity to redistribute high-quality supplies. For several years, Barnard has partnered with Grad Bag, a local non-profit, to collect bedding, rugs, and other supplies. Barnard has hosted Let's Get Ready's annual Transition Day program, where Grad Bag distributed cleaned and nicely packaged bedding to more than 250 first-generation and low-income (FLI) college students from around New York City. Similarly, we reserve a portion of dorm supplies for Barnard's FLI students. Our goal, of course, would be that eventually, *all* students would use repurposed dorm supplies every fall and that none of the annual mounds of puffy vests, colanders, rugs, fans, lamps, mini-fridges would go to landfill. No longer would the college pay for buses to go down to Bed Bath and Beyond on shopping trips; those resources would be deployed to better manage the flow and redistribution of gently used supplies. Students with the highest need are given first access to supplies, and reuse becomes the default for all.

This process demands the participation of many stakeholders, most importantly the FLI students who might benefit most from a Green Sale. A core principle of environmental justice is the importance of including affected communities in building programming. Accordingly, we have worked hard at Barnard to create initiatives in response to student needs and desires, rather than making assumptions. For example, a 2020 student summer cohort at the Digital Humanities Center (DHC) conducted a study "Examining Sustainability at Barnard" that explored concepts of circularity. The project makes clear recommendations, like a laptop lending library and revising the 4-year "expiration" timeline that leads to new computer purchases for faculty and staff. The DHC students underlined the links between circular practices and inclusion, pointing out that "many aspects of the core principles of circular economy and other popular sustainability initiatives have a lot in common with Indigenous sustainability practices," (Burton et al., n.d.) and that naming and acknowledging these similarities must be central to Barnard's approach.

Another instance where students were able to shape programming occurred as the result of a 2020 course called "Change and Climate Change." A student created a project exploring sustainability and access, examining ways that Barnard might ensure that sustainability initiatives are inclusive and shaped by the communities they impact. Regarding reuse, for example, the class discussed how some of them came from communities where reuse was seen as something to move away from as a family income increased, rather than something to embrace. The students themselves were largely supportive of reuse but were careful to note that programming needed to take into account a range of attitudes and backgrounds. The project, and the class discussions that ensued, led directly to a Campus Conversation on Sustainability and Access the following year, where students, faculty, and staff made recommendations on policy and programming. This process of engaging the community in defining priorities and policies is one key strategy for overcoming some of the all-too-familiar roadblocks on the way to a circular, sustainable campus.

20.9   Pathways to Circularity: Starting Small, Thinking Big

In 2018, faced with piles of trash on a tarp, a campus-wide holistic response felt very distant. And it's still true – we have a long way to go. But we have made progress, and, despite the many challenges, that progress is thanks to a few key strategies that can be implemented on any campus:

1. Build on what already exists
2. Find your allies
3. Link to other goals
4. Address inertia and psychological safety
5. Measure progress

At Barnard, we began building a circular campus by expanding or adapting systems that already existed. Our students, faculty, and staff, without using the term circular economy, had been championing the practices for years. "Barnard Buy Sell Trade" is a student Facebook group where students can exchange or sell everything from clothing to textbooks to kitchen supplies. Students led the Give and Go Green process. Faculty members placed used textbooks outside their office with paper signs saying "free books." Business Services staff tried to find new homes for office furniture, on-campus and off. Public safety regularly donated their "lost and found" items to local charities. The Art, Architecture, and Theatre departments, along with Barnard's Design Center, maintained (usually informal) inventories of free shared supplies and tools. Of course, the libraries hummed along as they always had, the original circular campus. And when the Sustainability Office was formed, the faculty and staff leadership team happened to have a passion for reuse, thrift, and repair. These formal and informal activities across campus, from sharing art supplies to working with waste haulers to signing food contracts, are found on many campuses, and when united, measured, and scaled under one ambitious umbrella, can have a significant impact.

Moving beyond the informal, adaptive networks of Facebook groups and cardboard boxes in the hall towards a campus-wide system and culture, however, requires conscious strategizing, including finding early allies, and linking initiatives to other core value programming. Most of the informal systems described above are maintained or supported by individual stakeholders on campus, either out of personal conviction or because the systems serve a real but otherwise unanswered need. For example, as noted above Barnard's Purchasing department undertakes a significant amount of "furniture rehoming," due to a desire to save money and reduce waste. These efforts are significant but largely voluntary and often under-resourced. Finding these champions, celebrating their work, and inviting them to join the process of naming and resourcing these informal systems is a necessary first step. For other campus partners, it may be more effective and is just as necessary to link the work of circularity to other core value initiatives. For example, at Barnard, adopting the Rheaply pilot was seen as a way to reduce waste and also to support FLI students.

Despite the many supporters of sustainable and circular new policies, and despite the perceived benefits for students, moving change through any organization is a significant challenge. The University of Manchester recommends creating "mixed teams of senior managers and operational staff" (Mendoza et al., 2019, p. 841) to ensure successful implementation. At Barnard, we have found that including faculty and students as well as staff of all levels is indeed critical, as is linking the work of circularity to other campus-wide goals and initiatives.

We have identified a number of practices designed to situate our approach to circularity within a larger net zero pathway and overarching social justice goals, and to involve many members of the community. First, the Office of Sustainability regularly convenes stakeholders to discuss challenges and identify goals. The Office was launched in 2017–2018 with a year-long Campus Conversation series, open to students, faculty, and staff. Students are active participants (and often a driving force) on the Sustainable Practices Committee. Partnerships with the Center for Engaged Pedagogy in 2020 and 2021 provided a forum for faculty from a range of disciplines to discuss climate in the curriculum. And in 2020, the Office of Sustainability hosted a Citizens' Assembly and Campus Conversation series. These regular convenings, plus many other related events, help to keep our community engaged and invested.

The at times creative chaos of this engagement strategy is balanced by a conscious process of specific goal setting, supported by outside consultants, and linked to global science-based targets. The 2017–2018 Campus Conversation series fed directly into the College's first comprehensive published Climate Action Vision in 2019, encompassing goals in Academics, Finance & Governance, and Campus Culture & Operations. Three key goals in the 2019 Vision highlight areas where circularity intersects with overall climate strategy:

1. Barnard achieves emissions reductions across all three scopes, in line with or exceeding the ambitious New York state and city standards.
2. Barnard's environmentally sound and socially ethical system of consumption on campus serves as a model for circular economy solutions.

3. Campus initiatives on sustainability, diversity and inclusion, wellness, and other mission-driven values are integrated and mutually supportive.

Circularity provides a tool kit for working at the micro and macro scale, toggling between the mounds of duvets and abandoned mini-fridges at Give and Go Green, and the big picture visions and goals. A key part of keeping the small and large-scale components moving in the right direction is figuring out how to measure progress.

20.10   How to Measure

As we have seen, many of the behaviors and associated emissions that fall under the main components of our circular campus framework fall under "Scope 3." This category of emissions is notoriously hard to measure, and is often, as a result, under-addressed. Conscious of this common gap in emissions accounting, in 2015 Barnard commissioned a small study from Gotham 360 to create a comprehensive carbon footprint. This may sound like an oxymoron – a small study for a comprehensive footprint? The College had been participating in the NYC Mayor's Carbon Challenge for several years, which required only very limited reporting on Scope 3 emissions, largely in terms of waste tonnage. No other Scope 3 emissions were being measured. In 2015, the Sustainable Practices Committee secured a small amount of funding for a study that would look more carefully at the whole emissions picture. We worked with consultants from Gotham 360 (including several Barnard alumnae) to create what we called a "hazy, but complete" snapshot of our emissions, as opposed to the more granular but incomplete data we were familiar with from the Mayor's Carbon Challenge. We felt it was more important to get an overview analysis of all our major sources of emissions than to worry about total precision. We didn't need to know the exact numbers to start making changes where the impact would be the greatest, and we were not satisfied with the more typical practice of only reviewing very limited and selective Scope 3 data. The results of the Gotham study, shown in Fig. 20.2, indicated that, indeed, Scope 3 emissions were a significant portion of our total emissions – as much as two-thirds.

This initial measurement provided a clear rationale for pursuing a strong Scope 3 emissions reduction strategy. Now, 6 years later, we are able to launch a more in-depth study as part of our larger pathway to net zero, which includes analysis of and strategies pertaining to all three Scopes (see Fig. 20.1). This roadmap to net zero, under development with the help of consultants from Energy Strategies, will assess our current Scope 1 and 2 emissions, and outline strategies and potential costs for reduction, from electrification to renewable energy procurement. In addition, this roadmap includes a 3-step approach to Scope 3 emissions:

1. Create a detailed inventory of Barnard's Scope 3 emissions, using global greenhouse gas protocols.
2. Determine focus categories for reductions in line with science-based global targets, including a weighted system for choosing areas of focus based on a number

Scope Type	Labels	Emissions Source	FY15 Metric Tons
Scope 1	A	Stationary Combustion	4,594.60
	B	Fleet Vehicles	29.7
Scope 2	C	Purchased Electricity	2,529.30
Scope 3	D	Commuting	913.7
	E	Directly Financed Travel	1,681.90
	F	Study Abroad Air Travel	872.3
	G	Solid Waste	1962.2
	H	Food	3,101.30
	I	Stuff	1,734.90
	J	Scope 2 Line Losses	156.30
		TOTAL	17,576.20

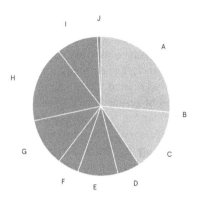

Fig. 20.2 Barnard Emission Sources, 2015

of factors, such as perceived importance to the College community or our ability to impact results.

3. Translate these goals into an operational language that meshes with Barnard's operational systems, e.g., by framing goals in terms of our existing chart of accounts, budgeting protocols, and business management systems.

The initial assessment tool will then be able to be used for internal tracking year over year. The College is also in the process of transitioning to a new enterprise management system called Workday, and we are working with that team to make collecting data on Scope 3 related expenditures much easier. For example, our current system includes airfare, hotel stays, and food under one label of "domestic or international travel." By coordinating with the Workday transition team, we will in the future be able to separate these elements, making data collection simpler and emissions calculations more accurate.

As a holistic framework, of course, a circular campus requires tracking more than greenhouse gas emissions. Measuring progress on a circular campus involves monitoring waste, diversion, purchases, reuse, food, and emissions associated with all of these activities. And, as a college campus, measuring success also requires tracking related courses and research. Since that first Gotham 360 carbon footprint, and including those waste audits on tarps in classrooms, gathering data has been a challenge, though we have made significant progress.

Outside of the NYC Mayor's Challenge reporting, the first circular campus data we ever collected was through those classroom waste audits. Students went out back to the dumpsters and started weighing bags while others sorted through samples. Our waste hauler did not provide us with monthly tonnage, so these student audits were the best source of data for our diversion rates. It wasn't much, but we had enough to convince our facilities team to let us create new signs, reposition waste collection stations, and implement limited organics collection. Even then, it wasn't until a new building was built with room for compactors to be installed that we could begin getting consistent and accurate data on waste. Other informal methods of collecting initial data included student surveys, and conversations with campus partners, like the Purchasing office, or individual departments with circular practices in place, like Architecture. Our student work-study team also spends a great deal of time collecting data manually when it is available, but not easily accessible. This includes combing through course catalogs using keywords to determine courses related to climate or sustainability, searching faculty profiles to determine areas of research, and creating inventories of waste stations on campus. It is not efficient, but it is a way to start the process, demonstrate need, and generate the buy-in necessary for building more robust and centralized tracking mechanisms.

Compiling the first round of data eventually brought us to the point where we felt we could participate in the Advancing Sustainability in Higher Education's (AASHE) Sustainability, Tracking, Assessment & Rating System (STARS). STARS is a holistic self-study that when completed, acts as a full baseline for all of the sustainability metrics the College reported. Before we started, we got permission from the President's office for this year-long undertaking that required sending surveys to the community and working with 25+ offices and individuals across campus. Working with so many offices gave us the opportunity to start discussions about sustainability with many of our colleagues and build a replicable process for ongoing tracking of metrics related to circularity.

Barnard's internal efforts to measure circularity are tailored to our organization and will evolve over time. For our campus, and for others, there is a range of external tools and frameworks to reference. For example, Circulytics, a measurement tool launched by the EMF, was created to support companies in the transition "towards the circular economy, regardless of industry, complexity, and size" (EMF, n.d.-e). This free, award-winning tool can identify what aspects of a company's operation enable circularity and what outcomes the company can accomplish by transitioning to a circular economy. Another tool to measure circularity is the Circular Transition Indicators (CTI) framework created by the World Business Council for Sustainable Development (WBCSD). This quantitative framework provides companies with a common language for internal decision-making and communication to key stakeholders (WBCSD, n.d.). Besides looking at processes or materials in which circularity is already present, this tool helps understand the company's material consumption and its relationship with successful business more widely; however, it does not measure wider environmental or social impacts (Lehtinen et al., 2020). Such tools support companies and institutions in making the decision to adopt the circular economy in their activities, measure progress, and increase the transparency of circular activities. Largely useful as references at this point, these tools might in the future be adapted to serve institutions of higher education.

20.11   Teaching and Research

Circularity is rooted in community engagement and radical collaboration, and therefore research, diversity of thought, and both theoretical and empirical experience are needed to become circular. To create innovative circular systems and make sustainable decisions we need research in the sciences, advances in design and technology, and a deep understanding of the historical, social, and political implications of any changes. We also need diversity of intellectual disciplines in order to develop new approaches, revive powerful practices from the past, or identify and resource existing but often unseen successful strategies. These requirements place teaching and research at the heart of any movement towards an inclusive and just circularity.

Barnard does not yet have a clearly identified body of courses and research aimed specifically at circularity. We have, however, identified 198 courses as "sustainability-related" or "sustainability-focused." These courses are taught across 64.4% of departments, including Architecture, Economics, Environmental Science, Theatre, and Women's Gender and Sexuality Studies. 45 faculty in 29 departments engage in sustainability or climate-related research. A number of these courses (and some research) include modules related to circularity, though we do not have an accurate count at this time.

In addition, teaching and research do not only happen in a physical classroom or a lab. Barnard students have been involved in every step of the climate action and circular campus planning, from serving on committees to participating in our 2019 Citizen Assembly to attending events like "Women, Clothing, and Climate," a 2018 used clothing sale, repair workshop, and panel discussion. In addition, Barnard's Climate Action Grant program allows students and staff to apply for funding for campus projects, and a significant portion of the initial cohort of grantees proposed initiatives related to circularity. Classroom projects, like our waste audits, and student-led initiatives help fuel lasting operational changes. For example, students at Barnard's Athena Center developed an internal carbon tax prototype that helped engage our Finance department in thinking through strategies to more accurately measure carbon-heavy activities like travel and purchasing, and eventually helped lead to the Workday partnership on gathering sustainability metrics. Most importantly, in the classroom and in the "campus as lab," the circular campus framework aims to provide students with skills and knowledge to empower them to address society's future challenges and make a positive impact in the world.

Many global peer institutions are expanding circularity in their curricula, performing research related to the subject, and creating outstanding initiatives to help us shift from a linear model to a circular one. The Interdisciplinary Circular Economy Centre for Circular Metal at Brunel University will explore how the reuse of metals can benefit the environment and the UK economy and will develop an understanding of how the country can best shift towards a carbon-neutral, circular economy by 2050. Over the next decade, this could generate more than £100 billion for the UK economy (The Interdisciplinary Centre for Circular Metals, n.d.). The Laurea University of Applied Sciences has a project called *Purater,* which provides information about the usability and safety of demolition materials reuse and analysis of suitable policy instruments (EMF, n.d.-f). Another example is the Campus Nature demo (2019–2020), a joint project of B2N Research Group, the University Properties of Finland Ltd., and Tampere University in which novel green areas around the city center campus were created together with students to enhance biodiversity and recreational opportunities (EMF, n.d.-g). Scientists at the University of Portsmouth have successfully developed a naturally occurring enzyme that can digest some of the most commonly polluting plastics (EMF, n.d.-b). Falmouth University is implementing a number of initiatives at the campus level, such as promoting free public transport and *Edible Pathways'* initiative in development, which will, in addition to planting new edible fruit trees and plants around campus, provide signage for areas of biodiversity and edible food plants (EMF, n.d.-c). African Leadership University's students have started circular economy ventures, such as Wastezon, an e-waste recycling company (EMF, n.d.-h). The Fashion Institute of Technology (FIT) has created a Natural Dye Garden as part of a student project. The plants – including sunflowers, coreopsis, and marigolds – can provide nontoxic fabric dyes used in campus textile research. Incorporating composting, the Dye Garden promotes more circular and eco-friendly dyeing practices and related research (EMF, n.d.-i). New York University (NYU) has an Urban Food lab, an aquaponic vertical farming class in which students learn about sustainability and farming by running their own projects connecting their fields to the farm (EMF, n.d.-j). These examples, and many more, demonstrate the power of an approach that utilizes the classroom and the lab to develop and test pathways to a just and sustainable circular campus.

20.12 Conclusion: Connecting the Dots Between Circularity, Net Zero, and Diversity, Equity, and Inclusion

Circularity is a key strategy in any global approach to climate change and environmental degradation. Similarly, Barnard's circular campus approach is a core component of our overall approach to sustainability and climate action, including our pathway to net zero emissions and our commitment to environmental justice and diversity, equity, and inclusion (DEI). As we have seen, Scope 3 emissions have historically been significantly undercounted on our campus, as has been the case for many other schools and organizations. The circular campus framework creates specific pathways for tackling those emissions, and the behaviors that fuel them; circular systems provide opportunities to link the College's emissions reductions work with the everyday life of our community. While addressing Scope 1 and 2 emissions is critical to our net zero roadmap, many of the necessary changes to address them can be made without significant participation from the majority of our students, faculty, and staff. But addressing patterns around purchasing, food, waste, green spaces, and course supplies by definition demands participation from everyone on campus. And, in building behavioral and cultural changes, the College has an opportunity and a responsibility to directly link our emissions reduction work with our commitment to social and environmental justice. Circularity can be a framework to reduce our environmental impact; it is an equally powerful tool for distributing resources more widely, both within our campus and with our neighbors. This simple act of sharing resources – and building the administrative and operational systems to do so purposefully and at scale – is a tangible and measurable way to support a more equitable and less environmentally destructive pattern of living.

References

Amed, I., Balchandani, A., Berg, A., Hedrich, S., Jensen, J. E., & Rölkens, F. (2021). *The state of fashion 2021: In search of promise in perilous times.* Retrieved from https://www.mckinsey.com/industries/retail/our-insights/state-of-fashion

Apte, A. (2020). *MIT startup wraps food in silk for better shelf life.* Retrieved from https://news.mit.edu/2020/mit-based-startup-cambridge-crops-wraps-food-in-silk-0605

Burton, E., Arredondo, G., & Jones-Davidis. (n.d.). *What is a circular economy?* Retrieved from https://dhcbarnard.org/scalar/dhclifecycle/what-is-a-circular-economy

Circle Economy. (2021). *The circularity gap report 2021.* Retrieved from https://www.circularity-gap.world/2021#downloads

Clayton, T. B. (2021). *Refocusing on diversity, equity, and inclusion during the pandemic and beyond: Lessons from a community of practice.* Retrieved from https://www.higheredtoday.org/2021/01/13/refocusing-diversity-equity-inclusion-pandemic-beyond-lessons-community-practice/

Columbia University. (n.d.). *Columbia University Sustainability Plan 2021–2030.* Retrieved from https://sustainable.columbia.edu/content/plan-2030

Cooper, G. (2020). *The sustainability industry needs sustainable change for racial justice.* Retrieved from https://solve.mit.edu/articles/the-sustainability-industry-needs-sustainable-change-for-racial-justice

Duncan, J. (2019). *Should I stay or should I go: The embodied carbon of buildings.* Institute for Market Transformation. Retrieved from https://www.imt.org/should-i-stay-or-should-i-go-the-embodied-carbon-of-new-and-existing-buildings/

Ellen MacArthur Foundation. (2020). *Designing out waste and driving a circular economy on a university campus.* Retrieved from http://sustainability.mit.edu/sites/default/files/resources/2020-10/mit-case-study.pdf

Ellen MacArthur Foundation. (n.d.-a). *What is a circular economy?* Retrieved from https://ellen-macarthurfoundation.org/topics/circular-economy-introduction/overview

Ellen MacArthur Foundation. (n.d.-b). *University of Portsmouth.* Retrieved from https://ellenma-carthurfoundation.org/university-of-portsmouth

Ellen MacArthur Foundation. (n.d.-c). *Falmouth University.* Retrieved from https://ellenmacar-thurfoundation.org/falmouth-university

Ellen MacArthur Foundation. (n.d.-d). *Stellenbosch University.* Retrieved from https://ellenmacar-thurfoundation.org/stellenbosch-university

Ellen MacArthur Foundation. (n.d.-e). *Measure business circularity: Circulytics.* Retrieved from https://ellenmacarthurfoundation.org/resources/circulytics/overview

Ellen MacArthur Foundation. (n.d.-f). *Laurea University of Applied Sciences.* Retrieved from https://ellenmacarthurfoundation.org/laurea-university-of-applied-sciences

Ellen MacArthur Foundation. (n.d.-g). *Tampere Universities (TAU & TAMK).* Retrieved from https://ellenmacarthurfoundation.org/tampere-universities-tau-and-tamk

Ellen MacArthur Foundation. (n.d.-h). *African Leadership University.* Retrieved from https://ellenmacarthurfoundation.org/african-leadership-university

Ellen MacArthur Foundation. (n.d.-i). *Fashion Institute of Technology.* Retrieved from https://ellenmacarthurfoundation.org/fashion-institute-of-technology

Ellen MacArthur Foundation. (n.d.-j). *New York University.* Retrieved from https://ellenmacarthur-foundation.org/new-york-university

Ellen MacArthur Foundation & Material Economics. (2019). *Completing the picture: How the circular economy tackles climate change.* Retrieved from https://www.ellenmacarthurfoundation. org/assets/downloads/Completing_The_Picture_How_The_Circular_Economy-_Tackles_ Climate_Change_V3_26_September.pdf

Fashion Revolution. (n.d.). *The women who make our clothes.* Retrieved from https://www.fashionrevolution.org/asia-vietnam-80-percent-exhibition/

Godelnik, R. (2020). *It's decision time: Reimagine or kill the circular economy.* Retrieved from https://razgo.medium.com/its-decision-time-reimagine-or-kill-the-circular-economy-ef9dca1646fd

Goldmark, S., & Purdum, K. (2021). In words and chairs: Making meaning of sustainability, equity, and circularity in American theatrical design and production. *Theatre and Performance Design, 7*(3–4), 152–162. https://doi.org/10.1080/23322551.2021.1996106

Gotham 360 Innovative Energy Solutions. (2017). *Some hero carbon footprint – Preliminary results.*

Heath, M. (2020). *Transform & reuse: Low carbon futures for existing buildings.* ARUP. Retrieved from https://www.arup.com/perspectives/publications/promotional-materials/section/transform-and-resuse-low-carbon-futures-for-existing-buildings

Humes, E. (2012). *Garbology: Our dirty love affair with trash* (p. 155). Avery.

Intergovernmental Science-Policy Platform on Biodiversity and Ecosystem Services (IPBES) & Intergovernmental Panel on Climate Change (IPCC). (2021). *Biodiversity and climate change.* Retrieved from https://ipbes.net/sites/default/files/2021-06/20210609_workshop_report_ embargo_3pm_CEST_10_june_0.pdf

Ishii, N., & van Houten, F. (2020). *To build a resilient world, we must go circular. Here's how to do it.* World Economic Forum. Received from https://www.weforum.org/agenda/2020/07/to-build-resilience-to-future-pandemics-and-climate-change-we-must-go-circular/

Ivanova, D., Stadler, K., Steen-Olsen, K., Wood, R., Vita, G., Tukker, A., & Hertwich, E. G. (2016). Environmental impact assessment of household consumption. *Journal of Industrial Ecology, 20,* 526–536. https://doi.org/10.1111/jiec.12371

Kimmerer, R. W. (2013). *The gift of strawberries. Braiding sweetgrass* (pp. 22–32). Milkweed Editions.

Labour Behind the Label. (n.d.). *A living wage is a human right.* Retrieved from https://labourbehindthelabel.org/living-wage/

Lehtinen, A., Leppänen, R., & Hughes, M. (2020). *Three tools for measuring and developing a company's circular economy performance.* Retrieved from https://www.sitra.fi/en/articles/three-tools-for-measuring-and-developing-a-companys-circular-economy-performance/

Leonard, A. (2010). *The story of stuff: How our obsession with stuff is trashing the planet, our communities, and our health – And a vision for change* (p. 233). Free Press.

MacGregor, N. (2011). *A history of the world in 100 objects* (pp. xv–xvi). Viking.

Meadows, D. H. (1999). *Leverage points: Places to intervene in a system.* Retrieved from https://donellameadows.org/wp-content/userfiles/Leverage_Points.pdf

Mendoza, J. M. F., Gallego-Schmid, A., & Azapagic, A. (2019). A methodological framework for the implementation of circular economy thinking in higher education institutions: Towards sustainable campus management. *Journal of Cleaner Production, 226,* 831–844. https://doi.org/10.1016/j.jclepro.2019.04.060

Mori. (n.d.). *Mori: Naturally fresher for longer.* Retrieved from https://www.mori.com/

New York City Environmental Justice Alliance. (n.d.). *Waste equity.* Retrieved from https://nyc-eja.org/campaigns/waste-equity/

Palese, B. (2021). COP26 will host the top men in climate policy. Now, where are the women? *Climate and Capital Media.* Retrieved from https://www.climateandcapitalmedia.com/ahead-of-cop26-activists-ask-where-are-the-women/

Scientific American. (2012). *Use it and lose it: The outside effects of U.S. consumption on the environment.* Retrieved from https://www.scientificamerican.com/article/american-consumption-habits/

The Interdisciplinary Centre for Circular Metals. (n.d.). *About the centre*. Retrieved from https://www.circularmetal.co.uk/

The University of Manchester. (n.d.). *The University of Manchester: Living Campus Plan: A living campus where we work alongside nature and nature works alongside us*. Retrieved from https://documents.manchester.ac.uk/display.aspx?DocID=33154%20

United Nations Environment Programme & International Resource Panel. (2020). *Global resources outlook 2019: Natural resources for the future we want*. Retrieved from https://www.resourcepanel.org/reports/global-resources-outlook

World Business Council for Sustainable Development. (n.d.). *Circular transition indicators (CTI)*. Retrieved from https://www.wbcsd.org/Programs/Circular-Economy/Factor-10/Metrics-Measurement/Circular-transition-indicators

Woynarski, L. (2020). *Ecodramaturgies: Theater, performance, and climate change* (Vol. 56). Springer. https://doi.org/10.1007/978-3-030-55853-6

Zero Waste International Alliance. (n.d.). *Policies*. Retrieved from https://zwia.org/policies/

Sandra Goldmark is a designer, teacher, and entrepreneur whose work focuses on circular economy solutions to overconsumption and climate change. She is an Associate Professor of Professional Practice and Director of Campus Sustainability and Climate Action at Barnard College, and Senior Assistant Dean for Interdisciplinary Engagement at the Columbia Climate School. Sandra is the founder of Fixup, a co-creator of the Sustainable Production Toolkit, and the author of *Fixation: How to Have Stuff without Breaking the Planet*.

Leslie Raucher is Barnard's Associate Director for Campus Sustainability and Climate Action. She works to realize Barnard's Climate Action Vision while implementing policies and initiatives that promote equity, access, and sustainability. She has a BA in Geology and Environmental Geoscience from Lafayette College, MA in Sustainability Studies from Ramapo College of New Jersey, and Certification of Professional Achievement in Sustainability Analytics from Columbia University.

Ana Cardenas is an undergraduate student at Columbia University majoring in Sustainable Development. She works for the Department of Sustainability and Climate Action at Barnard College as the student lead for the Circular Campus. Ana is interested in the circular economy, issues of equity, and climate change adaptation.

Part V
Sustaining Curricula 2021 Panel Discussions at Barnard

Chapter 21
Sustaining Curricula: Multidisciplinary Faculty Dialogues for Ecological and Pedagogical Renewal

Jennifer Rosales

21.1 Sustaining Curricula 2021

Sustaining Curricula 2021 was a series of four interdisciplinary faculty panels at Barnard College that took place virtually in the Spring 2021 semester. Barnard's Center for Engaged Pedagogy (CEP) planned this series with the Office of Sustainability and Climate Action and faculty to encourage professors to integrate climate action, sustainability, and the environment into their courses in authentic, inclusive, and justice-oriented ways and to cultivate a culture of interdisciplinary pedagogical exchange and collaboration.

To involve multidisciplinary faculty, the planning committee chose four themes that could be researched and taught from different and intersecting disciplines: Theorizing the Environment, Climate Change and the Anthropocene, Environmental Justice, and Design Decarbonization. Each panel consisted of three to four faculty who spoke to how they interpreted the theme and integrated it into their courses. We asked each panel a set of questions, including: How have you changed your course design? How do you teach sustainability within your subject and field? How does sustainability connect to your research and work outside of the classroom? Although the questions remained fairly standardized, the conversations varied based on the panelists and theme. Faculty members in the audience contributed questions as well as their own perspectives and expertise.

We aimed to cast a wide net of participants to attract faculty who might be drawn to one or two themes as well as faculty who were eager to contribute to every conversation. For faculty who were only able to join us for one panel, we hoped to spark interest and motivation to relate the themes, ideas and practices discussed to their own teaching. For the faculty who attended repeatedly, we hoped to foster

J. Rosales (✉)
Center for Engaged Pedagogy, Barnard College, New York, NY, USA
e-mail: jrosales@barnard.edu

© The Author(s) 2023
M. S. Rivera Maulucci et al. (eds.), *Transforming Education for Sustainability*,
Environmental Discourses in Science Education 7,
https://doi.org/10.1007/978-3-031-13536-1_21

connections not only to the material but to one another by building a community around interdisciplinary pedagogical approaches that center the environment. We invited faculty attendees to stay after the panel for a workshop session that enabled them to brainstorm with colleagues how to overcome challenges and concerns for incorporating the environment into their courses and how to collaborate with one another across disciplines and curriculum. We invited students to attend the panel portion of the event to listen and ask questions about the inclusion of the environment in their curriculum. Students also moderated each of the panels. The student moderators came from the CEP and the Student Government Association's (SGA) Sustainability Representatives and Committee. Since many of our students have grown up with the effects of climate change and will inherit the world we leave behind, their voices were crucial to the conversation.

This series was a response to Barnard's Climate Action Vision (2019) which was the result of a campus-wide effort from 2016 2018 that outlined the college's 360-degree approach to climate leadership in academics, finance and governance, and campus culture and operations while prioritizing the role of women, people of color and low-income communities. The academic vision laid out in this document states two goals: (1) "All students graduating from Barnard engage with climate and sustainability from multiple perspectives in coursework and research across the curriculum; and (2) Barnard's interdisciplinary approach to climate and sustainability supports innovative research, fosters civic engagement, and builds practical solutions." This series was also a response to the recommendations that came out of Barnard's 2020 Citizen's Assembly. Barnard faculty, staff, students, and alumnae came together to create a set of recommendations to advance the goals articulated in Barnard's Climate Action Vision. It specifically addressed the recommendation that Barnard create pedagogical support for faculty to create new courses or adapt existing ones in order that sustainability encompass all disciplines, with the "goal of 100% of the College's students engaging with climate and sustainability before they graduate, in courses across the curriculum."[1]

In order to achieve this vision and carry out the recommendations, the CEP along with the Office of Sustainability and Climate Action created the series to inspire and foster the development of more classes and curricular pathways for students to engage with the environment. This aligned with the CEP's mission to strengthen Barnard's deep academic engagement and support for student and community well-being by creating diverse learning contexts, developing and sharing scholarship, building and sustaining relationships, and providing tools and resources. This series combined all of the CEP's strategies for the dual purpose of supporting the integration of the environment into more courses and creating the space for faculty to partner, collaborate and build community in interdisciplinary ways.

Last year, the CEP partnered with the Office of Sustainability to create two communities of practice made up of faculty from different disciplines with a common interest in integrating sustainability, climate change, and the environment into their

---

[1] https://barnard.edu/sites/default/files/inline-files/ClimateActionVision-2019-final-5.pdf

courses. Some of the faculty in these groups already had whole courses dedicated to studying the environment in fields like Environmental Science and Ecology, while others were intrigued to integrate it into their courses in English, Psychology, Education, History or deepening/broadening the focus of the environment in their courses within Urban Studies, Theater, and Anthropology. The exchange of scholarship, ideas, and pedagogical approaches that occurred in these two communities was matched by the community and relationship building that was taking place among colleagues from different departments. Faculty were guest lecturing and guest critiquing, sharing source recommendations, and helping construct learning activities in one another's courses. As multidisciplinary communities, they helped review and conceptualize the Environmental Humanities Minor/Concentration proposal to create a new curricular pathway for students. These communities made practical changes to their own courses, those of their colleagues, and the curriculum at large while also pushing the bounds of their disciplinary understandings and approaches to the environment.

This year, we wanted to build on last year's work by shifting from learning communities of a few committed participants to a series open to all Barnard faculty and students. The faculty panels, each of which included at least one member of last year's learning communities, showcased the critical inquiry and creativity that took place in those communities of practice and inspired further thinking and innovating for the subsequent workshops. During the workshop sessions, faculty discussed how to integrate climate, environment, and sustainability into course content and requirements; generated different forms of faculty synergies; offered possible interventions for challenges that arise when integrating the environment into a course or trying to design collaborative teaching practices; and acknowledged the emotional intelligence and responsibility that is required for teaching these topics and teaching them well.

Before we introduced each panel, Yuval Dinoor ('21), CEP student worker and advisory member, read a land acknowledgment that she wrote especially for the series. The following statement she created recognizes Indigenous people and their land rights and reinforces the importance of land pedagogy and Indigenous knowledge in conversations about sustainability and climate action.

We may all be joining from various locations, but we would like to acknowledge that our university campus exists as a result of violent territorial displacement. In the making and expansion of our campus, Barnard College is occupying Lenapehoking, unceded land stewarded by the Lenni Lenape people, and gentrifying Harlem, a historically Black and Latinx neighborhood. As we discuss sustainability at Barnard during these sessions, we must also take seriously the responsibilities that come with taking institutional accountability for the ongoing violence we perpetuate in these communities, including redistributing institutional resources to Indigenous organizations, hiring and fairly compensating BIPOC workers for their labor, advocating for holistic community safety practices that decenter policing, and using Barnard's academic spaces to reflect honestly on the devastating impacts of colonization.

This land acknowledgment she created signifies some of the intersecting themes that were addressed in the panel discussions, specifically colonization and environmental justice and, more broadly, the relationship between land and people.

I would like to thank the Center for Engaged Pedagogy team including Alex Pittman, Joscelyn Jurich, Annebelle Tseng, and Hana Rivers in partnership with Sandra Goldmark and Leslie Raucher from the Office of Sustainability and Climate Action and Maria Rivera Maulucci, Professor of Education and Severin Fowles, Associate Professor of Anthropology for conceptualizing, implementing and sharing the series with the Barnard Community.

The following sections provide brief summaries of the four panels, which feature faculty from History, Spanish and Latin American Cultures, Biology, Chemistry, Education, Sociology, Environmental Science, American Studies, Theater, and Architecture.

21.2 Theorizing the Environment

"I am convinced that climate change and the environmental crisis is not something that is only practical. It is also ideological," says Orlando Bentacor, panelist in the session "Theorizing the Environment" (Chap. 22). The theme for this session came from the conversations we had in the communities of practice from the previous year, in which we discussed the ideological implications of the current climate crisis from multidisciplinary perspectives. It is further explored in the panel conversation between students, Yuval Dinoor and Rachel Elkis and faculty panelists, Orlando Bentancor, Associate Professor of Spanish and Latin American Cultures, Hilary Callahan, Professor of Biology, and Carl Wennerlind, Professor of History.

The panelists critique the dominant Western imperial and capitalist ways of conceptualizing the environment and the practical ramifications of these ideologies on human interactions with nature. Wennerlind and Bentancor both explore how hegemonic capitalist and imperial ideologies have impacted assumptions about nature and the ways people have engaged with it for centuries. Wennerlind argues that dominant ways of historical economic thinking that presume human desires and nature are infinite and expandable are partially responsible for our current climate crisis and lack of policies for present and future sustainability.

As an ecologist, Callahan adds that while she uses a cross-scale approach to teaching ecology from different spatial and temporal circumstances and vantage points, she includes some focus on sociopolitical ideology throughout her courses because she understands the implications that ecology has on law and policy and wants students to understand the relationship between ecology and the social, health and political spheres. She does not teach science as ideology-free or *pure*, because that was not how she was trained, and because she engages with scholarship outside of the white-male-dominated canon of historical texts from the United States and the United Kingdom. She seeks out ecologist publications from authors with different identities to provide a more inclusive and diversified learning experience and approach to studying ecology. Engaging with works from diverse authors and talking to colleagues outside of her discipline like Bentancor and Wennerlind help her become more aware of the hegemonic economic language that has been adopted by ecologists, such as "abundance," "scarcity," and "tradeoffs;" this work has also inspired her to seek out diverse voices and varied conceptual understandings of the field.

The panelists also demonstrate how their own thinking and theorizing about the environment has evolved over time as they continue to research the relationship between humans and the environment. Bentancor explains how his original post-structuralist theorizing of the environment blamed technology for the climate crisis. However, after reading a few different texts, named by Bentacor in the subsequent discussion, he shifted blame to capitalism for the climate crisis. Betancor's research and teaching connect to Wennerlind's because they both critically examine historical narratives of power, expansion, and economics, which have led to devastating material ramifications, as exemplified in Bentancor's focus on the mining industry in South America.

In the discussion, the panelists emphasized the value of critically examining the dominant ways of theorizing the environment for the last 500 years and the detrimental practical implications of this thinking so that educators and students can change the narrative and transform how they think about and engage with the environment.

21.3   Climate Change and the Anthropocene

"How we define the problem is going to define where we look for solutions to the problem. If climate change isn't happening, then there is nothing we should be doing differently. If humans are not responsible, then there is nothing we really can do to fix it," warns Maria Rivera Maulucci, one of the panelists in "Climate Change and the Anthropocene" (Chap. 23). She then defines the Anthropocene for participants as the current geological age during which human activity is most responsible for climate and the environmental conditions and explains its significance in understanding the climate crisis and human responsibility.

This panel focuses on different pedagogical approaches to teaching about humans and the environment. Maria Rivera Maulucci, Professor of Education, is joined by Anthony Cagliotti, Assistant Professor of History, Andrew Crowther, Assistant Professor of Chemistry, and Jon Snow, Associate Professor of Biology and students, Aastha Jain and Yuval Dinoor.

The panelists summarize how their teaching and research connect to the theme. Snow, who teaches cell and molecular biology, acknowledges that most students come to this subject with an interest in human health and that he has slowly shifted the focus from human health to the health of other organisms, specifically pollinators. During the Anthropocene period, pollinators such as bees, which are important to the agricultural and natural ecosystems, are suffering and Snow's hope is that while students may continue on their path into the medical and health fields, they will have a better understanding of ecological issues connected to human activity. Cagliotti acknowledges that history may not be the most obvious discipline to turn to when confronting the current climate crisis and future sustainability efforts because, like molecular and cell biology, people tend to think of history as human-centered and focused on the past. However, for Cagliotti, the focus on the Anthropocene, the part of history that connects human activity to climate change, must be understood in order to describe the devastating consequences of industrialization and farming on the environment as well as to contextualize scientific concepts like "climate." Crowther teaches the physical chemistry and mechanisms that cause global warming, revealing inefficiencies in power plants and other modes of human-generated energy. He addresses climate change as a physical chemist with a very specialized and precise area of expertise, while Rivera Maulucci addresses climate change and environmental sustainability in courses on teaching science more broadly. She relies on standards for education for sustainability that can be used for K-12 curriculum development, offering her college students multiple pathways to create lessons on sustainability.

Though approaches and disciplinary knowledge differ between the panelists, all of them create authentic opportunities for students to actively engage in the learning process with meaningful real-world applications or connections. As Rivera Maulucci emphasizes, it's important for students to *do* the work. She has students teach the lesson plans they create in her course in the K-12 classrooms. On the first day of class, Snow takes students outside to meet bees and assigns students their own independent projects at the end of term. Crowther applies concepts students learned in foundational courses to more specific cases like how solar cells and batteries work, how power plants waste energy, and how specific mechanisms cause global warming in his physical chemistry course. Cagliotti partnered with the Computing Science Center to offer a student data visualization workshop on how scientists visualize climate change in maps and graphs to emphasize the importance of modeling in understanding and predicting climate change as well as comprehending how knowledge is produced based on these visualizations.

The panelists also connect the theme to issues of justice and inclusion. Cagliotti builds on Rivera Maulucci's discussion on how the concept of the Anthropocene has been contested by climate deniers and those who do not believe humans are to blame for the current climate crisis. According to him, the concept has also been considered controversial by those who argue that it conceals which humans, societies, and countries have predominantly contributed to global warming and the intersection of human inequities and environmental concerns. Similarly, Snow's research on bees has led him to consider the connection between pollinators and food justice because bees are pivotal to the agricultural systems that reinforce uneven food distribution. Rivera Maulucci uses culturally sustaining pedagogy (CSP) as an umbrella approach to sustaining "rather than eradicating" the environment and culture of marginalized communities experiencing oppression. The Anthropocene becomes an important way to frame the relationship between humans and their diverse environments. At the same time, the panelists acknowledge that there are disparities in power dynamics between humans that impact how they engage with those environments.

In the discussion, the panelists share what and how they teach as well as the interdisciplinarity of teaching climate change, enabling educators and students to understand the Anthropocene as a multifaceted pedagogical topic that can be explored from various disciplines and areas of expertise.

21.4   Environmental Justice

"Environmental justice, in many ways, is just an extension of looking at unjust relationships. The relationship between people and the relationship between the earth..." says one of the panelists, Angela Simms, in "Environmental Justice" (Chap. 25). This discussion expands upon the comments made by the previous panelists by focusing on the intersections of human inequities, such as structural racism and wealth disparity with hazardous environmental conditions and unjust distributions of natural resources. Simms, Assistant Professor of Sociology, is joined by Elizabeth Cook, Assistant Professor of Environmental Science and Manu Karuka, Assistant Professor of American Studies and students, Rachel Elkis and Yuval Dinoor.

The panelists describe the evolution of incorporating environmental justice in their teaching. For example, Cook admits that while it is a major part of her research, she has only recently dedicated more time to it in her courses. As an ecologist, she explains that environmental justice, once focused specifically on economic disparity, now includes intersecting systems of oppression, such as structural racism. There is also a shift in concentration from the availability of green space and biodiversity in different communities to evolutionary trends in cities based on aspects of green spaces and species interactions with one another. For Karuka, his students' interest in current events like Standing Rock and the Black Lives Matter movement has motivated him to include environmental justice cases within his courses. For example, he introduced the concept of slow death in his "Profits on Race" course in connection to Freddie Gray's murder because before Gray was killed in the back of a police van, he was a victim of lead poisoning as a baby. Karuka seeks out collective questions from students to connect course material to current happenings in the world. Simm's policy work in the federal government before becoming a sociology professor motivates her coursework. She discusses how federalism helps explain how different nestings of power and authority interact to impact geographies of opportunities for people and she provides students with the sociohistorical context in which to do environmental justice work.

Engaging in this material can feel very personal for students, and Simms tries to frame their roles in a sociohistorical context. She says,

> You're standing on the shoulders of giants—people have been resisting oppression since people have existed—so you now have the baton. I am equipping you to think through how to be more effective. I have to be a good steward of your energy, your time, your resources—but think about this as a process of negotiating with yourself and with others about how we engage each other on the earth. This is ongoing and it is going to be a lifelong process, so I want you to feel empowered to engage effectively.

She encourages students to think about their social locations and spheres of influence through formal and informal relationships and the different power dynamics at play. She assigns students to write op-eds to share their opinions on policies. Karuka has students write letters to themselves about the skills they have learned at Barnard and their assessment of racism in society. He says that many of those letters address environmental justice. His second writing assignment builds off the first and asks them to answer what they will do to fight racism or war. By the end of the semester, he hopes they are aware of their own agency and abilities to make change. Environmental justice is at the forefront of conversations in Cook's classes, in which she prompts students to apply what they discuss in the classroom to their own communities, as she does in her participatory research with community members in Harlem and at the Mayor's Office.

Partially because environmental justice can be so personal, an audience participant asked the panelists how they address eco-anxiety and grief from students that may feel overwhelmed by the current state of structural oppression and the environmental crisis. Karuka suggests that for those of us who are non-native, we can learn from Indigenous communities and philosophies on "how to survive catastrophe," which Indigenous people have been doing for over 500 years. Likewise, Simms refers to a "reservoir of wisdom" from people of color who have been dealing with devastation for centuries. She recognizes that there are still serious emotional and mental challenges in doing this work and processing the pain and trauma of racism and climate crisis but her advice to students includes learning different histories and forms of wisdom.

In the face of the immense intersecting injustices discussed, the panelists model a collective agency and eagerness to work together, and with others, to make change and motivate those in their spheres of influence to do the same. The discussion is a call to action for educators and students to be agents in the collective struggle for environmental justice.

21.5 Design Decarbonization

"So what we are experiencing right now is enormous increases in CO_2 levels compared to historical levels. The last time we had that high of a level was 3 million years ago and the earth was much, much warmer than it is today" cautions Martin Stute, one of the panelists of "Design Decarbonization" (Chap. 26). He presents a few visual graphs to demonstrate the urgency of reducing greenhouse emissions such as CO_2. This process is known as decarbonization and requires individual, collective, and systematic changes to reach net zero emissions in order to prevent further climate change. The panelists share how they design decarbonization from their fields and in their classrooms. Stute, Professor of Environmental Science is joined by Kadambari Baxi, Professor of Professional Practice in Architecture, and Sandra Goldmark, Associate Professor of Professional Practice in Theatre and Director of Sustainability and Climate Action and students Linda Chen and Olivia Wang.

This panel uses data visualizations to reinforce the immediacy and seriousness of climate change and also to emphasize the role of design in conveying information and making change. Baxi explains that designing decarbonization requires collective authorship that includes scientific knowledge, global governance, worldwide collaboration, and activism. Goldmark shares her own experience learning about greenhouse emissions as a theater artist, trying to make connections between the global crisis and her own work. She started by decarbonizing her own design practices, which eventually led to redesigning her teaching practices as well.

The panelists reveal some of their pedagogical approaches to teaching and participating in decarbonization design. Goldmark uses circular design and production principles in her teaching and assignments. For example, after a student builds one model, they exchange their model with a peer and must design the second model from the material they find in their peer's first model. In one of Baxi's courses, she asks students to explore the relationship of museums to climate change and has them create exhibits that "make visible" the complex issues connected to climate change through a practice she has coined as "Climatorium." In Stute's Sustainable Development workshop course, groups of students act as mini consulting firms, and they work with outside clients on real problems that often involve climate change. These panelists employ pedagogical approaches that require students to work together and reinforce the "collective authorship" that Baxi mentions as essential for designing decarbonization.

In answering how to address the gravity of climate change authentically in everyday lives, the panelists address it at the individual level in which people can practice more sustainable daily routines and make change within their spheres of influence. They also emphasize the importance of systematic and structural change to tackle this global crisis. Stute advocates for voting as a small but essential practice and Baxi suggests collective organizing. Both practices connect individual action to structural change. Goldmark agrees that action should happen at the individual, collective and structural levels. She hopes this session will get participants to ask the question, "Okay, given where I am and where I am working, what can I do?" The panelists share what they are doing, given where they are, where they work, and what they do to reinforce the severity of our climate crisis and inspire educators and students to engage in their own efforts to design decarbonization.

Jennifer Rosales is the inaugural Executive Director of Barnard College's Center for Engaged Pedagogy. She provides academic leadership and vision for teaching and learning at Barnard, guiding and expanding its commitment to inclusive and imaginative pedagogy. Jennifer received her Ph.D., MA and BA from the University of Southern California.

Chapter 22
Panel One: Theorizing the Environment, March 10, 2021

Orlando Betancor, Hilary S. Callahan, Carl Wennerlind, Yuval Dinoor, and Rachel Elkis

22.1   Panel Discussion

Yuval Dinoor Before we dive into the wonderful panel discussion, I want to offer a brief land acknowledgment… (See Rosales Introduction.) Thank you very much for listening. I will pass it over to Rachel, who will introduce herself and begin to introduce our panelists today.

Rachel Elkis As Jennifer mentioned, after our climate conversations in response to the Climate Action Vision Statement last year, I am representing the student voice and encouragement for incorporating sustainability in the curriculum. So without further ado, I would love for our faculty panelists to introduce themselves, their department, courses they are currently teaching, and then their role in the Barnard community, both inside and outside of the classroom.

Panelists: Orlando Betancor, Hilary S. Callahan, and Carl Wennerlind
Student Moderators: Yuval Dinoor and Rachel Elkis

O. Betancor
Department of Spanish and Latin American Cultures, Barnard College, New York, NY, USA

H. S. Callahan (✉)
Department of Biology, Barnard College, New York, NY, USA
e-mail: hcallaha@barnard.edu

C. Wennerlind
Department of History, Barnard College, New York, NY, USA

Y. Dinoor · R. Elkis
Barnard College, New York, NY, USA

© The Author(s) 2023
M. S. Rivera Maulucci et al. (eds.), *Transforming Education for Sustainability*,
Environmental Discourses in Science Education 7,
https://doi.org/10.1007/978-3-031-13536-1_22

Carl Wennerlind I teach in the History department. I focus on seventeenth and eighteenth century Europe and am mostly interested in economic ideas, concepts, and institutions. But lately, I have gravitated towards the relationship between the economy, economic thought, and the environment. I am going to talk to you today about my course on the history of the idea of scarcity and introduce some aspects about how different ways of thinking about the world, can be viewed as instrumental in generating a culture in which the environment is treated and exploited in ways that have landed us in the predicament that we are now facing.

Orlando Betancor Hello, my name is Orlando Bentancor. I teach in the Spanish department. I am also the chair of the Spanish department. I teach a course, Introduction to Hispanic Cultures I, which is a survey from the Romans to the Iberian Empire, and I always gave it a material/ecological twist. So I was always interested in ideas of imperialism, domination of nature, imperial expansion, and early modern globalization. Now I am also teaching a class called, Between Science Fiction and Climate Fiction, where I am teaching authors that are all females except one, and it is centered around literature written in the last five years. The course pays a lot of attention to environmental consequences, the agro-industry, and extractivism, and basically the consequences of this corporate exploitation of nature in Latin America.

Hilary S. Callahan I am a botanist by training, specifically in plant ecology, and in addition to being a professor and the chair in the biology department, I do a lot of things around campus. I am the faculty director of the Ross Greenhouse and I work a lot on mentoring, research, and teaching experiences, like the Beckman's Scholars Program and the Barnard Noyce Teacher Scholars program. I love teaching and also working with students on their journeys at Barnard, looking ahead to what they are going to do with their lives, especially students starting to discover an avocation as a botanist or plant ecologist. I am the only person in the Biology department who teaches plant courses. My main focus today is about ecology and how to teach ecological thinking. It is a complex discipline. It is imperfectly unified, in part because it is integrated with other aspects of biology, like genetics and genomics, while also being integrated with the social world, the economic world, and the worlds of public health and environmental health. I feel privileged to have a career studying these things, especially how it allows me to integrate what I do for my job with everything else in my life and my environment.

Yuval Dinoor Thank you for sharing a little bit about what you do in the Barnard community. I think it is so exciting delving into our conversation that each of you comes from such different disciplinary backgrounds to this conversation about theorizing the environment in our classes. So to kick this off, I would love to hear more about how the courses that you teach right now connect to the theme, theorizing the environment. What does it look like to take that kind of focus in the different classes that you each teach?

Orlando Betancor I integrate environmental discourses by focusing on classical philosophy, meaning Greek philosophy and medieval philosophy, how the dichotomy between humans and nature paved the way to the official ideology of the Spanish empire, and how that contributed to what Carl was talking about today. The Spaniards conceived nature in an instrumental way and that had a great impact on the environment. The main topic of my research was mining.[1] What I tried to do is to criticize the hegemonic perspective. My other class, the one I am teaching now on climate fiction, is very inclusive because I tried to bring in as many different voices as I can. When I teach this other class, I emphasize how important it is to understand the ideas behind dominant notions of nature, and how the ideas about climate change brought us here to this current situation. I am convinced that climate change and the environmental crisis are not something that is only practical. They are also ideological. Abstractions in a way, rule the world, and these presuppositions are extremely important. That is why sometimes the students complain about Hispanic Cultures I. Why don't we have more diverse voices? First, it is very diffi-cult, and second, I really want to place more emphasis on these dominant ideas, on these ideas that drove sixteenth-century mercantilism and capitalism.

Hilary S. Callahan I teach a science course, and I will talk a little later about what I call cross-scale thinking, across a lot of different spatial scales from the local to the global, and also temporal scales. I am honored to be with people who have such strong history training, who emphasize history in their scholarship. I do aspire, in my science course, to keep close to the foreground the soft and ideological aspects of what we are studying, and the history of ideas. I do that in an untidy way. It is scattered throughout the course because ecology has so many implications for law, politics, and policy.[2] Ecology is a professional practice. People hire ecologists when they want to build a road or open a mine. There is a lot of interaction between the global and colonizing north and how their cultures and their impacts have spread all over the world throughout the diverse tropics, diverse in terms of human diversity and biological diversity. I am also thinking about the role of historically marginal-ized people who are ecologists—a lot of white men built this relatively young field of ecology.[3] But the field was drawing from much older knowledge systems that women and Indigenous people had developed to understand the world, and continue to develop.

The way I sell ecology to science students is that it is Biology. And people love biology. It is the study of life and people. The cell is fascinating, and the body is also fascinating, and I do show students how ecologists do experiments with these enti-ties, at the laboratory bench with test tubes and pipettes and instruments. The depar-ture of ecology is doing experiments out in the world, in forests and lakes or in the ocean. Students immediately grasp that ecology is richly informed by

---

[1] Betancor 2017.

[2] Neff 2011.

[3] Courchamp and Bradshaw 2018.

nature-observing and being out in nature, but what my courses also stress is how ecology and ecologists are adamant about being theory-driven, operating as physicists do, from first principles about matter and energy. Ecologists also develop formal quantitative models, mathematically and computer-based. Ultimately, ecology is also an applied science, informing land-use, energy use, agriculture, water and fire management, and many other activities, including human health. For those aiming for medical school, we explore how being well-versed in ecology is likely to help in health-related STEM careers. And of course, ecology knowledge is going to be empowering to them intellectually not just in their jobs, but beyond their jobs, as citizens in the world.

Carl Wennerlind The course that I am teaching on the history of scarcity shares a lot of common ground with Orlando's teaching, thus suggesting a nice kind of synergy. The course is based on a book that I am in the process of finishing, *Scarcity: Economy and Nature in the Age of Capitalism*. It is a co-authored book with a colleague at the University of Chicago, Fredrik Albritton Jonsson, and it explores the economic thinking of infinite economic growth, which we regard as an important underlying feature of the Anthropocene. Economics tends to think of human desires as endless and nature as infinitely expandable or infinitely substitutable. We are interested in investigating how such a bizarre idea took hold and offering a critique of why this way of thinking about the world is not appropriate for the development of sustainable economic processes. The course traces these ideas back to the seventeenth century and then moves forward in time, eventually ending up discussing contemporary challenges and opportunities.

Premodern writers had a tendency to think of both economic desires and nature as bounded. They viewed desires as limited by spiritual, moral, and political concerns and they thought of nature as scant, incapable of yielding ever-increasing amounts of goods. Matters began to change towards the end of the seventeenth century. Natural philosophers, such as Francis Bacon and Samuel Hartlib, gained confidence in their capacity to decipher nature's source code and thus operate on it, in an almost godlike manner. They viewed nature as God's gift to humanity and insisted that it was humanity's responsibility to fully utilize all of nature's resources. Along the same time, moral philosophers, such as Nicholas Barbon and Bernard Mandeville, began theorizing human desires as unlimited. Unlike their predecessors, who saw unlimited desires as unnatural and a danger to social stability, Barbon and Mandeville argued that never-ending desires for more consumption infuse society with a progressive energy that over time contributes to greater affluence. As such, they promoted a radically different way of thinking about the relationship between nature and human desires, a double helix of infinities.

The class then moves through the centuries, examining how the conceptualization of nature and the economy evolved. Most of it is not a linear but a dialectical story. People like Gerrard Winstanley, Jean-Jacques Rousseau, and Karl Marx posited alternative grand narratives designed to reshape economic and environmental thinking. In the end, however, the notion of infinite growth prevailed and became entrenched in western cultures. Although my course is intrinsically historical, it is based upon a critique of contemporary thinking, a critique of contemporary neoclassical economics, which has become an almost hegemonic way of looking at the world. This way of thinking has seeped into the social fabric and now exercises a great deal of influence over how people think about the world and its future. If we are to successfully develop more sustainable economic patterns, new ways of thinking about the nature-economy nexus are required.

Rachel Elkis Extremely interesting and working off of this historical context and how we think about these topics through time, I am wondering how this integration of environmentalism and sustainability has evolved within your courses over time?

Carl Wennerlind I have always taught courses on capitalism. Most of the time, my focus has been on a critique of capitalism through a historical lens, but the criticism has mostly been about things like alienation, monopoly, exploitation, slavery, and colonization. Earlier, nature and the environment didn't figure that much into my way of thinking about capitalism. Now that I have seen the light, so to speak, I recognize that you cannot think about capitalism without nature, nor can you think about nature without capitalism. So from now on, I will always explore prisms and readings that are integrating these two perspectives.

Hilary S. Callahan One thing I clearly have in common with Carl is a focus on economic ideas. If you look up the definition of ecology, it is typically phrased as the study of the distribution, and it often refers to abundance, such as population abundance or species abundance. Abundance, like scarcity, is an inherently economic idea. The word is not really needed to talk about distribution. Clearly, there is clumpiness rather than uniformity, but that can be referred to as low and high density, or low and high frequencies, not using such loaded words. Another really dominant language that you hear in ecology is trade-offs, a balance of costs and benefits. And ecology constantly harks back to Thomas Malthus and the idea of overpopulation and scarcity, overlooking the many moral dimensions of such arguments. Concepts of abundance and trade-offs and over-population remain as important ideas in textbook ecology, typically with little if any ideological critique. Students learn it, and then if they become teachers, they continue to teach it and never critically dismantle it.

Being at Barnard has been such a great privilege because I am always in conversation with people like Carl and Orlando and I have come to completely change my perspective and approach, noticing and calling attention to words like "abundance" or "dominance" and seeking more neutral terminology and frameworks. Sometimes that comes from another motivation, which is to try to read more papers by women, including a notable single-authored paper by a woman in conservation biology, entitled Seven Types of Rarity. Its author, Rabinowitz, set up three dichotomies to identify eight ways that organisms can be found in the world. Most of the species that make up the world's biodiversity are confined to seven types of rarity, being narrowly endemic in terms of geography rather than widespread, being ecological specialists with a narrow niche rather than broad generalists, or living in low densities within a habitat area. Or some combination of these. The eighth way is to be common in all three ways, as humans are—widespread globally, generalist and often at high densities, as in cities. Rabinowitz's paper is just one example of gaining key insights by adding more papers by women. Over time, I also have started to include papers that are less neutral than hers, and may go beyond ecology into other non-ecological ways of thinking about biodiversity, even going as far as decentering capitalist economic thinking or other tacit ideologies that are inherent to ecological ideas or terms.

Orlando Betancor Hilary, what you said, reminds me of Donna Haraway, and her 2016 book, *Staying with the Trouble.* The notion is simple: losing yourself with other things. But it is really beautiful. I was going to say something different about how my thinking about the environment changed over time, given the texts and the imperial ideologies I started with at the time. Since this is theorizing the environment, the theory I used relied heavily on fashionable French theory and this post-Heideggerian way of thinking, post-structuralism, Derrida, Deleuze, etc. But the basic Heideggerian topic is that technology is the problem. Technology is a way of considering nature in terms of a standing reserve as passive material and thus violent to nature. I was fully within that paradigm until I read two books. One of them was Alfred Sohn-Rethel's *Intellectual and Manual Labour: A Critique of Epistemology* and the other one was Jason Moore's *Capitalism in the Web of Life.* Then, I saw that technology was not the problem, and I saw clearly, in historical terms, that the problem was capitalism. That served my purposes, the purposes of reading these texts, way better than presupposing abstract ideas, such as the homogenous idea of humanity, and anthropocentrism, the technology. It was more a story of power, a story of an expansion, one of power and economics, one closer to what Carl was talking about, and that is why I appreciate his work so much.

Ideas such as the Aristotelian notions of desire were not just some kind of expression of a technological will to power that is threatening to engulf everything and eat the world. They were contingent, historically and geographically situated processes that could be changed. There is no necessity. It did not have to be that way. And that is how environmentalism changed in understanding over time, finding bits and pieces of theory here and there that somehow change the way you think about these topics.

Yuval Dinoor I think it is so interesting to hear from all three of you about the wide array of voices that you are all bringing into your course material to related stories in all of your different course disciplines. I'm curious about beyond the theoretical component, how this turns into practice. Are there ways that you designed or planned your course so that it encourages students to think about environmental theory beyond the classrooms? What do you think has worked well in that regard as you have experimented with this in the past?

Hilary S. Callahan There are two things that I use in developing a framework for the course and a reading list. One is crossed-scale thinking, which somewhat connects with Barnard's Foundations Curriculum, for example our Modes of Thinking— Thinking Globally and Thinking Locally. In my course, I try to cover things that are local. I love it when we can study something ecological about New York City that is very local. But I do not want to be restricted to just this specific city. I want to be able to go to other parts of the world. In Spring 2020, I was teaching Ecology and I made the decision to read up on the fires that were happening in Australia and to talk about them at the opening of the course. Ironically, in my previous notes from teaching quite a few years ago in 2011, I had opened the course with a discussion of disease ecology, because of the Ebola epidemic. So in December 2019, I replaced that opening focus with new material about Australia, because I thought it would be interesting to open the class up to a more global perspective, to focus on another part of the world and to think about the climate crisis and the ongoing wildfires. There are interesting conflicts between European settlers in Australia and Indigenous traditions in managing fire and managing land resources. It was great. But then, less than a month into the course, I realized I needed to use that epidemic and disease ecology material, because of the Covid-19 pandemic.

Regardless of current events topics in any given semester, I do try to look at all scales. Ecologists do go all the way from experiments that can be done in a test tubes in the lab or in the greenhouse, to studying tiny insects and mosses in the field, and also doing investigations requiring satellites to gather data remotely, to focus on whole swathes of the earth, if not the entire earth.

Also, I try to bring the canonical list of important ecology articles into dialogue with other voices. Ecology has a strong White, male-dominated list of influential papers, and there is rampant argument about it.[4] I bring that argument right into the classroom. That type of critique is such a Barnard thing. Reading women authors, and more authors from Latin America or Africa, goes hand-in-glove with having geographic diversity, so useful in ecology, and also a diversity of conceptual perspectives.

Along with diversity all around the globe, I add in diversity through time, and not just evolutionary time but also more recent changes through time. I am honored to be talking with these two distinguished people who know so much about the history of the last 500 years. Trying to understand the ecology of five hundred years ago is clearly relevant to ecology today, but perhaps less understood and less studied is the world as compared to studies that last just a few growing seasons, or that use the evolutionary perspective to think about what happened, say, 10 or 20 million years ago. This vast range of scales is important to comprehend the mess that we are now in with the climate. Modern evolutionary biology and ecology have existed for just a little over a century, and only in the last few decades have ecologists developed ecological lenses for this intermediate time scale, allowing us to investigate globalization, which has been happening over 500–600 years, and also the impact of the industrial revolution and fossil-fuel burning.

---

[4] Mayer and Wellstead 2018.

Orlando Betancor I was going to say something briefly about going beyond the classroom at this point. In the last few weeks. I have been talking to Chilean scholars who work on communication media studies on climate change. These scholars are interested in bringing feminine perspectives, Indigenous perspectives, and making a critique of extractivism. One of them specifically is making a critique of how corporations invent their own notions of sustainability, sustainability as a corporate invention, basically, which is something maybe we could pay more attention to in the United States. These mining corporations are the ones who are moving the strings when it comes to reinventing this notion of sustainability and the impact of the mining practices on the environment. We decided to engage in an interdisciplinary project where I can bring my own text-based approach to other perspectives open to the experience of communities that are suffering right now and the impact of destructiveness in these communities.

Carl Wennerlind Let me also give you an example. When I taught the course last time, we started by reading a glossy insert in The Financial Times that documented all the things that the corporate community is doing for sustainability and for greening capitalism. It was quite the study in self-congratulatory rhetoric. It almost felt as though The Financial Times exists in an alternate reality. The students compared the choice of grand narratives to the film, *The Matrix*. If you take the blue pill you can continue living as though nothing is wrong, but if you take the red pill and start learning about the coproduction of capitalism and the environment, you can never look back. You then bring that perspective with you into all of your other classes and into life in general, and it is almost impossible to unlearn the inconvenient truth. For me, it took a long time to get there, but luckily my students are taking in these lessons early in life.

Rachel Elkis Really interesting. We want to get to the audience's questions.

Student I wanted to ask a little bit more about how the way your disciplines are structured informs the way that we perceive or conceptualize sustainability. I think you touched on this a little bit in terms of at least for the economics example, talking about how the discipline carries these assumptions about human behavior as inherently wanting to maximize consumption and the implication that says about our behaviors surrounding consumption versus sustainability. I am curious to know if there are other things you have noticed about the way knowledge is produced in each of these disciplines that maybe reproduces these assumptions surrounding sustainability.

Carl Wennerlind Economics is an inherently ahistorical discipline. It is based on a series of rather crude assumptions about human psychology. One such assumption is that people always prefer more consumption to less. This assumption might make sense in a world of consumer capitalism, say after the 1870s in the western world, but it would be difficult to maintain that this is the aim that characterizes people in earlier periods or in non-capitalist societies. The notion that people approach the world as a set of consumption choices only makes sense in a consumer society. The assumption that people have insatiable desires leads economists to conclude that it is in people's interest for the economy to grow as much as possible. The problem, as this panel has discussed, is that economic growth can no longer continue unchecked. Therefore, we need to rethink the purpose of the economy. We might not have to dedicate ourselves to no-growth, but we definitely have to find ways to grow that improve our lives without destroying earth systems. In writing our book on the history of scarcity and in teaching this course, I view it as essential to deconstruct the notion of scarcity and point out that it is based on assumptions made in the 1870s, grounded in limited psychological or anthropological research. It was just analytically convenient to assume that human desires are insatiable and that people are mostly focused on consumption. While this way of thinking fostered behavior and policies conducive to economic growth, our present challenge is to build a social science that is conducive to more sustainable practices.

Severin Fowles (Anthropology)   I am interested to hear what the others have to say about that question as well. This has just been great. Thank you all for doing this. Carl, what you were just saying made me suddenly leap back to readings of Marshall Sahlins and the Zen Road to Affluence, which is being resuscitated in so many communities I circle into, that are all about degrowth right now. The question that I have actually builds off of the fascinating comment that Hilary made about the incorporation of a certain kind of economic language in ecological theory. What interests me is that ecological language and ecological discourses have been taken up in the humanities as the core metaphor now. That started with Bateson and in the ecology of mind. But now, the new materialist language is all ecological language. So suddenly, I had this vision of the laundering of certain kinds of economic principles through Ecology back into Social Sciences. I am curious about how all three of you think about the language that we use now, which draws upon so many ecological metaphors, but maybe too loosely and maybe in ways that are denying that complicated history you were just pointing towards.

Hilary S. Callahan   There has been a lot of work done to critique the overemphasis on competition as the most important type of biotic interaction among organisms, rather than other ways that organisms interact.[5] Additionally, for years, there has been this tension and hand-wringing among ecologists and in teaching ecology— the pressure to keep ecology pure, to keep it scientific and therefore untainted by ideology or politics.[6] I was fortunate to have had mentors who were on the other end of the spectrum. My dissertation adviser, while he was mentoring me and running our lab, was suing the United States Forest Service, alleging that it was negligent in protecting biodiversity. I learned so much by having a mentor who is not afraid of the soft, more social, and very political side of ecology.

---

[5] e.g., Raerinne 2020.
[6] Also see Pfirman and Winkler, this volume.

Severin Fowles I am just going to follow that because it is so fascinating: if ecological discourse is being mobilized in anti-capitalist theory in the humanities right now, what you are just pointing to really concerns me, because its grounding is also in conservative, and capitalist economic theory.

Hilary S. Callahan Yes, and some of the neo-Malthusian stuff is especially troubling. For years, people who teach about ecology have used a "Tragedy of the Commons" metaphor to discuss resource management, citing the opinion piece by Hardin (1968) in the prestigious journal *Science*. Over the years, it is more and more recognized that the idea and the person who advanced it was questionable, and also associated with white nationalism (see also Ghoche & Udoh, this volume, chapter 5). Yet his ideas are written into many ecology textbooks.[7]

Yuval Dinoor That is such a rich starting point for so many conversations. We are going to transition to breakout rooms now.

References

Betancor, O. (2017). *The matter of empire*. University of Pittsburgh Press.

Courchamp, F., & Bradshaw, C. A. J. (2018). 100 articles every ecologist should read. *Nature Ecology and Evolution, 2*, 395–401. https://doi.org/10.1038/s41559-017-0370-9

Ghoche, R., & Udoh, U. (this volume). A commons for whom? Racism and the environmental movement. In M. S. Rivera Maulucci, S. Pfirman, & H. S. Callahan (Eds.), *Education for sustainability: Discourses on authenticity, inclusion, and justice*. Springer.

Haraway, D. (2016). *Staying with the trouble*. Duke University Press.

Hardin, G. (1968). The tragedy of the commons. *Science, 162*, 1243–1238. https://doi.org/10.1126/science.162.3859.1243

Mayer, A. L., & Wellstead, A. M. (2018). Questionable survey methods generate a questionable list of recommended articles. *Nature Ecology and Evolution, 2*, 1336–1337. https://doi.org/10.1038/s41559-018-0637-9

Moore, J. A. (2015). *Capitalism in the web of life: Ecology and the accumulation of capital*. Verso Press.

Neff, M. (2011). What research should be done and why? Four competing visions among ecologists. *Frontiers in Ecology and the Environment, 9*(8), 462–469. https://doi.org/10.1890/100035

Pfirman, S., & Winckler, G. (this volume). Perspectives on teaching climate change: Two decades of evolving approaches. In M. S. Rivera Maulucci, S. Pfirman, & H. S. Callahan (Eds.), *Education for sustainability: Discourses on authenticity, inclusion, and justice*. Springer.

Rabinowitz, D. (1981). Seven forms of rarity. In H. Synge (Ed.), *The biological aspects of rare plant conservation*. Wiley.

Raerinne, J. (2020). Ghosts of competition and predation past: Why ecologists value negative over positive interactions. *Bulletin of the Ecological Society of America, 101*(4), e01766. https://doi.org/10.1002/bes2.1766

Sahlins, M. (1972/2017). *Stone age economics*. Routledge.

Sohn-Rethel, A. (2020). *Intellectual and manual labour: A critique of epistemology*. Academic.

Wennerlind, C., & Jonsson, F. A. (2023). *Scarcity: Economy and nature in the age of capitalism*. Harvard University Press.

---

[7] See Ghoche and Udoh, this volume.

Orlando Betancor is an Associate Professor of Spanish and Latin American Cultures at Barnard College. His research interests include Colonial Latin American literature and intellectual history; speculative fiction; ecomarxism.

Hilary S. Callahan is the Ann Whitney Olin Professor of Biology. Her integrative research examines many different features of plants—from roots to flowers to seeds. She seeks to understand how plants and their traits function in nature, and how external factors affect trait expression and how rapidly traits evolve.

Carl Wennerlind is a Professor of History at Barnard College. He specializes in seventeenth- and eighteenth-century Europe, with a focus on intellectual history and political economy. He is particularly interested in the historical development of money and credit, as well as attempts to theorize these phenomena.

Yuval Dinoor '21 holds a BA from Barnard College in American Studies concentrating in Education, Civic Engagement, and Community. Her research focuses on creating experimental learning spaces to build politically active communities; recently, she aired a podcast series exploring the Highlander Research and Education Center's pedagogical approaches to community organizing.

Rachel Elkis '22, is a senior at Barnard College studying Economics and Environmental Science. She is the former Representative for Sustainable Initiatives for Barnard's Student Government Association and remains involved in sustainability programming and activism. She is eager to leverage her Barnard experience as a Power & Renewables Investment Banking Analyst at JP Morgan.

Chapter 23
Panel Two: Climate Change and Anthropocene, March 19, 2021

Angelo Caglioti, Jonathan Snow, Andrew Crowther,
María S. Rivera Maulucci, Yuval Dinoor, and Aastha Jain

23.1   Panel Discussion

Yuval Dinoor   Before we get our conversation started, I would like to offer a brief land acknowledgment (See Rosales Introduction to this panel series.). I will pass it on to Aastha to introduce herself after introductions from our panelists.

Angelo Caglioti   Hello, everyone. It is a pleasure to be here with you. And thank you so much for inviting me to join this series. My name is Angelo Caglioti, pronouns he/him/his, and I am in the History department. I am currently teaching a class that fits perfectly in the conversation we have today. It is an intensive lecture titled "Climate and History: Intersecting Science, Environment and Society." I am

---

Panelists: Angelo Caglioti, Jonathan Snow, Andrew Crowther, María S. Rivera Maulucci
Student Moderators: Yuval Dinoor, Aastha Jain

A. Caglioti
Department of History, Barnard College, New York, USA

J. Snow
Department of Biology, Barnard College, New York, USA

A. Crowther
Department of Chemistry, Barnard College, New York, USA

M. S. Rivera Maulucci (✉)
Education Program, Barnard College, New York, NY, USA
e-mail: mriveram@barnard.edu

Y. Dinoor · A. Jain
Barnard College, New York, NY, USA

401
M. S. Rivera Maulucci et al. (eds.), *Transforming Education for Sustainability*,
Environmental Discourses in Science Education 7,
https://doi.org/10.1007/978-3-031-13536-1_23

also teaching a senior thesis seminar throughout the year. To share something about myself and my role in the Barnard community, I am very excited to have finally moved to New York. The pandemic delayed my arrival, unfortunately, but I am really happy to be here. I am particularly grateful that here I have the intellectual freedom to wear different hats at the same time: my interests in environmental history and history of science and my expertise in European history. I can see how they can intersect very well and serve the Barnard community at large. So, I am very excited to join this conversation.

Jonathan Snow I am Jon Snow, my pronouns are he/his/him, and I am in the Biology department. I exclusively teach cell and molecular biology courses. Specifically, I teach a very large introductory cell and molecular biology course. I also teach several upper-level cell biology courses which are not automatically focused on sustainability. But I am excited to tell you about some of the things I have tried to do in those courses.

Andrew Crowther My name is Andrew Crowther. I am a physical chemist, which is a discipline that is at the intersection of physics and chemistry, but more on the chemistry side. My research is focused on the fundamental properties of different nanomaterials. Pronouns are he/him/his. I am in the Department of Chemistry, and I mainly teach physical chemistry courses, both lecture and laboratory courses, in the department. At present, I am teaching Thermodynamics and Kinetics and also an advanced laboratory course primarily focused on physical chemistry topics.

María Rivera Maulucci Hi, my name is María Rivera Maulucci, I am a Professor of Education, and I am really excited to talk to you today about how I am integrating sustainability into my courses. My background is in science education, so sustainability is an obvious fit, although I also teach classes across education, including arts and humanities and all subject areas. I have supported the sustainability initiatives in the Barnard community and was involved in the conversations early on. I helped out with the faculty sustainability workshops last year. The other hat that I love to wear is around diversity and inclusion. I have served on the Faculty Diversity and Development Committee pretty much my entire time at Barnard up until last year. But certainly, that is a hat that I love to wear. And pronouns she/her/hers.

Aastha Jain Thank you for introducing yourselves. As always, it is so exciting that we have folks on our panel from so many different Barnard departments teaching and doing so many different things. Today's theme is Climate Change and the Anthropocene. I would love to hear what that theme means to each of you and how you connect your courses to that?

María Rivera Maulucci First, I think it is helpful to define the Anthropocene, which means relating to or denoting the current geological age, viewed as the period during which human activity has been the dominant influence on climate and the environment. And I think this definition is important because this idea is disputed. We know that climate change deniers come in at least two forms. Those that deny it is happening at all and ask, "Can we really say climate change is even happening?" So, they deny the science behind it. And then there are those that recognize that maybe climate change is happening, but they deny that humans are responsible. And so, this basic controversy that is at the root of this issue is really unfortunate, because how we define the problem is going to define where we look for solutions to the problem. If climate change isn't happening, then there is nothing we should be doing differently. If humans are not responsible, then there is nothing we really can do to fix it.

The courses that connect most directly to sustainability are the Science in the City courses. These are pedagogical electives offered by the Education Program that focus on the methods of teaching science. In those courses, we use the standards for education for sustainability[1] as the basis for the curriculum development that my students do when they work with K-12 students in their field placements. There are nine core standards and those include cultural preservation and transformation, responsible local and global citizenship, the dynamics of systems and change, sustainable economics, healthy commons, natural laws, and ecological principles, inventing and affecting the future, multiple perspectives, and a strong sense of place. What I really appreciate about this framework is the idea that there are many pathways to sustainability. You can come to sustainability from the perspective of cultural preservation and transformation, or you can come to it from a strong sense of place.

In education, we have an approach to pedagogy called culturally sustaining pedagogy. Django Paris and Samuel Alim explained culturally sustaining pedagogy as an approach that:

> seeks to perpetuate and foster—to sustain—linguistic, literate, and cultural pluralism as part of schooling for positive social transformation and revitalization. CSP positions dynamic cultural dexterity as a necessary good, and sees the outcome of learning as additive, rather than subtractive, as remaining whole rather than framed as broken, as critically enriching strengths rather than replacing deficits. Culturally sustaining pedagogy exists wherever education sustains the lifeways of communities who have been and continue to be damaged and erased through schooling. As such, CSP explicitly calls for schooling to be a site for sustaining—rather than eradicating—the cultural ways of being of communities of color.[2]

What this means is that the umbrella for what counts as sustainability is quite broad. Many things can be included, and when you are working with local communities or even a college campus, the many questions about what needs to be sustained versus what needs to be changed can fall under this umbrella. So whether you are working to ensure the city has more green roofs or clean water or the campus has more green spaces, or the curriculum has more conversations about sustainability, or we are trying to document or reclaim the rich cultural history of Black and Latinx neighbors in Harlem, or working to help students sustain their native languages and cultures as they engage with western educational systems, or creating opportunities for civic

---

[1] The Cloud Institute 2010.
[2] Paris et al. 2017.

engagement for our students, whether it's in K-12 classrooms as my students are, or in other spaces, we can all begin to recognize the work we are doing as informed by tenets of sustainability, or we can all begin to see how we might elevate the work by bringing those tenets in.

Jonathan Snow I study honey bees, and I just love bees. But I did not start out that way, and this comes to what I have tried to do with my classes. My background is in biomedical sciences and using cell and molecular approaches to understand disease in humans. I switched to bees about ten years ago. What I have found as someone who teaches cell and molecular biology classes is that the vast majority of the students are focused on human health, as I was earlier in my career. Most students really want to use their biological training to go into disciplines that are focused on human health, such as medicine, public health, or biomedical research. I have tried to broaden their horizons in terms of how cell and molecular biology can play a role in asking biological questions beyond human health and then help them appreciate how a broader number of organisms on this planet are worth studying and thinking about. During this Anthropocene period, one of the things that we see is a lot of important organisms facing difficult times, and pollinators, those animals that help to pollinate plants both in agricultural ecosystems but also natural ecosystems, are really facing the brunt of those difficulties. One goal of mine is to bring those ideas to students who are more focused on human health than environmental issues or ecology issues.

The class where I have been able to do that most easily is an upper-level cell biology lab class where I have been slowly shifting the focus so that we can frame it in the context of how organisms, especially pollinators, and especially bee pollinators, are dealing with stress at the molecular and cellular level.[3] The focus of the class has changed from one on human health to one on how disease happens in other organisms like pollinators and how we are playing a role in that. My belief is that most of these students are going to continue with their dreams of working in the healthcare industry, but they will have the knowledge about bees from this class and a greater appreciation for ecological issues when they move on into their lives. We have been able to expand our focus from looking at honey bees, which I call the gateway bee, to all these other different bee species. There are actually around 20,000 bee species on the planet,[4] and most students are not aware of that when they start the course. These other bee species are so important in so many different ecological contexts, and they do not have the same level of visibility and appreciation that honeybees have. Broadening the horizons of students who love molecular biology but are only thinking about it in a certain context is our main goal in that class. Still, it only reaches a few students every year. We have begun using honeybees in some of the introductory lab classes to reach more students.

Angelo Caglioti In terms of history, you might be wondering, what does history have to do with climate change and the Anthropocene? Probably most people think of history as a very human-centered study of the past. Whereas climate change, of

---

[3] Snow this volume.
[4] Michener 2007.

course, seems like a threat. It is a threat to our future. Yet history is connected to the Anthropocene – and the question why it is a useful concept – because of the long-term perspective that it offers to think about it over time, especially over the historical perspective of geological time. Geological time is so vast that it is really hard for us to understand, grasp, and conceptualize. So, if you compare it to an hour, then of that, we experience the appearance of humans taking place more or less in the last minute. Within the last minute, the Anthropocene occupies the very last seconds. Everything that took the entire hour to put together – the earth, the climate – has been changing in the very last seconds of geological time much more rapidly and dramatically than over the rest of the hour. For me, this is the best answer to the question: why do we need history? We need to study and understand what has been happening in these last minutes.

The way I teach this point in class is by offering a short global history of the relationship between society, humanity, and the climate from two perspectives. The first one is the perspective of environmental history, which shows the material changes that brought us here, such as farming, industrialization, and the so-called "Great Acceleration" in the twentieth century. The second perspective is from the history of science. This approach contributes to our understanding of the concepts in use to describe and talk about climate. What is the "climate"? How did we come to think about the climate as just one global phenomenon? Etymologically, "climate" comes from the plural of the Greek word, *klima, klimata*, which means inclinations or latitudes, a word used to describe how different regions of the Earth at varying latitudes enjoyed different temperatures and climatic characteristics. So how did we come to think about the "climate" not as a plurality of regional phenomena, but as a global scientific idea? This cultural and intellectual transformation pertains to the history of science. Even a concept like the Anthropocene is not neutral. As María pointed out, it can be critiqued from the angle of climate change denialists. But academics and activists have also criticized it for not capturing enough of the process that produced the Anthropocene. If you look at how we got here, the history of global warming is another side of the history of global inequalities. Not all countries and societies contributed in the same way to the pollution levels that we experience today. That historical disparity is a huge issue for the success of international negotiations to address climate change. History as a discipline sheds light on the origins of these challenges. Climate change is not just an environmental, scientific, or policy issue. Its history needs to take into account all these different aspects and their relationship. Thus, my history lecture class trains students coming from different disciplines and going into different fields to analyze the history of how we got here. In short, history helps to connect the local and the global challenges of climate change.

Andrew Crowther I teach an upper-level physical chemistry elective course that is focused on the physical chemistry of climate and energy. For a little context, the two classic physical chemistry lecture courses are quantum chemistry and thermodynamics. Quantum chemistry and thermodynamics are usually semester-long courses. We lay a lot of conceptual foundations in these courses, along with introducing applications of these concepts relevant to chemical research. In the physical chemistry elective course, I build on these introductory courses by applying the previously learned foundations to the topics of climate and energy. Physical chemistry is integral to fundamental climate science and to transferring energy between different forms. In the elective course, we talk about the mechanism of global warming, which involves the vibrations of molecules in the atmosphere absorbing light. We also talk about how solar cells and batteries work. We can take a lot of the foundations that students have developed in earlier courses and apply them to these different topics.

Aastha Jain Fantastic. Thank you all so much for that overview. I would love to now talk about how your approach to this theme extends beyond the classroom and how you integrate this theme beyond just the classroom space?

Andrew Crowther My elective course was designed to focus on climate and energy from the beginning, but I have incorporated some of these ideas beyond this course over the years, primarily into other courses. For example, I often teach Thermodynamics and Kinetics. Thermodynamics places fundamental limits on how efficiently you can convert energy from one form to another. When I teach this topic, I now make specific points about how power plants or other modes of generating energy have an inefficiency to them that is inherent. There are fundamental limits on the efficiencies of these processes. So, we talk about how there is wasted energy that is not wasted on purpose. There is actually just a fundamental limit on how efficient power plants can be. That is one example of how I have started to bring some of these ideas into earlier courses.

María Rivera Maulucci The evolution in my courses has been about trying to sustain conversations about sustainability across the semester. In the beginning, I would introduce the education for sustainability standards, and I actually had a different framework that I used until I found the one from the Cloud Institute,[5] which is more comprehensive. But when I first started bringing sustainability in a couple of years ago, we would have one class where I would introduce the sustainability standards. We would talk about them and have a great discussion about it but that discussion was not really connected with what came next. What I found was that if you really want to sustain these conversations across the semester, it has to be part of what I am asking students to do. That meant redesigning the assignments or the reflections that students had to complete so that it was part of what my students were doing.

So, I started incorporating sustainability into what my students had to do when they were placed in schools. Right now, they are teaching over Zoom, but they are usually placed in schools and they work with classroom teachers to develop science curricula. So, in one course, I said, you have to develop a sustainability science unit. My other course focuses on integrating engineering with the science curriculum. So, I said, you have to develop engineering design units that have a sustainability

---

[5] The Cloud Institute 2010.

theme. For example, one team was teaching a unit on the water cycle, and they worked with the teacher to develop a unit on flooding. The students had to design flood mitigation methods, which they tested to see if they worked. Another team looked at water quality, and students had to design water treatment devices, which they tested to see how clean the water was. Their work was embedded in a unit on earth materials.

By requiring a sustainability theme, my students were not just having this great discussion at the beginning of the semester as to why sustainability connects to science education. Now, they actually had to do something with it. And that is the key. When we ask students to do something with these ideas, that is when they start to own the ideas, and start to really think about them, and start to incorporate them into their practice, and start to think that sustainability is really important. I need to know more about it. And then, the ways that they incorporate sustainability are incredibly creative. So, for me, it needs to be a theme that we talk about at the beginning of the course because it is important, but then we have to keep returning to it, and it has to be part of the work that my students are doing across the semester in their field placements.

Angelo Caglioti It is easy to feel powerless in the face of such a global and complex issue as climate change. After learning about it, students naturally wonder: "Okay, so what can I do?" My goal in teaching this subject is to empower students. They should feel that they can always make an impact regardless of their major, regardless of what they will be doing after college, and regardless of where they are situated and where they come from, precisely because this is such a global issue. "Think globally and act locally", to use an expression that is very common among environmentalists.

Regarding the theme of sustainability, some of my own research deals with the crucial question of what is sustainable development. I got into this topic while working on the history of colonial meteorology in Africa. It quickly became very clear to me that studying only colonial science, which, of course, was political by itself, was insufficient. Studying only science would miss the environmental aspects of European imperialism and how it affected the rest of the world. Sustainability is a vital goal not just in the West, but also in developing countries. In fact, it is an even bigger challenge for them. For example, in 2019 I was in Munich, Germany, as a fellow at the Rachel Carson Center for Environment and Society. Munich is one of the most environmentally-friendly cities I have ever lived in. It is also one of the wealthiest cities in Germany. In the same year, I traveled to Ethiopia for research purposes, where you need to always drink bottled water in order to avoid water-transmittable diseases. This practice produces a great amount of waste, but often environmentalism is perceived as a white-man's problem. To conclude, we need to think about sustainability not just in our backyard here, but in a larger global perspective, especially where local socio-economic conditions make it a more challenging target.

Jonathan Snow In the beginning, my class was really focused on trying to bring sustainability ideas to students that were maybe not as familiar with ecological questions and pollinators. What I found is that they just took to it immediately. So, I kept putting more and more of that into the class. Also, it is pretty cool to be in a cell biology class where we have to go outside the first day and try to learn about some bees, see some bees, and then also work with actual bees before we dive into the cell and molecular way of thinking about biology. I think the students really enjoy that. And to María's point, we try to frame the class with those ideas of pollinator health and how the Anthropocene is impacting that and then end up with an independent project so that they are able to bring it all back together at the end of the class to understand the big picture. The independent project lasts for two or three weeks, and the students are able to apply what they have learned in terms of cell biology lab techniques and ways of thinking to try to answer a question about pollinator health. Not only do they really enjoy that, but it helps them go back to the framing of the class.

In terms of going outside the class and bringing this to a broader community, I was a part of the Sustainable Practices Committee, and we were able to start thinking about how our campus is bee-friendly. One thing people can do is try to create environments that are friendly for other organisms, plants, invertebrates, and insects especially. So we tried to do that, initially just on Barnard's campus. We worked with facilities and others to transform our landscaping protocol into more organic and sustainable practices. Then, collaborating with Terryanne Maenza-Gmelch (Environmental Science) and Hilary Callahan (Biology) and some others, and Angie Patterson at Black Rock Forest we began to think about how we might do educational outreach beyond just Barnard students to include other groups. This is sort of in the infancy stage right now, but I think those are different ways of trying to help sustainability go outside of the college classroom.

Yuval Dinoor I think your last answer there is actually a great transition to our last question for you all, which is, what might be the opportunities that you've identified for using climate change and the Anthropocene to build connections between different disciplines at the university? And what are you imagining as other opportunities for different disciplines to come together to do this work going forward?

Angelo Caglioti I hope that conversations like this one will lead to more opportunities for interdisciplinary collaboration. For example, I have started partnering with the Computer Science Department thanks to Rebecca Wright (Computer Science), who involved me in the Computing Fellows program. In the class titled "Climate and History" that I am teaching this semester, we included a visualization workshop about climate mapping and modeling. In it, Bryn Stecher, a student from the Computing Fellows program, demonstrated how scientists turn climate data into maps and graphs. This visualization workshop shows the importance of modeling to understand past climate, and especially to predict climate change in the future. It clarifies that climate data visualization does not just produce aesthetically pleasing maps. Rather, graphs and maps have an important epistemological role in how climate science is made. I was grateful for the opportunity to run this workshop this semester and I very much welcome other opportunities to collaborate in the future with other departments. Because it is possible to apply a historical approach to any subject, there are endless opportunities of collaboration on the history of the environment and climate sciences.

Jonathan Snow When you think about pollinators, one of the things that you can think about ecologically is that an environment that is healthy for pollinators is also healthy for people. So you automatically get into aspects of climate justice, and where people live, and how the environment they are in is going to impact their health. But there are also obvious links to issues like food justice because bees are a huge part of the agricultural system. The agricultural system, such as it is today, is not necessarily providing food for everybody in the same way. So there are lots of opportunities to try to think about food justice as a part of social justice. Thinking about pollinators is a really obvious connection to that work, and that is something that also is happening in the Barnard community and the broader communities of Morningside Heights and Harlem.

Andrew Crowther The science of climate and energy is extremely interdisciplinary. It includes chemists, physicists, engineers, environmental scientists, biologists, and many others. It is really a huge, broad range of people. We read a paper from the literature each week in my elective course, so we can start to think about having guest lectures or guests to help guide discussions in areas that are a little bit more in their scientific discipline, rather than the physical chemistry topics that I am more familiar with. This is one way that I could imagine bringing other disciplines into the courses that I have taught.

María Rivera Maulucci For me, the connections are, in a way, super easy because my discipline is education. There is a tremendous amount of scholarship in, for example, environmental education. Historically, that is where you could say sustainability has its roots. So, everything from reducing, reusing, recycling of the 1970s, moving onward, to the current time where we are now, environmental education is certainly an easy connection to make. More recently, in science education, there has also been an emphasis on socio-scientific issues or the idea that there are issues that cross borders with other disciplines that have to do with society but also have strong science components. These issues do not just sit in the realm of science. I had the pleasure of collaborating with Jon Snow, and we wrote an article together about bees, looking at pollinators as a socio-scientific issue.[6] Yes, they are pollinators, and so they are essential to our agricultural system, but they are also an introduced species. So, from an ecological perspective, the honey bees came from Europe. The Europeans brought the bees with them when they came here, but they also are out-competing many native pollinators. So how do we make sure that we have a vast supply of honey bees to pollinate our major crops without eliminating all the native pollinators? This brings in questions about industrial agriculture. Also, now that we have them here and we need them to pollinate our crops, how can we treat them better? How can we make sure that they are healthier? How can we prevent things like colony collapse disorder? Some of Jon's research looks at specific ways to identify the pathogens affecting the bees. If we can have a diagnostic assay for that, we can actually treat the bees only when we know the colony has the dis-

---

[6] Snow and Rivera Maulucci 2017.

ease, as opposed to prophylactically throwing antifungal or antibacterial agents into their colonies that we know are also toxic for the bees, and those toxins can get into our food supply. So, it is interesting to think about the interdisciplinarity of it. Going back to those pathways of sustainability that I mentioned earlier, there are just so many pathways to sustainability, and so there are so many pathways to interdisciplinarity. The connections are sort of endless for us.

Yuval Dinoor Thank you all so much for sharing. I think it's so inspiring to hear about all of the ways that you are incorporating different knowledge and different resources in our campus community to do this work. I can speak for myself as a student, thinking about how I engage with these issues and imagining all the different possibilities that all the professors here can take on in their coursework.

23.2 Question and Answer Session

Yuval Dinoor Before we get further into those conversations in the breakout rooms, we are going to take a little bit of time to hear any questions that folks have that they'd like to ask in front of the entire group.

Severin Fowles (Anthropology) Jon, I wanted to pick up on something you mentioned in regards to a Black Rock project that you were developing. I am really interested in how people are mobilizing that space in their research or teaching. Are there other initiatives like that that are place-based projects, perhaps in Manhattan itself or somewhere that we could incorporate into our classes during the academic year? I imagine Black Rock might be singular in some ways, but perhaps there are other examples of place space work in the region.

Jonathan Snow Terryanne or Hilary might be the best people to talk about Black Rock specifically because it is definitely in the early phases.

Terryanne Maenza-Gmelch (Environmental Science): First, I would like to remind everyone that Black Rock Forest is available for just about anything from art to archeology to architecture. So, if anyone wants to try and get up there, just please email me and I will make it happen for you. I go up for my own work, but I have also gone up with faculty like María Rivera Maulucci and Lisa Northrop and Jessica Goldstein, and Hilary Callahan, where we hike around and generate ideas and it is fabulous. So if anyone is interested in doing that, contact me, please. The project that Jon is talking about is a mentoring project between undergraduates, high school students, and professors trying to increase awareness about insect conservation, and that is very much in the infant stage. We have not really moved much on that yet because of pandemic reasons.

María Rivera Maulucci   What I will say to Angelo is that the Black Rock Forest has this incredible land-use history that could be really interesting to incorporate into some of the work that you are doing on how the land use shifted over time and how it became a forest. But of course, there were other things happening on that land prior to becoming what it is today. So, it is really interesting.

Aastha Jain   We have a question in the chat for Angelo. Can we hear your thoughts on teaching and studying the history of discourse and research about climate?

Angelo Caglioti   Before answering this question, I would like to add a point building on María's intervention. Many economic historians turned to environmental history because they started raising interesting questions about land, property rights, and how they were used. However, environmental history shifted the focus of their analysis. Environmental historians examine not just human beings and their economic activities, but also – for example – the history of the forest, of the animals living in it, or of the species of plants that make up a forest. Environmental history is centered on the relationship between society and nature as it developed over time, including species other than just humans.

Regarding your question about the study of the history of discourse and research on climate, some of the key questions I would ask are: Which concepts were used in the past to describe the climate? How did they develop? Studying the history of research and discourse about climate requires examining how the climate was described in the past. I like to tell my students, especially those that do come from the environmental sciences, that before the 1800s they would have been called "natural historians", because natural history was the discipline describing the history of the earth, the history of nature. This point, I believe, creates a stronger common ground between what scientists and humanists do. It emphasizes the analogies rather than the differences between the "two cultures." The development of meteorology and climate science is hard to pigeonhole in the history of a single scientific discipline. It seems very recent, but it actually belonged to the broader scientific inquiries of natural history for many centuries.

Aastha Jain   We have another question in the chat from Sandra, who is curious about what types of interdisciplinary courses the panelists teach or would like to teach and what other disciplines would you envision co-building a course with, and the benefits or disadvantages of a co-taught course in your opinions?

María Rivera Maulucci   I have a background in forestry, my master's degree is in forest science. So, I would love to teach a class that goes to the forest, and we can do science methods there and really get into sustainability and work with my friends Terryanne and Hilary, and other folks that do work in the forest. To me, that would be a wonderful class. Of course, there are the issues of getting there, getting our students there. And then, I think challenges around interdisciplinary courses come when we talk about, how do you get credit for a course? So, does a team-taught course count as one? Or does it count as half? How do we count it? Because I think team-taught courses, actually, in a lot of ways require more work. It requires a collaborative piece. There are more conversations happening around what we are doing and why we are doing it and how we are doing it.

Hilary S. Callahan (Biology): I just want to ask a question about the urban topic, which none of you really brought up. The Anthropocene has a specific meaning and we are talking about it in terms of its climate impact, which is its global impact, for example, thinking about it in terms of ocean sediments, which is a globalized phenomenon. But another thing that is happening globally in a multiplicity of specific locations is the transition to an urban world. So we have to acknowledge the fact that more than half of the human population is now living in cities and the importance of cities. One of the topics that now we are dwelling on and that I do want to encourage is taking advantage of Black Rock Forest and taking advantage of some reasonable level of well-thought-out travel where the impact of the travel in terms of its carbon footprint is weighed against the benefit for learning and for research and new knowledge. I think there is this tendency to sometimes overlook, in thinking about the Anthropocene, that we have all these cities, but that does not mean that there is no more nature in the cities. The cities are just as much part of the solution as they are part of the problem that we are encountering in the Anthropocene. So, I am just wondering if anybody would like to comment about the urban context of our campus and how the concept of cities might enter into what you're teaching.

Jonathan Snow I think that what you are talking about definitely impacts things like food justice, where people in urban settings have vastly different access to nutritious food than people in other settings. I was lucky enough to have a colony of bees with Harlem Grown, which is an organization that tries to produce food locally and also educate people about growing their own food. That was a really fun collaboration. I think those kinds of collaborations can just as easily be done here as somewhere in a suburban or rural context.

Aastha Jain Amazing. For the sake of time, I want to thank all our panelists so much for all of their insights.

References

Michener, C. D. (2007). *The bees of the world* (2nd ed.). John Hopkins University Press.
Paris, D., Alim, S., & Ferlazzo, L. (2017). Author interview: 'Culturally sustaining pedagogies'. *Education Week.* Accessed from: https://www.edweek.org/teaching-learning/opinion-author-interview-culturally-sustaining-pedagogies/2017/07
Snow, J. (this volume). What does cell biology have to do with saving pollinators? In M. S. Rivera Maulucci, S. Pfirman, & H. S. Callahan (Eds.), *Transforming sustainability research and teaching: Discourses on justice, inclusion, and authenticity.* Springer.

Snow, J., & Rivera Maulucci, M. S. (2017). You can give a bee some water, but you cannot make her drink: A socioscientific approach to honey bees in science education. In M. P. Mueller, D. Tippins, & A. J. Stewart (Eds.), *Animals in science education* (pp. 15–28). Springer.

The Cloud Institute. (2010). *Cloud education for sustainability standards*. Accessed from: http://www.cloudinstitute.org/cloud-efs-standards

Angelo Caglioti is an Assistant Professor of History at Barnard College. His research focuses on environmental history, history of science, and late modern European history, with a particular emphasis on nineteenth and twentieth-century Italy. He teaches courses about climate history and the history of climate science.

Jonathan Snow is an Associate Professor of Biology at Barnard College. His research focuses on the cellular stress responses of the honey bee, a species that is crucial to agricultural and ecological systems and the infection of honey bees by the microsporidia species *Nosema ceranae*. He teaches courses in molecular and cellular biology.

Andrew Crowther is Associate Professor of Chemistry at Barnard College. His research focuses on using Raman spectroscopy to study fundamental properties of nanomaterials, including the vibrational structure of nanocrystals and the chemistry of carbon-based nanostructures. He teaches physical chemistry lecture and laboratory courses.

María S. Rivera Maulucci is the Ann Whitney Olin Professor of Education at Barnard College. Her research focuses on the role of language, identity, and emotions in learning to teach science for social justice and sustainability. She teaches courses in science and mathematics pedagogy and teacher education for social justice.

Yuval Dinoor '21, holds a BA from Barnard College in American Studies concentrating in Education, Civic Engagement and Community. Her research focuses on creating experimental learning spaces to build politically active communities; recently, she aired a podcast series exploring the Highlander Research and Education Center's pedagogical approaches to community organizing.

Aastha Jain '21, graduated from Barnard College in April 2021 where she studied Economics with a concentration in Environmental Science. She is passionate about intersectional climate activism and animal rights, and hopes to combine her interests to explore and pursue new macroeconomic models that consider people and the planet ahead of profit.

Chapter 24
Panel Three: Environmental Justice, March 26, 2021

Elizabeth M. Cook, Manu Karuka, Angela M. Simms, Joscelyn Jurich, Yuval Dinoor, and Rachel Elkis

24.1 Panel Discussion

Yuval Dinoor (She/Her) I am co-moderating this session with Rachel, who will introduce herself shortly. Before we get started, I would like to take a moment to offer a land acknowledgment that we here at the CEP put together… (See Rosales Introduction). Thank you for listening. I will pass it to Rachel to begin our panel conversation.

Rachel Elkis Thank you. As Yuval mentioned, I am Rachel. I am a junior at Barnard studying Economics, and I am the former SGA representative for sustainability. I am excited to be here. I would love to get started by having our faculty

---

Panelists: Elizabeth M. Cook, Manu Karuka, Angela M. Simms
Joscelyn Jurich (Ed.)
Student Moderators: Yuval Dinoor, Rachel Elkis

E. M. Cook (✉)
Environmental Science, Barnard College, New York, NY, USA
e-mail: ecook@barnard.edu

M. Karuka
Department of American Studies, Barnard College, New York, NY, USA

A. M. Simms
Departments of Sociology and Urban Studies, Barnard College, New York, NY, USA

J. Jurich
Center for Engaged Pedagogy, Barnard College, New York, NY, USA

Y. Dinoor · R. Elkis
Barnard College, New York, NY, USA

© The Author(s) 2023
M. S. Rivera Maulucci et al. (eds.), *Transforming Education for Sustainability*,
Environmental Discourses in Science Education 7,
https://doi.org/10.1007/978-3-031-13536-1_24

panelists introduce themselves with name, pronouns, department, courses you are teaching, and your role within and outside the Barnard community.

Elizabeth Cook Hi everyone, I can start us off. I am Elizabeth Cook. I use she/her pronouns, and I am an Assistant Professor in the Environmental Science department. I teach a couple of classes right now, including our senior seminar, in which several students are researching environmental justice and injustices in their senior theses. I also teach an *Environmental Data Analysis* class where environmental justice is incorporated into case studies of data analyses using New York City-related data. Finally, I teach *Urban Ecosystems,* which is in the Environmental Science department, but it is an interdisciplinary course, and we discuss environmental justice topics in that as well. I am relatively new to Barnard. Last year was my first year, and I was only on campus for a short time before we were swept away by Covid, so I am still figuring out my role, but I am really happy to be here.

Manu Karuka I am Manu Karuka. I use he/him pronouns. I am an Assistant Professor in American Studies, and none of my courses have an explicit focus on environmental justice. But my courses do focus on imperialism, racism, and colonialism. That is the entry point in ways that strongly resonate with the powerful land acknowledgment that Yuval read. When we think about these themes, it impels us to think about environmental justice. It comes into the conversation in a first-year seminar called *Liberation.* I am teaching an *Introduction to American Studies* course where environmental justice comes into the conversation as well. In addition to these broader overarching themes of imperialism, racism, and colonialism, I think environmental justice is also a way to think about place and our relationships and responsibilities to place.

I have learned a great deal in our preparations for this conversation, just thinking across our divisions in the college. One of the things I have learned is how all of us have thought about Harlem, Barnard's location in Harlem, and how Harlem connects to our disciplines, and with our teaching and research approaches. We found that we even discuss some of the same campaigns in our courses. For example, WE ACT for Environmental Justice,[1] a community organization based in Harlem, comes up in our courses in different ways, so that has been exciting and enriching for me to have that conversation with my colleagues on this panel. In terms of my role, I have been helping my colleagues to build the department of American Studies and I feel very committed to helping to build the Critical Consortium for Interdisciplinary Studies, which I think is a really unique formation at Barnard. It brings together American Studies, Africana Studies, and Women's, Gender and Sexuality Studies. This is a group of scholars and teachers for whom questions of environmental justice are really at the center of how our conversations have been shaped in the past and are shaping our visions for the future.

---

[1] WE ACT for Environmental Justice (2021).https://www.weact.org/

Angela Simms Thank you, Manu, for setting me up so well. My courses dovetail well with his. But first, my name is Angie Simms. For Elizabeth and me, this is our second year, so I had a whopping nine months in Harlem and I am now working remotely from Maryland. I use she/her pronouns. I teach two lectures: *Race, Ethnicity & Society* and *Metropolitics of Race and Place,* and two seminars: *Suburbs, Racism, and the United States Opportunity Structure* and *Advanced Topics in Race.* In these courses, we think about the socio-historical context of the development of race and racism and examine the interactions between capitalism and racism. One of the big takeaways of my classes is that racism comes from the needs of capitalists. We also investigate how racism shapes metropolitan space, including relationships within and between cities and suburbs, centering historical and social processes in regions. Environmental justice is one of many aspects of unjust power dynamics we cover. Social connections between people and their resource exchanges, particularly in the Global North, are not sustainable because they distribute benefits and burdens unevenly among people and diminish the Earth's capacity to maintain equilibrium – this leads to social and natural catastrophes. My courses equip students with language for understanding environmental justice concerns as embedded in power structures across levels and types of social organization. We also address cumulative effects over time. Additionally, I am working on the exhibit "Undesign the Redline," which highlights how racialized capitalism has led to different kinds of investment and divestment in Black, Latinx, and White neighborhoods in New York, resulting in the racial inequities we see today.

Outside of Barnard, I am a member of Renaissance Church. I have been there for about a year. As a Christian, I think about the earth in terms of stewarding what we have been bequeathed by the Lord. When Genesis says, we have "dominion" over the Earth, that indicates we have responsibility for the planet's wellbeing. It's not an invitation to manipulate the Earth's resources purely for our short-term gain. In addition, I have been doing a series of anti-racism workshops with churches and civic groups about what racism is and how it's connected to unjust social relationships. I want all people, whether academics, Christians or otherwise, to be socio-historically grounded and participate skillfully in shaping society in ways increasing humans,' other creatures,' and our natural environments' capacity to flourish.

Yuval Dinoor Thank you all for introducing yourselves. I am so excited to hear the rest of this conversation. To get started, I am really curious to hear how your integration of environmental justice in your coursework or in your external work as well, has evolved over time. Manu's *What is American Studies* course is the first place I learned about environmental racism back in 2018, which is part of what got me so excited to commit to the American Studies major. I am curious how, as current events and different world transformations have progressed over the past few years, you may have changed how you speak about environmental justice in your classrooms.

Elizabeth Cook I will start by putting myself out there a little bit and being honest. When I started teaching the *Urban Ecosystems* class, which I have taught now for several years, including at other institutions, at first, I only dedicated a tiny portion of one class, one individual class, to environmental justice. Even though it is something that I care about and have been considering in my work and research for a long time, I thought other urban ecologists didn't value it very much as a critical part of the field, and I think, to some extent, that might have been true. But now, I dedicate much more class time to discussing environmental justice, so that is one obvious evolution.

The field of and scholarship on environmental justice has also evolved. Early work on the environmental and ecological framework of environmental justice was very much done in terms of distributional justice. Distributional justice frameworks ask which communities are impacted disproportionately by either environmental disamenities or hazards. For example, which communities are more impacted by higher heavy metal concentrations in different locations? Or which communities have greater access to particular amenities like green space and parks? Early on, when I was teaching about environmental justice, we discussed the distributional aspect of environmental justice, largely dominated by analyses focusing on classism and wealth. That framing has evolved quite a bit in the last several years in the ecology field, and I know it has already evolved quite a bit in other fields. In terms of ecology, recently, there has been much more discussion around the role of structural racism in driving those differences and taking a procedural justice framing. With this framing, we ask, what are the drivers of those inequitable distributions of amenities and disamenities in different communities? I try to bring both of these perspectives and framings into my work now.

One other point is that early environmental justice work from an ecology perspective focused on topics such as access to green space and distribution of toxins or heavy metals. Now, the ecology work on environmental justice is not just focusing on green space and species richness in different communities but also considering how environmental injustices can drive evolutionary trends in cities differently. For example, the size of the green space can impact how species interact, which may lead to different trends and evolutionary patterns.[2] So there is really cool scholarship happening on these topics, and I think that is a huge advancement for the urban ecological community as well.

Manu Karuka I will jump in next. Thanks, and Yuval, thanks for sharing that about the intro classes; that makes me really happy to hear. One of the ways that I teach this and other topics is through dialogue with students. It is something I have taught on a number of campuses, and there is something I have felt that is distinct about teaching at Barnard, the dialogical relationship between instructor and students even with a big class like, "*What is American Studies?*" When I first started teaching at Barnard, it was during the first wave of Black Lives Matter. Students wanted to start class by giving reports to each other. I was teaching a class called the "*Profits of Race,*" about the political economy of racism, and the uprising in

---

[2] Schell et al. 2020, Des Roches et al. 2021.

Baltimore was happening. In that context, activists and scholars were pointing out that Freddie Gray was killed by the Baltimore police, but he had been a victim of lead poisoning as an infant. Activists and intellectuals were asking questions about the slow death that someone like Freddie Gray had been subjected to, even before that horrific violence that he suffered in the back of a police van. When the #NoDAPL camps at Standing Rock were happening, every day at the beginning of class, students wanted to spend about 10 minutes sharing the news from the day before, and so there was an immediacy to these questions. It is harder for me right now on Zoom to get a sense of what the immediacy is. Of course, we have the pandemic and the lockdown. It is staring us in the face, but it is harder to have that space where we begin sharing information, which gives me a sense of seeing where students are getting information, and what questions they are asking of the information they are getting. Last week, I was talking about the news about Pfizer, that they have been negotiating with the government of Argentina and demanding collateral on infrastructure and national debt to purchase vaccines. There are countries around the world that have been asked to put their roads and bridges up as collateral to buy vaccines. It is harder for me to gauge, but I think part of it is really trying to understand what is happening right now and what the students are thinking about and asking them how to respond. What are the collective questions that are arising for us?

Angela Simms I will take a step back to discuss what motivates my coursework and my research. Before Barnard, I had a mini-career in public policy. I was a legislative analyst at the U.S. Office of Management and Budget during the George W. Bush and Barack Obama Administrations. My scholarship is at the intersection of race and class and thinking about the policy implications for reducing social inequities. Specifically, I study political economy through the lens of the Black middle class. The Black middle class is a window into understanding how class and race interact because we see a group with class resources still facing legacy and contemporary headwinds from racist policies and market practices. One of the points I highlight for students is the importance of shared authority between levels of government. I tell them their new favorite "f word" is federalism. Federal, state, and local government action are at the heart of the geography of opportunity. There are nestings, clusters, of public and private authority and resources that elite Whites are disproportionately able to leverage to pursue their interests at others' expense. One of the historical events I highlight regarding non-Whites resisting White dominance in terms of access to habitable neighborhoods is the 1991 National People of Color Environmental Leadership Summit. Here we see Indigenous, Black, and Latinx folk coming together to discuss the common burdens they bear and how to pursue policies and other social processes that correct harms and envision a future where people and the natural environment are whole.

My course material has evolved over the past two years at Barnard. Ruth Wilson Gilmore's new book (2022), *Changing Everything* is on my reading list. My syllabus includes, among others, works by Bob Bullard, a professor at Texas Southern. He's the self-styled "father of environmental justice" and has written several books about what environmental justice entails.[3] I invite students to think about the options racial groups have for creating life-sustaining spaces in a context where they have differential amounts of political and economic power. When was the EPA established? In the wake of the Modern Women's Rights Movement. While People of Color have been petitioning for environmental justice for decades, it was not until White women took on the issue that significant political change occurred. Even then, it was focused on issues that affected them and their families. Also important is how incremental wins can lead to major breakthroughs. In 1994, the Clinton Administration established an environmental justice office at the EPA. While it did not correct past harms, it was charged with equitable distribution of negative amenities, such as trash dumps. Thus, the justice movements of People of Color and Women led to a shift in policy focus.

I also use Daniel Faber's chapter in *Environmental Justice and Environmentalism: The Social Justice Challenge to the Environmental Movement* (2007), called "A More Productive Environmental Justice Politics." He offers three great principles for thinking through what sound environmental stewardship looks like. The first is "the precautionary principle" of risk assessment. The onus is on the potential polluter to demonstrate to us that their product is not harmful. If we have a history of pollution, and we have a history of certain people bearing the weight of that, then why is it that people have to sue after they realize they have been poisoned? Second is the "substitution principle," replacing toxic chemicals with non-toxic alternatives. Finally, there is the "clean production principle," which means our default is to work in concert with the Earth's processes to achieve economic goals and improve quality of life. Again, we see the theme of stewardship, as opposed to commodification at any cost. Faber's principles are not anti-capitalist per se, but they do challenge capitalist forms to work in concert with values we believe lead to thriving people and a healthy planet.

In doing my fieldwork in Prince George's County, Maryland, I became familiar with the School of Public Health at the University of Maryland. Sacoby Wilson directs the Community Engagement, Environmental Justice, and Health (CEEJH) Laboratory.[4] This project integrates the public in tracking pollution and devising remedies to reduce negative fallout. Wilson and his team equip people to be citizen scientists who gather data and deploy that data to engage decision-makers. CEEHJ models how different stakeholders can work together – from health professionals to teachers and students, advocacy groups, and policymakers.

Of course, the horizon on environmental justice and climate change issues is always moving out ahead of us. Therefore, I hope I empower students with transferable skills and a core vocabulary for thinking critically about how to intervene in natural processes to sustain a high quality of life while also achieving equity across social groups and a planet capable of renewing itself over the long run. Still, this set of topics often leaves students discouraged. "How are we going to fix things?" they

---

[3] i.e., Bullard 1994.

[4] University of Maryland 2021a, b.

ask. I tell them: "You're standing on the shoulders of giants." People have been resisting oppression since people have existed. Now you have the baton, and you can learn lessons from your elders, empowering you to be more effective. This is a life-long commitment. Engage strategically with your time, talent, and material resources. Press forward alongside like-minded others in organizations. My courses involve real talk about what is happening, alongside skill and knowledge building for being effective as we negotiate with each other about our future.

Rachel Elkis That just led us perfectly into our next question: how we can extend discussions about current events outside of the classroom through our work and how do you encourage your students to do so.

Elizabeth Cook I can speak about that and build on what Angie just said. My research uses participatory methods to engage communities and different stakeholders in thinking about the future. Through participatory engagement processes, such as the development of positive future visions, we explore innovative solutions to address some of the challenges that we are facing, particularly extreme weather-related events. This work lends itself to thinking about climate justice and the impacts of climate change on communities. When we talk to communities about the changing climate and the future, they are often thinking about these topics in addition to meeting their basic needs, which tends to go hand-in-hand with considering environmental justice and climate justice. So, as we develop scenarios or visions for the future, equity and addressing environmental justice is really front and center, whether it is in conversations with community members who are living in Harlem or whether it is with decision-makers in the Mayor's Office of Climate Resiliency in New York City. Through these conversations outside of the classroom and this framing of the future, I try to engage students, in the same way, to ensure that they are thinking about how the topics apply to their own lives, in their own communities where they live. For example, in the *Urban Ecosystems* class, the students engage in a visioning exercise to imagine a positive, equitable future for New York City in 2080. It is a creative and important exercise to discuss how we address environmental justice, and it is easy to do because most of the students are living in New York City, and it feels directly applicable.

Manu Karuka I also try to address some of these questions in my assignments. The *Intro to American Studies* class that I am teaching right now is organized around Dr. King's speech, "A Time to Break Silence," which he gave at Riverside Memorial Church. He spoke out against the Vietnam War, and in that speech, he spoke about the giant triplets of racism, militarism, and extreme materialism. So the class is organized into three sections: racism, war, and poverty. There are three assignments, and the first essay, which students are writing now, is, "What will I do to fight racism?" They have definitions of racism that I have given them and some material that I want them to have, and there is a range of ways students respond to it. Some take it as an academic exercise, and some students take this very personally. I encourage them to do so. I tell them that their graders and I are just there to encourage them through the process. Some write it as a letter to themselves, something that will hopefully be meaningful to them later on. For those students who take this really personally, it is a very, very difficult assignment to do. And it becomes a way for them to think about what skills they are developing at Barnard: "What is my major? What am I learning in this major? Where am I from? What have I learned and how am I assessing racism in this society, and how can I contribute to fighting it in really concrete ways?" Not all of the students, but many of them, bring in questions of environmental justice. The second paper is about what I will do to fight racism and war. So they are accumulating their analysis and thinking.

My hope is, by the end of the semester, in a purely personal way, they have given themselves a sense of what they can actually do. I think a lot of students wrestle with this question, and then, when they think about it, they think about the resources they have at their disposal. They start to see, "Oh, actually, I could do this; there are specific things that I can do." They think about participating in collective struggles. It has been a productive space and a moving space to see students work through. When they are in person, I have them do collaborative research projects. Small groups of students are assigned to a state, and they research poverty in a particular state and specific modes of poverty, like hunger and education. One of the modes of poverty that they are looking at is environmental justice, broadly speaking, which is one of the ways that we can see premature death, as Ruth Wilson Gilmore talks about in her definition of racism. That is new research about actual poverty in these different parts of the United States.

Angela Simms Well, I was taking notes on what Manu and Elizabeth discussed because I need to think more concretely about how to expand my assignments. I encourage students to engage in the material and apply it in their own lives. We talked about my favorite "F word," federalism, but in thinking about policy, one of the trends regarding the post-Modern Civil Rights Movement backlash is the issue of disparate impact. The 2001 U.S. Supreme Court case *Alexander v. Sandoval*[5] established the principle that disparate impact does not in itself indicate racism. So if a policy is not explicitly racially-biased, meaning that it is not targeting racial groups, the policy is not unconstitutional. Elizabeth brought up different types of justice. Procedural justice is what we often focus on, due process of the law. Due process is important, but it's not sufficient. The full meaning of justice involves reckoning with power as historically and presently wielded and how resources flow to individuals and groups, and locations. We must account for the sedimentation of history and the roles of government and market institutions in it – and thus how they're situated for helping to create a more equitable society. Whose interests matter to these institutions? Who has the time, money, energy, and skill to navigate and reform them?

In my courses, we also think about intersectional social statuses. What is your race, your class, your gender? One of the things I encourage students to do is to understand their social statuses. For many of my students who are White and who come from privileged backgrounds, there is a sort of eye-opening moment of recognition about what has been happening in their communities. For my students who

---

[5] Legal Law Institute 2001.

have come from backgrounds where they have not had those same resources, say those from the Bronx, they feel validated: "Now I understand why a freeway cuts through our neighborhood and why asthma rates are higher here than in the Upper West Side." One of the exercises students do is write an op-ed. It is an opportunity to take a stand on a policy. This assignment dovetails fairly well with what Manu described. It's not a letter to yourself, but it does ask students to think about what they value and the specific kinds of social change they seek. Another thing I tell students who are White is that there are spaces you are going to be in that I am not and things people will say around you that they will not say around a Black woman. You have a unique opportunity to champion justice when this happens.

All of us have to own our spheres of influence. I mentioned, for example, that I'm a Christian. So when I talk about environmental justice and racism, I use the language of theology when I'm with other believers. We're all made in the image of God. If we are all image-bearers of God, we all ought to have human dignity, and our social systems should reflect that. If some people are treated like trash and other people like humans, that's a profound problem. You are calling God a liar. Certainly, these are uncomfortable conversations, really sitting with the emotional, intellectual, and spiritual weight of this. But the discomfort is unevenly distributed. I, as a Black woman, have to stand up, again and again, to assert that I am as fully human as everyone else. It's exhausting and demonstrates how deeply distorted our social systems are.

The convergence of climate change and environmental justice is critical for racial justice. For too long, they have been separate movements. We know that the burdens of climate change are likely to be borne by people of color and poor people because of the history of capitalism. The people who have benefited the least from these capitalist practices, the people whose bodies have absorbed the negative externalities, are experiencing the effects disproportionately. So we need to think about how we can have a system that really honors these disparate distributions.

Yuval Dinoor   Thank you so much for that. I think that, on that same note of what kind of language and argument engages these conversations in different spaces, I am curious how all of you are seeing environmental justice as an opportunity to come into contact with other disciplines and do cross-departmental work or make these connections across the university that you might not get to really act upon in other kinds of work. What do you imagine those collaborations to be?

Elizabeth Cook   I can start us off. Angie, Manu, and I already have fun ideas about how we can integrate our work and think about next steps. So it is already happening, and thank you for bringing us together to do that. In terms of urban ecology, the field of urban ecology in and of itself is interdisciplinary. So already, there is a lot of cross-talk and integration with social sciences and certainly with planning, urban studies, and architecture, and I feel very grateful to be working with and learning from people who are experts in those disciplines. From an urban ecological perspective, many sub-disciplines are important in understanding the ecological interactions happening within a city. For example, community ecologists are examining how species distributions are different among different neighborhoods, and the environmental justice implications, or, for example, some ecologists study the ecosystem services or the benefits that people get from the environment in those different communities where ecological disparities may exist. As I mentioned at the beginning, there is more recent work by evolutionary biologists and ecologists in cities thinking about how structural racism has driven evolutionary processes in different parts of cities, impacting communities differently – including the benefits that people might get from the environment in their community. So I appreciate that question, and my answer is that urban ecology is already a very interdisciplinary field. I think by bringing in the environmental justice focus, there is room for additional collaborations with other departments and disciplines on campus.

Manu Karuka I want to echo that I appreciate the question, and I think part of the excitement for me just being part of this conversation was to think across our divisions. So when the three of us met before the panel, I talked about teaching *Silent Spring* in the first-year seminar class. We had an interesting conversation about that because it is an old text. It is not current in terms of research, yet it spoke to the students. For me, there is a question about how we read across our literatures in a deep sense – I like to teach texts that I think are foundational – but then, how can we collaborate to build a sense of literacy about the cutting-edge directions various disciplines are taking. I think environmental justice is an opening, potentially, to think across our divisions. So, for example, water is becoming increasingly important in my classes in the last few years. There are approaches to thinking about water that are firmly rooted in Black Studies, Indigenous Studies, and new work in those fields, and also in Latin American Studies. I am thinking about struggles, for example, in Bolivia. It becomes a way to think across geographies, but I do not yet feel literate enough to say what the cutting-edge analysis of water in conversations in Biology is, for example. One of the concrete takeaways the three of us had from our earlier conversation was thinking about doing collaborative or guest lectureships in each other's classes, which I think is an exciting way to go forward. I have an *Intro to American Studies* course and I can bring in someone from a division that seems very far removed in terms of the teaching and the methods to talk to an American Studies course about water. Maybe it could be a mini-lecture, maybe not for the whole class, but I think there are opportunities for us to collaborate as teachers and build those relationships with each other, as colleagues. And I think environmental justice seems like a vibrant site to imagine those conversations across our academic divisions.

Angela Simms Certainly, the lecture swaps intrigue me! Sometimes the best interdisciplinary work is really about grounding yourself in your field and then recognizing your limits – and from there seeking growth by partnering with an expert in another subject, not trying to master another field of research. Pairing with an expert in another discipline means you have a mentor to guide you through the literature. One of my growth edges is knowledge in the biological sciences regarding carbon dioxide and how it affects our planet. Elizabeth and I have talked about microclimates within New York City, but I do not consider myself an expert on microclimates. This is potentially an area where I could partner with Elizabeth, so I can ask more "literate questions," to use Manu's phrase. Elizabeth could share her expertise in my class, and I could offer to her class a deeper dive into the socio-historical context of racism and how environmental justice is situated within it.

We need to think about professors' and students' growth edges and how to pivot out strategically, respecting where we're based while discerning where we are headed next. For instance, many students in my classes are pre-med. Thus, I make sure we talk about environmental justice and climate change in relation to public health, not just the social determinants of health. How are racial groups living in fundamentally different social worlds? Lead contamination is much more likely in Black neighborhoods, relative to White communities. Why is that? In Flint, it took pediatrician Hanna-Attisha's sound of the alarm as she noticed a pattern in Black children's health outcomes.[6] I am excited to hear what others are thinking about and how we might collaborate. Let's start with the low-hanging fruit – the course material we've already prepared and how to share it more broadly across campus.

Rachel Elkis This is fantastic! I am very eager to see how these collaborations come to fruition. Environmental justice is such a unique topic to be able to do that. So that is extremely exciting, and we would love to open up to questions from the audience. If you would like to raise your hand, you can unmute and then just say a question or put it in the chat and we will get started.

Yuval Dinoor One from Leslie.

Leslie Raucher (Assistant Director of Sustainability) Thank you all so much! This was really great. One of the questions that I have, I do not know if anyone is prepared to answer, but eco-anxiety and environmental and ecological grief are starting to get talked about. We are beginning to hear about this from students, and it was touched upon a little bit with the hopelessness and what to do next. I was wondering if you are taking a conscious approach to thinking about this or dealing with it, and what we could do to help provide resources if it is something you are noticing. I am hearing about it from students, but I do not experience it in the classroom, so I was just hoping to get a little bit more of your perspectives.

Manu Karuka I can jump in. My mind immediately goes to some of the lessons I have learned from Indigenous Studies. I learned that there are climate scientists researching carbon dioxide levels in the Earth's atmosphere who found that in the decades after Europeans started coming to the Americas, there was a change in the atmosphere itself. Some interpret this as reflecting the rapid drop of the human population on the planet because of the earliest phases of European colonization in the Americas. It is devastating to think about that, to understand that this violence was registered on the planet's atmosphere. But this is something that Indigenous communities and Indigenous philosophies have been thinking through for a long time, which is: how do you survive catastrophe? So it is something that is being voiced in different ways in different Indigenous movements around the world. The

---

[6] Hanna-Attischa (2019).

apocalypse already happened centuries ago, and part of what it is to be Indigenous now, Indigenous presence today, is living after the apocalypse. So there are many lessons I think that Indigenous philosophies are trying to propose, not just to Indigenous peoples, but to humanity. That is one entry point for all of us who are non-native is to begin thinking, "Okay, what are the lessons that we can commit ourselves to studying. This whole area of thought that goes back 500 years, and what can we learn from that to think about the choices we are making now?"

Angela Simms It is a provocative question, and I appreciate you asking us to think it through. In my course, I support students in developing their capacity to push past anxieties arising from confronting that America's history is not one of linear progress and society is not a meritocracy. I tell my students that there is no app to cure racism. There is no hack for poverty. These are fundamental relationships. As Manu shared, there's a reservoir of wisdom we can tap into. For Black people, their ancestors have already survived the Middle Passage. Indigenous people have survived genocide. These groups have had to define their humanity outside of a White gaze to survive, live lives of meaning, and have hope up to this point. So while there is trauma, there is resilience and a language for processing devastating experiences and imagining new futures.

To move forward, we must honor that violence underpins the political, economic, and other social relationships we have. Manu just centered violence too. None of what he described could occur without fierce use of force. Whether implicit or explicit, people are experiencing slow death through violence. Toni Morrison, and other people in my wisdom tradition, have already been contemplating what it means to move beyond horrific experiences. Resmaa Menakem in *My Grandmother's Hands: Racialized Trauma and the Pathway to Mending Our Hearts and Bodies* (2017) discusses how flight, fight, freeze, and annihilate responses override the sophisticated thinking of the prefrontal cortex and thus influence our social relationships more than we expect. We feel threatened by conversations that cut to the core of our identities and worldviews, and we react to this threat the way we would if we were to encounter a wolf in the wild. Emotional and spiritual processing is critical. As a Christian, prayer – reaching for a higher wisdom and power than myself, that of the Creator and sustainer of the world – is crucial. People of Color have many communal processes, from churches to dance halls, for managing social distress. Circling back to *My Grandmother's Hands*, Menakem said that his book's name comes from noticing the calluses on his grandmother's hands. His grandmother picked cotton starting at age six. She combed through tiny, razor-sharp seeds, and her hands were left bloody and raw. Eventually, the skin responded by developing calluses to protect her.

Let us honor those traditions that reflect the social calluses we have. Yet let us also move through the pain to get to the other side. As people in positions of power at Barnard, we are responsible for amplifying the voices of people who do not hold our platforms. I hold class privilege and have access to an institution that many of my brothers and sisters do not, even as a Black woman experiencing my own forms of discrimination.

Yuval Dinoor   Thank you so much for sharing. I am so sad that we have to wrap up the panel portion of the event, but I know that these will be amazing springboards for conversations in the breakout rooms. So another huge thank you to all of our wonderful panelists for everything you shared with us today. Thank you all for having Rachel and me here with you for this conversation. It was an honor. Thank you.

References

Bullard, R. D. (1994). *Dumping in Dixie: Race, class, and environmental quality*. Westview Press.

Des Roches, S., Brans, K. I., Lambert, M. R., Rivkin, L. R., Savage, A. M., Schell, C. J., Correa, C., De Meester, L., Diamond, S. E., Grimm, N. B., Harris, N. C., Govaert, L., Hendry, A. P., Johnson, M. T. J., Munshi-South, J., Palkovacs, E. P., Szulkin, M., Urban, M. C., Verrelli, B. C., & Alberti, M. (2021). Socio-eco-evolutionary dynamics in cities. *Evolutionary Applications, 14*(1), 248–267. https://doi.org/10.1111/eva.13065

Faber, D. (2007). A more "productive" environmental justice politics. In R. C. Sandler & P. C. Pezzullo (Eds.), *Environmental justice and environmentalism: The social justice challenge to the environmental movement*. MIT Press.

Gilmore, R. W. (2022). *Changing everything: Racial capitalism and the case for abolition*. Haymarket Books.

Hanna-Attisha, M. (2019). *What the eyes don't see. A story of crisis, resistance and hope in an American city*. One World, Random House.

Legal Law Institute. (2001). *ALEXANDER V. SANDOVAL (99-1908) 532 U.S. 275, 197 F.3d 484, reversed*. https://www.law.cornell.edu/supct/html/99-1908.ZS.html. Accessed 7 Dec 2021.

Menakem, R. (2017). *My grandmother's hands*. Central Recovery Press.

Schell, C. J., Dyson, K., Fuentes, T. L., Des Roches, S., Harris, N. C., Miller, D. S., Woelfle-Erskine, C. A., & Lambert, M. R. (2020). The ecological and evolutionary consequences of systemic racism in urban environments. *Science, 369*(6510), eaay4497. https://doi.org/10.1126/science.aay4497

University of Maryland. (2021a). School of Public Health: Sacoby Wilson. https://sph.umd.edu/people/sacoby-wilson. Accessed 7 Dec 2021.

University of Maryland. (2021b). Community Engagement, Environmental Justice, & Health (CEEJH). https://www.ceejh.center/our-mission. Accessed 7 Dec 2021.

WE ACT for Environmental Justice. (2021). https://www.weact.org/. Accessed 7 Dec 2021.

Elizabeth M. Cook   is an Assistant Professor of Environmental Science at Barnard College and an urban ecosystem scientist with interdisciplinary expertise in ecology, cultural geography, and sustainability sciences. Her research focuses on future urban sustainability and human-environment feedback using mixed methods and a comparative approach. She teaches Urban Ecosystems and Environmental Data Analysis, both of which focus on urban environmental justice.

Manu Karuka is an Assistant Professor of American Studies at Barnard College. He is the author of Empire's Tracks: Indigenous Nations, Chinese Workers, and the Transcontinental Railroad (University of California Press, 2019). He teaches courses on the political economy of racism, U.S. imperialism and radical internationalism, Indigenous critiques of political economy, and liberation.

Angela M. Simms is an Assistant Professor of Sociology and Urban Studies at Barnard College. Her research examines how legacy and contemporary market and government processes in metropolitan areas shape racial inequity. She teaches the courses Race, Ethnicity & Society and Metropolitics of Race & Place, among others. Prior to Barnard, she was a Presidential Management Fellow and legislative analyst for the U.S. Office of Management and Budget.

Joscelyn Jurich is a graduate assistant at the Center for Engaged Pedagogy and a Ph.D. candidate in Communications at Columbia University. At the CEP, she focuses on faculty support, pedagogy research, and media projects. She has taught media studies and journalism at NYU, CUNY, and CELSA-Sorbonne and she researches visual culture.

Yuval Dinoor '21 holds a BA from Barnard College in American Studies concentrating in Education, Civic Engagement and Community. Her research focuses on creating experimental learning spaces to build politically active communities; recently, she aired a podcast series exploring the Highlander Research and Education Center's pedagogical approaches to community organizing.

Rachel Elkis '22 is a senior at Barnard College studying Economics and Environmental Science. She is the former Representative for Sustainable Initiatives for Barnard's Student Government Association and remains involved in sustainability programming and activism. She is eager to leverage her Barnard experience as a Power & Renewables Investment Banking Analyst at JP Morgan.

Chapter 25
Panel Four: Design Decarbonization, March 30, 2021

Kadambari Baxi, Sandra Goldmark, Martin Stute, Joscelyn Jurich, Linda Chen, and Olivia Wang

25.1 Panel Discussion

Linda Chen Hi everyone, welcome to today's session. Before we get started, I would like to read a brief land acknowledgment. (See Rosales Introduction).

Olivia Wang And now we will do introductions of the faculty panelists: please share your department teaching and research scholarship and your role in the Barnard community.

Kadambari Baxi Hi, everyone, and thank you to the CEP for organizing this panel. I am Kadambari Baxi, she/her. I am a Professor of Practice in Architecture, and I teach design studios at the introductory and senior levels. Occasionally, I also

---

Panelists: Kadambari Baxi, Sandra Goldmark, Martin Stute
Joscelyn Jurich (Ed.)
Student Moderators: Linda Chen, Olivia Wang

K. Baxi
Department of Architecture, Barnard College, New York, NY, USA

S. Goldmark (✉)
Department of Theatre, Barnard College, New York, NY, USA
e-mail: sgoldmar@barnard.edu

M. Stute
Department of Environmental Science, Barnard College, New York, NY, USA

J. Jurich
Center for Engaged Pedagogy, Barnard College, New York, NY, USA

L. Chen · O. Wang
Barnard College, New York, NY, USA

M. S. Rivera Maulucci et al. (eds.), *Transforming Education for Sustainability*,
Environmental Discourses in Science Education 7,
https://doi.org/10.1007/978-3-031-13536-1_25

teach seminars on special topics, such as Environmental Visualizations. My practice centers on design research projects that circulate as exhibitions in art, design, and architecture forums. I will mention one project briefly, as it may link to this panel's theme. This is a work in progress, a spatial film montage that uses two sites: the United Nations headquarters in New York and the Supreme Court in Washington DC. The film excerpts climate actions at UN COP Conferences over the last ten years in ten different cities worldwide. This montage is paired with another one, where I use the live court footage of a recent youth climate case in the US: "Juliana v USA,"[1] projected on the Supreme Court plaza. The film juxtaposes two different kinds of climate actions: activists' disruptions at international climate negotiations and climate litigation demanding the constitutional rights of future generations.

Sandra Goldmark My name is Sandra Goldmark, and I teach in the Theatre department. I am a designer by training, and a lot of my work over the past years has been a transition from designing shows to looking at the theatre industry and the performing arts at the organizational or institutional levels, and to begin thinking about how we can incorporate climate impacts, climate responses, and climate justice into our practice as theatre artists. Everything from our design practices, to our budgeting processes, to how we can integrate and understand the overlap between questions of climate change and social justice, which are very much alive, of course, in the American theatre. That is my teaching and research thread in terms of my role in the community. I am also the Director of Sustainability for the campus, which means that I have been helping to move our campus forward in how we think about climate change across academics, operations, campus culture, the student experience—all of these different facets of our campus and our campus life. How can we think about this question of climate change in all that we do and all the decisions that we make? I hope today's conversation will connect to these themes.

Martin Stute Martin Stute is my name. I am a Professor in the Environmental Science Department at Barnard. I am also a faculty member in the Department of Earth & Environmental Sciences at Columbia and in the Lenfest Center for Sustainable Energy that is working on batteries, carbon sequestration, and various other energy-related things. My teaching is mostly focused currently on the senior seminar, which is a joint Columbia/Barnard course where students work on their research and write a senior thesis. Every year, we have a lot of students, but maybe half or so of the students have a project that is related in some way or another to climate change. We actually have a carbon sequestration project at the moment. I also teach classes in Hydrology and the Workshops in Sustainable Development, which maybe I can come back to later. More interesting perhaps is the research that I am doing because it is very closely related to the topic today. I guess my expertise is in water on land, mostly. I actually have studied water on all continents of the planet, and one of my main contributions to science has been to reconstruct past climate conditions, using groundwater as an archive of past climate conditions—

---

[1] https://harvardlawreview.org/2021/03/juliana-v-united-states/

how cold or warm it was in the past, how wet or dry it was in the past. For decades, I have been working on this topic.

More recently, I have been moving to more applied topics that also relate to water, including hydraulic fracturing and its health effects—except it is very hard to raise money for it, so I have not been able to do as much as I wanted to, and then finally carbon mineralization, which is basically trying to find a safe way to store CO_2 collected at sources or from the air directly in the subsurface. I was very much involved for eight years in this study in Iceland, where CO_2 is taken from a geothermal power plant and then injected into the ground. We mineralized the CO_2, so it's stored, not as CO_2 but as carbonites in the subsurface. It was surprisingly successful, and it was the first application of this approach in a field site.

Linda Chen Thank you so much for the brief introduction and moving on to our first question, What is decarbonization and why is it so urgent, and what exactly does the phrase "design decarbonization" mean?

Martin Stute The panel has been thinking about those questions, and we felt that it might be a good idea to explain what decarbonization is as a starting point. We have prepared a few slides. In Fig. 25.1, I am showing you a graph of temperature and carbon dioxide (CO_2) concentration from an ice core in Antarctica covering the past 350,000 years. Temperature is reconstructed from the isotopic composition of the ice, and the CO_2 is actually measured in small air bubbles that form in the ice when it comes to the surface—so it is a measure of atmospheric CO_2 concentrations. For about the last 350,000 years, you can see how uniform the correlation between them has been. So CO_2 and temperature are very closely related, although the relationship is somewhat complicated.

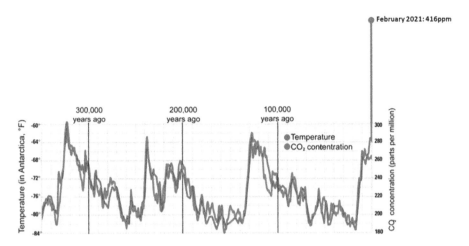

Fig. 25.1 CO_2 and temperature over the last 350,000 years from the Vostok ice core and atmospheric measurements. (Source: Adapted from the Marian Koshland Science Museum of the National Academy of Sciences)

Clearly, there is some connection between the two, and you can show that in a laboratory, when you have CO_2 in a glass box and shine a light on it, the temperature rises, more so than in a box that does not contain CO_2. What is really scary is the right part of that graph that shows you temperature and CO_2 in the last 50 years or so. We currently have a CO_2 concentration of 416 parts per million (ppm), and the temperature in Antarctica has risen by about five or six degrees Fahrenheit in the same time period. So what we are experiencing right now is enormous increases in CO_2 levels compared to historical levels. The last time we had that high a level was 3 million years ago, and then the earth was much, much warmer than it is today.

If you wanted to project this to 2100, the blue curve would reach far beyond the top of the figure, and we are exceeding any precedent in the last few million years. These high CO_2 concentrations will cause all sorts of havoc. We see the beginning of it: massive storms, flooding, droughts, wildfires, massive migrations, death at a very high level, it will trigger wars—so it's really a disaster that is in front of us.

The concept of decarbonization: I wanted to show you this diagram and that is my last one (See Fig. 25.2). It is a little complicated, so let me guide you through it. In a nutshell, decarbonization means we are trying to do whatever we are doing right now in terms of standard of living, providing food for most people, living a safe life, but without these enormous CO_2 emissions. This graph shows how challenging this is. We are currently emitting about 50 gigatons of CO_2 (equivalent) per year, as a combination of CO_2 and other greenhouse gases. Emissions will likely rise in the future, and then in mid-century sort of drop off a little bit. Climate scientists tell us that if we want to limit climate change to about two degrees warming global average, we have to actually follow the red curve in the diagram. We have to reach zero net emissions of greenhouse gases in this scenario, by 2090. We can accomplish this by avoiding the CO_2 emissions in the green area of the diagram

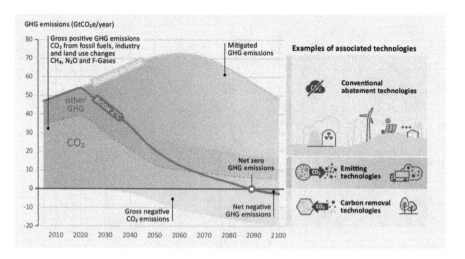

Fig. 25.2 Potential future carbon dioxide emissions for 'business as usual' as well as what would be required to limit warming below 2 °C. (Sources: UNEP, 2017; NAS, 2019)

using wind energy, renewable solar energy, and perhaps nuclear energy. Now, the problem is that there are considerable CO_2 and methane emissions that are very, very difficult to avoid. This includes airplanes, cement production, and agriculture that emit a lot of methane. It is very hard to completely get rid of those emissions. So in order to get to zero, we actually have to have negative emissions. We have to remove CO_2 from the atmosphere by growing trees, increasing uptake of CO_2 by the oceans, or taking CO_2 out of the atmosphere and putting it into the subsurface and storing it there. A lot has to happen in order to get to zero and fast. My final point is this graph is originally from 2017 and has been publicized by the National Academy of Sciences in 2019, but it is outdated already. Newer studies say that we have to reach carbon neutrality already by 2050 because we need to limit global warming to not 2 but 1.5 °C. So our window of opportunity is actually even more compressed. Hopefully, that is enough background for defining this term.

Sandra Goldmark Given what Martin just said, laying out in the plainest terms the urgency and the scale of the problem, there are a number of reactions we might have. One reaction might be to be totally overwhelmed and say, "I don't know what I can do about this." Another might be and what I hope that this workshop is inviting us to do, is to say, "Okay, given where I am and where I am working, what can I do?" One of the things I wanted to talk about today is how—and in a way, this is going to kind of make it seem worse, but in a way, hopefully, it'll get us somewhere that might feel better in the end—this question of decarbonization is also linked to so many other issues and questions that we are living with and grappling with. I use this visual exercise in one of my classes to begin to open up a conversation about the intersections between the specific metric of carbon in the atmosphere and some of the other disciplines, challenges, and human quandaries that are actually all inter-connected. Figure 25.3 is like Martin's graph, but just zoomed-in—you are looking at 1000 years of carbon dioxide emissions growth. Temperature follows the same "hockey stick" pattern.[2] Figure 25.4 represents World Population growth over the last 12,000 years; Fig. 25.5 is species extinction over the last 200 years. Figure 25.6, hard drive capacity by year, follows the same pattern. I could continue with the hockey sticks—sugar consumption per capita, extraction of resources per capita, storage unit capacity in the U.S.—it goes on and on. But I am going to pause on this next one: Fig. 25.7 is the reverse shape, interestingly enough visually from a design point of view, because this is the distribution of emissions relative to the percentage of the population living in extreme poverty. The richest 10% emit at least 50% of emissions. These multiple hockey sticks hopefully help illustrate, visually, that there are a number of artistic and disciplinary ways to think about the question of decarbonization and the related challenges we face. And furthermore, as we see with the sugar consumption graph, it can't all be tied back to population growth. It's actually about the fact that, as time moves on, we've begun extracting more resources per capita on top of the actual growth in population, so in that way, it's twofold.

---

[2] Term used by Mahlman to describe the reconstructed 1000-year temperature pattern shown by Mann et al. (1999).

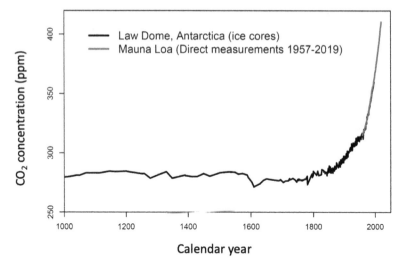

Fig. 25.3 1000 years of atmospheric carbon dioxide concentration showing dramatic recent increase from anthropogenic emissions. (Source: Earth Institute of Columbia University)

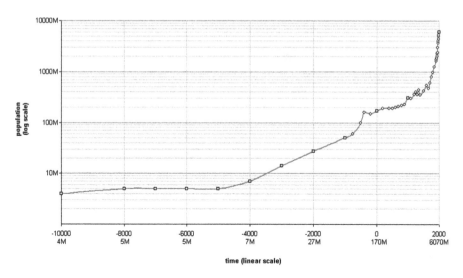

Fig. 25.4 World Population growth over the last 12,000 years. (Source: Wikimedia Commons)

Kadambari Baxi Thank you, both Martin and Sandra. Sandra compares visualizations of data in compelling ways. Martin talks about scientific data that is also visualized very simply, and very impactfully. I think the message is very straightforward, clear, and urgent. We must also acknowledge that decarbonization is a debate, and often, different angles may be highlighted with the same data. Data can be parsed in many different ways. Right now we're at this moment where–a kind of complex conversation between past, present, and future that also leads to some controversial

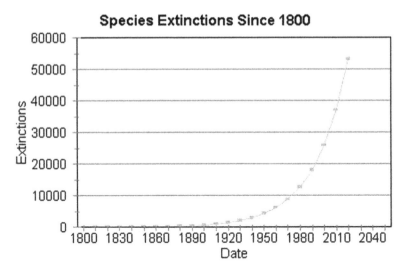

Fig. 25.5 Species extinction over the last 200 years. (Source: Extinction Symbol)

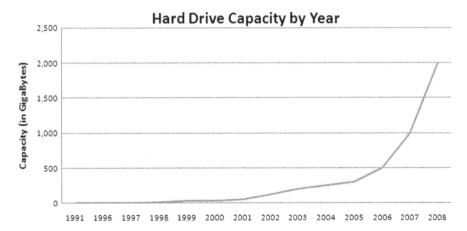

Fig. 25.6 Hard drive capacity by year. (Source: UCLA)

questions, such as: who governs decarbonization? Who funds it? To emphasize these types of questions: I want to share three images: Fig. 25.9 from "Our World in Data" shows: "who has contributed most to the global CO_2 emissions?" You can see USA, EU, China, and India are among the major contributors. Note that these figures are based on production-based emissions, which means that they represent CO_2 produced domestically from fossil fuel combustion and cement, and do not correct for international trade. Excluded here are any consumption-based emissions. This inclusion, most likely, will significantly increase the numbers for the USA and EU. Seeing this scale of comparative accumulations as country-based comparisons, to me, suggests obvious implications. Figure 25.8 shows per capita

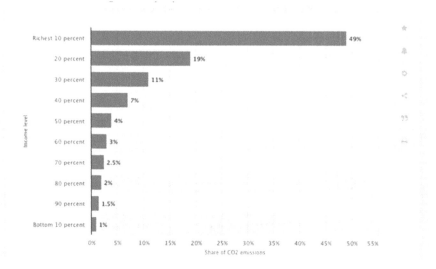

Fig. 25.7 The distribution of global population's carbon dioxide emissions in 2015, by income level. (Source: Statista)

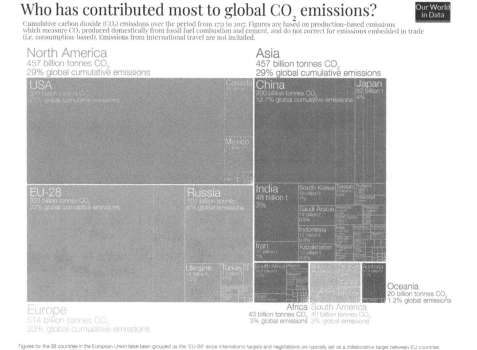

Fig. 25.8 Global distribution of per capita CO_2 emissions from the burning of fossil fuels for energy and cement production, 2019. Land use change is not included. (Source: "Our World in Data," Creative Commons)

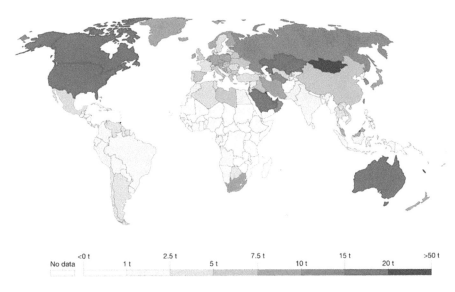

No data   <0 t   2.5 t   7.5 t   15 t   >50 t
              1 t   5 t   10 t   20 t

Fig. 25.9 Contributions to global CO2 emissions. (Source: "Our World in Data," Creative Commons)

CO_2 emissions, and this world-view of rich countries versus poor countries is probably not surprising for anyone. Any decarbonization questions, and debates, will need negotiations and solutions at many scales: the individual scale, the national scale, per capita scale—there are so many different ways to look at the urgency around such questions.

The last Fig. 25.10, simply breaks down the emission sources and what remains in the atmosphere: 59%. That is a very large number! But what I want to emphasize are also the different sources noted on the left side. They are shown as separate sectors, but are, of course, also interdependent. In academia, for teaching and learning purposes, the different sectors suggest interdisciplinary conversations and research around how they shape many aspects, for example the built environment, our society, etc. But what we really want to highlight here is the urgency represented by the 59% number and the emissions that remain in the global atmosphere.

Designing decarbonization, thus, is an urgent discussion. It includes issues of global governance, scientific knowledge as well as worldwide collaborations, and further, must also include locally specific activism that can also lead to large impacts. We should consider design as a kind of collaborative authorship. Our title includes "Design" in the most expanded sense: we design curricula here at Barnard. We design our lives to a certain extent.

Sandra Goldmark To echo what Kadambari said, and make it a little more personalized, looking at Martin's Fig. 25.1, as a theatre designer, many years ago I

EMISSIONS SOURCES & NATURAL SINKS

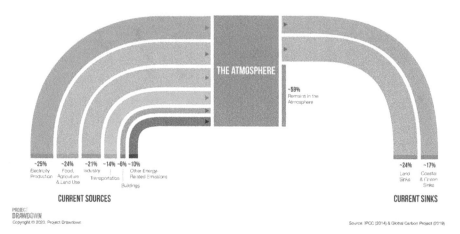

Fig. 25.10 Sources of current global emissions and natural sinks of greenhouse gases. Sources data from IPCC Fifth Assessment Report, Working Group Three. Sinks data on carbon dioxide from the Global Carbon Project (adjusted here for whole greenhouse gas mixes). (Graphic by Project Drawdown © 2021)

would have said, "Well, what do I have to do with this? What does a theatre artist have to do with decarbonization?" I started with what Kadambari indicated as the first thing, trying to sort of decarbonize my own design practice or my own designs in theatre. And then I realized over the years that what Kadambari is inviting us to do, i.e. all participate in the process of designing decarbonization, is incredibly valuable and necessary. If we are going to be thinking about emissions along with species extinction, and social justice, and equity, we are all going to have to be part of the process of designing a decarbonized future. It's not just about emissions, it's about the whole complex package.

Olivia Wang Thank you for sharing, we will move on to the next question, which is how have you integrated this theme in your courses so far or any thoughts on future courses?

Kadambari Baxi Last year, I taught a studio course for seniors, *Architectural Design III,* that explored the role of museums in exhibiting issues related to climate change, and in providing climate forums for the general public. Students designed what we termed a "Climatorium," at two natural history museums in New York and Berlin. "Climatorium" is a word that is made up, similar to a planetarium that models deep space and atmospheres, students designed spatial installations that "made-visible" complex issues related to climate change. Another course, *Environmental Visualizations of NYC*, is one that I co-taught with Karen Fairbanks. We used Newtown Creek in Brooklyn and Gowanus Canal as sites, both of these are Superfund sites. We were looking at mitigation strategies and also how city and state planning agencies are conducting cleanup efforts, but that also led to large-scale planning and zoning initiatives.

Sandra Goldmark   For my classes, as I mentioned, the first step is decarbonizing my design practices, and the way I teach design, and the way we practice in our classes. So, in terms of set and costume design classes at Barnard, I have started incorporating practices built around circularity, because embracing circular design and production principles is one of the easiest ways to slash your emissions from design and production. In the classroom, I have a series of exercises to explore circularity and how students can approach design with those principles in place. In one specific exercise, for example, students build a model. For the next project, they are given the model of one of their classmates, and they can only design using the materials they find in the first model. It is a way to simply and clearly physicalize and actualize the concepts of circularity in three dimensions. As another example, in my costume design classes, I have started incorporating modules around labor and environment, i.e., looking at the environmental impacts of our design practices and the social impacts. Students learn about what textiles are made of, who makes our garments and apparel, how much they are paid, and how that knowledge can influence our design decisions. We also have modules on budgeting, which include thinking about how to track and budget sustainably in a theatrical production. The second part of the question, designing decarbonization, is a much newer thing for me to be trying to teach in my courses because most of my courses are within theatre. I have really been playing with the question in a first-year seminar and in a course I developed last year during the pandemic. It was called *Change and Climate Change*, and that is where I am really trying to unpack some of the questions that point to how we might "design decarbonization" by first understanding change itself. How are we going to change all of these systems we live within? What are different theories of change, what are different approaches to it? For that course, I brought in several fellow faculty members to do guest visits to physicalize and embody this idea of interdisciplinarity and collaboration as a necessary part of designing our decarbonized future.

Martin Stute   So I wanted to start with a bit of history. Of course, climate change and sustainability, in general, are the heart and center of whatever we do in the Environmental Science Department. But we did play a role in putting climate front and center for undergraduates when I started at Barnard in 1995. Supported by a grant from the National Science Foundation, we built a sequence of classes that deal with Earth systems. I was involved in developing the course about the Earth's Climate System (see Pfirman and Winckler, Chap. 19). This was the first comprehensive climate change class on the Columbia/Barnard campus. It was also the first class that I am aware of that was served on the internet in 1995, which is pretty early in the process. That class still exists. It is our basic introductory course for the major and also has always been co-taught, which is a concept that I think we are pushing a little bit in this context. So it works—it has worked for like 25 years, so we should keep pushing along those lines.

I teach the environmental science senior seminar; we have a lot of topics that deal with research on climate change. I am an advisor of a project that uses rocks and water to capture CO_2 from the atmosphere and concentrate it to sequester it automatically. I teach hydrology. Water is my field, and of course, there are many interactions between water and CO_2. We use water as a carrier for CO_2, we need water for chemical reactions that involve CO_2, and water and climate change are very, very tightly connected. The effects of climate change on the water cycle might be more important than just temperature changes, for example, because we all want to survive and we need to drink water, eat food, etc., so it is front and center in that class.

I did want to mention one more thing. I teach a workshop in sustainable development, which is a concept, maybe of broad interest. Students basically form a consulting firm in their class and they work for an outside client on a real world problem, and often those projects involve climate change. We were once tasked, for example, by another campus to figure out electricity consumption and how that could be reduced, and the students found out that in one giant laboratory building, just by having motion sensors in there, the institution would save $40,000 a year of electricity costs. It is an interesting concept. Students learn how to work as a team, which is actually something that most of us do for the rest of our lives. We are rarely working alone on anything and yet we are not prepared for it. We do not learn how to work in a team, we are not educated about how to do this, students aren't and faculty aren't. So that is a skill that students learn and I, as an instructor, step back, more and more. I am really only actively involved in the first couple of weeks or so, and then they pretty much take over and run the whole show themselves. And then, finally, I do give a lot of guest lectures about various climate related topics in various departments, not just the sciences. This Iceland study in my background here I have talked about 20–25 times already in various forms.

Linda Chen Looking at these slides, we can kind of understand the gravity of the situation. Professor Goldmark, earlier, you mentioned that we should incorporate decarbonization design into our daily lives. Do you have any tips on what we could or should be doing about this?

Sandra Goldmark That is such a big question because who is "we?" Is it me as an individual, is it faculty and students, is it Barnard as an institution, is it New York City, is it the federal government? There is every level of "we" embedded in that question. I am a believer—given the urgency of what Martin laid out in that first slide—in looking at the action at each level. I get a lot of questions because of my work on sustainable consumption along the lines of, "Who should be doing this? Is it the individual consumer? Is it the businesses? Is it the policymakers?" and I say, "Yes! All of the above! We do not have any time to wait for somebody else to start." There is a lot of finger-pointing in the world of sustainability. There are a lot of people saying, "The government should make policies," or "It's all the corporation's fault," and then the corporations say, "Well, there's no consumer demand, so we can't do this." Everybody at every level can—and must—take action now. That might sound kind of huge but it is not, because I am not saying everyone can do this work and be done right now, but I am just saying we can and have to start. One way to start is to say, "I am teaching a course next semester. Is there a module or reading I might include?" That is one of the things this workshop is hoping to open up. Or to say on an individual level, "Is there a practice that I am engaged in that does not feel right for me?" I did not try to go 100% green when I started. I just started realizing that my theatrical designs were kind of wasteful and nasty, and I began changing the way I designed one show. I bought less stuff. The next show, I bought no stuff. And before I knew it, ten years later, I had committed my whole career to climate action.

But of course, let's not forget the policy level. In addition to looking at our own practices, our own spheres of influence, we need to push for policies that will help move things at scale. Most importantly, we need to approach all of this work with a justice and equity lens. That's something we are in an excellent position to do at Barnard. So perhaps I should not even say, "Just start," because we have started. I feel like everyone at Barnard has either started this work or is way down the road, as in the case of Martin, who has been showing the way for many years. So, it is more a question of evaluating where you are and continuing to turn up the volume.

Martin Stute I am very much on board with what Sandra just said, but to rank the efforts, a little bit, I would say if we want to have any hope in addressing this problem, we need a very large-scale shift, and that can only happen on a global scale. We do not have control over the globe, but we do have some control of what is happening in this country, and you saw earlier how important we are. So if I ranked the efforts, I would say the number one thing is, make a difference on the political level: vote. You already see how much of a difference one year made compared to where we were last spring to where we are this spring in terms of climate change discussions and what is actually on the agenda. It's a huge difference. So that is a requirement without which we will not have a clue to address this problem. I am not against all the personal measures you put in place—that is part of the big picture. But the other one is the most important one in my mind, so we have to move people's minds, we have to generate some groundswell of support for these massive changes we have to see in the future of our lives.

Sandra Goldmark I totally agree, but I just want to jump in and say, I think the way you get the political will, the way you get the groundswell is by everybody clicking in and pushing for change in their homes, their community, their sphere of influence because people vote where their hearts are, people vote where their communities are, people vote where their belief systems are. So political change doesn't happen in a vacuum; politicians need to see the political will from everybody, and for me, that can start at a very personal, individual level. It just can't stop there.

Kadambari Baxi I absolutely agree with both Martin and Sandra, on all of the above, of course. I would echo and emphasize that what we really need is systematic change, and we need structural change. So how do we achieve that? Certainly, it was on my list to say to vote and to participate in political action. But we also have to recognize that for some people, voting may be a privilege. Moreover, one is often not able to vote on global issues. Electoral democracy is based on nation-states and the power of nation-states. I would argue that to generate any kind of structural and systematic change, in terms of individual action, organizing collectively is essential. We can organize to produce collective action. We need to learn how to organize. As Martin said, we have seen over the last year that activism matters. Given the urgency, I think we need cultural activism, we need imagination that facilitates collective collaboration. We are not all going to agree, but we need processes and forums, participatory processes where these issues can be debated and some decisions can be made. The last thing I would say is that we must also think of the role of universities in this. This is our realm. We are knowledge producers, and we are also cultural producers, and I think that in terms of these discussions, we also need to think about our places in universities and what universities can do. As a quick example, consider the EPA (Environmental Protection Agency) over the last four years. When most of the climate data was removed from its websites during the Trump administrations, a university consortium (Environmental Data and Governance Initiative: EDGI, https://envirodatagov.org/) was formed to quickly download the data before it was lost for public use. Universities can become open sources for this kind of important data.

Olivia Wang Thank you everyone. We will now be doing a Q & A session so you may unmute to ask your question, or you can also put your question in the chat.

Ralph Ghoche (Architecture) I have a question. I am in a design department as well, and I wonder to what extent this word "design" is used, I think, very broadly here, but also very specifically. There is an example Sandra gave of really heavy engagement with her students and design practices in the Theatre department. But I wonder to what extent we can shift the focus back on Barnard with our design practices. I wonder to what extent we can actually engage with design practices on campus. I think Sandra you mentioned that we need to act more locally. Do you think there's a desire on the part of students to focus more on the very local institutions of Barnard and Columbia and do you think we could target design much more in the very institutions which we are part of?

Sandra Goldmark I feel like we have gotten into this whole sustainability debate of the individual versus the collective and I just do not see it as an either/or. I do not think you go out and vote and then you do whatever the hell you want in your own life. To me, they are two tracks that need to move forward at the same time. And I am not saying anybody on the panel is saying that. This is a debate that has been happening for years in the climate change world. I think that we need to look at our design practices at Barnard, not only to reduce our own emissions, but as Kadambari was indicating, to pioneer and to serve as a model to train the next generation. And so I think there are huge opportunities coming down the pike for the college.

Martin Stute Construction is going to happen again on the Barnard campus. Altschul is going to be very heavily renovated and decarbonization is very high on the agenda, looking at architects and making choices there. Several of us have been involved in those discussions and are going to push very hard that this will be as carbon neutral as possible. That is definitely a campus opportunity and we should definitely have done more of this when the previous two buildings were built, but we lost that opportunity to a large extent. There are many student organizations that help with various activities on campus, so students can be active there. That is one thing. Then, by building a real-world impact into our classes, such design studios and workshops where students work with outside entities, they can really make a difference. These are real projects, the students produce something that can be and often is implemented. So there is a real-world impact that students can make within those courses.

Kadambari Baxi I think that it is very interesting to see what happened last year, with the pandemic. There were so many things we thought that could never happen, like working from home at such a large scale, that had to become a quick reality. Going forward, we should really assess what we have learned from this last year. I remember Sandra and I had a very quick conversation when I ran into her when the campus was closed and nobody was around. None of us were allowed to use our offices on campus but we noted how all the lights were on in Diana and in Milstein. I hope we can challenge our business as usual practices and learn something different from this entire pandemic year.

Linda Chen There is a quick question in the chat. How important is divestment from fossil fuel companies as a practical and or symbolic step?

Martin Stute I think it is important. I was not involved in Barnard deciding to divest from fossil fuel, but there was a strong movement some years ago. I think Sandra was involved. It is a good thing. How much difference it really makes in the end, I do not know because at this point, not a tremendous amount of people are willing to divest on the national level. So if that grows, I think there is a chance that it will make an impact. I am not sure if it has yet, but it is worth doing and certainly has value, but Sandra, you are the expert here.

Sandra Goldmark I think of divestment as part of a portfolio of climate action. Just like I think the Theatre department should figure out how they will take action, I think the financial management of the College needs to figure out what their role is. So, if you are working in Theatre, you need to put that house in order. If you work on investments and endowment you need to put that house in order. Of course, people pay more attention to certain arms of the College than others. But for me, it is just part of a philosophy where every decision-making process needs to take climate, impact, and justice into account. Every single one. So yes, investments for sure, but also purchasing, design, renovating Altschul, you name it, this needs to be a lens that we use when we make decisions, and so absolutely investment should be part of it.

The only thing about divestment that I don't like is when it becomes seen as the only thing we should do. Then, I personally do not agree, because I think it is not enough. I am not looking for just the one symbolic act. I am looking for consistent and coherent actions across all spheres and operational levels. For me, that is why it is not only what endowments do, or what theatre does, or what individuals do, or what corporations or policymakers do. It's about decision making processes, how we all make our decisions. How do we use this lens when we make those decisions? I think that logic can apply to the smallest decision, like, "Am I going to buy this or am I going to buy that?" and to the biggest decisions. For example, the US Government is considering how to invest $2–$3 trillion in infrastructure over the next six years. Obviously, infrastructure investment is more important than one individual purchase. But underneath it all, the most important question is, what is our decision-making process and what kind of lenses do we apply? And that applies to all categories, including investments.

Kadambari Baxi is an architect and Professor of Professional Practice at Barnard + Columbia Architecture Department. Her practice produces multimedia design projects and exhibitions. Recent work focuses on climate research and environmental visualizations and highlights human rights, climate justice, design ethics and activism. Project excerpts at www.kbaxi.com.

Sandra Goldmark an Associate Professor of Professional Practice and Director of Campus Sustainability and Climate Action at Barnard College and the Senior Assistant Dean for Interdisciplinary Engagement at the Columbia Climate School. She is the founder of Fixup, a circular economy social enterprise, a co-creator of the Sustainable Production Toolkit, and the author of *Fixation: How to Have Stuff without Breaking the Planet.*

Martin Stute is Co-Chair of the Department of Environmental Science at Barnard College. His research interests include water resources, contaminant transport in groundwater, carbon sequestration, unconventional gas production, paleoclimate, mathematical modeling of environmental phenomena, and the social and economic impact of global environmental change.

Joscelyn Jurich is a graduate assistant at the Center for Engaged Pedagogy and a Ph.D. candidate in Communications at Columbia University. At the CEP, she focuses on faculty support, pedagogy research, and media projects. She has taught media studies and journalism at NYU, CUNY, and CELSA-Sorbonne and she researches visual culture.

Linda Chen, a native New Yorker, is a junior at Barnard College studying political science and human rights. She is an active member of the Student Government Association Committee on Sustainability and the Athena Center's Student Advisory Board, Managing Editor of the Barnard Bulletin, and secretary for Barnard's Alliance for Humanitarian Aid.

Olivia Wang is a Sophomore at Barnard College studying International Relations. In her free time, she enjoys spending time outdoors and learning new languages.

Printed in the USA
CPSIA information can be obtained
at www.ICGtesting.com
LVHW020558170924
791293LV00001B/24